Institutions for Enviro

Institutions for Environmental Aid

Pitfalls and Promise

Robert O. Keohane and Marc A. Levy, editors

The MIT Press
Cambridge, Massachusetts
London, England

This book was set in Sabon by Graphic Composition, Inc.
Printed and bound in the United States of America.

Library of Congress Cataloging-in-Publication Data

Institutions for environmental aid: pitfalls and promise/Robert O.
  Keohane and Marc A. Levy, editors.
        p.   cm.—(Global environmental accords. Strategies for
  sustainability)
    ISBN 0-262-11213-2 (alk. paper).—ISBN 0-262-61120-1 (pbk.:
  alk. paper)
    1. Economic development—Environmental aspects.   2. Economic
  assistance.   I. Keohane, Robert O. (Robert Owen), 1941–   .
  II. Levy, Marc A.   III. Series: Global environmental accords series.
  HD75.6.I573   1996.
  363.7—dc20   96-2486
                                                                    CIP

Contents

Contributors

Hildegard Bedarff is research fellow at the Center for Transatlantic, Foreign and Security Studies at the Free University, Berlin, Germany.

Thomas Bernauer is senior lecturer in international relations at the University of Zurich, Switzerland. He is the author of *The Chemistry of Regime Formation* (1993) and co-editor of *Towards the New Europe* (1996).

Barbara Connolly is assistant professor of political science at Tufts University, and is currently writing about institutional governance of international cooperation.

Elizabeth R. DeSombre is assistant professor of government and environmental studies at Colby College.

David Fairman is a doctoral candidate in political science at MIT. His dissertation analyzes the politics of community forestry in developing countries.

Wendy E. Franz is a doctoral candidate in government at Harvard University and a graduate student associate at the Center for International Affairs.

Tamar Gutner is a doctoral candidate in political science at MIT. She is co-editor of *Comrades Go Private: Privatization in Eastern Europe* (1992) and *Investing in Reform: Doing Business in a Changing Soviet Union* (1991).

Cord Jakobeit is a postdoctoral fellow of the German Research Foundation (DFG) at Hamburg University, Germany.

Joanne Kauffman is a doctoral candidate in political science at MIT and a former administrator in the OECD Environment Directorate.

Robert O. Keohane is Stanfield Professor of International Peace, Harvard University, author of *After Hegemony: Cooperation and Discord in the World Political Economy* (1984), and co-editor of *Institutions for the Earth: Sources of Effective Environmental Protection* (1993) and *Local Commons and Global Interdependence* (1994).

Marc A. Levy is assistant professor of politics at Princeton University and co-editor of *Institutions for the Earth: Sources of Effective Environmental Protection* (1993).

Martin List is assistant professor of international relations and comparative politics at Hagen University, Germany. He is the author of *Umweltschutz in zwei Meeren* (1991) and co-author of *Internationale Politik* (1995).

Michael Ross is a doctoral candidate in politics at Princeton University, and is working on the political economy of commercial logging in Southeast Asia.

Preface

This book has its origins in an earlier project at the Center for International Affairs, supported by the Rockefeller Brothers Fund, which culminated in the publication of *Institutions for the Earth: Sources of Effective International Environmental Protection* (Cambridge: MIT Press, 1993). That volume, which we edited with Peter M. Haas, discussed how international institutions have affected environmental policy on issues ranging from depletion of the stratospheric ozone layer to pollution of the Baltic and North Seas. Its conclusions were relatively optimistic: that international environmental institutions could exert significant influence by affecting the degree of public and governmental concern about environmental issues, the nature of contractual arrangements, and national capacity—what we called the "three C's."

Showing that international environmental institutions could make a difference was a necessary condition for further investigation of this subject, but it did not answer some key questions, either for social science or for policy. From a social scientific standpoint, *Institutions for the Earth* was flawed because we selected only prominent institutions for study, making it likely that we had oversampled successful cases. It would not have been valid to generalize our findings about these cases to international environmental issues as a whole. Furthermore, *Institutions for the Earth* covered such a broad domain that it could not compare different institutional mechanisms on roughly similar issues, controlling to some extent for the type of issue in order to examine the impact of choices of institutional design. And from a policy perspective, our previous analysis, however valuable in providing relevant context, lacked the sharpness of focus on a given set of issues that policymakers value.

Our pride in our earlier work was thus tempered by our awareness of how much more there was to learn about international environmental institutions. Discussions with Professor William C. Clark of the Kennedy School of Government at Harvard revealed common concerns, and we therefore returned to the Rockefeller Brothers Fund with a proposal for further work in this area. The president of the fund, Colin Campbell, and his assistant, Michael Northrop, recommended that we sharpen our focus on contemporary policy dilemmas. They came to Cambridge for a long session, at which we proposed the idea of studying what we called "financial transfer mechanisms"—institutional arrangements for transferring resources for environmental protection. We hoped that such a focus would have clear policy relevance and that it would also provide a set of comparable cases for analyzing the sources of institutional success and failure. The fund generously agreed to support this work over a two-year period, with Professors Clark and Keohane as co-principal investigators and Professor Levy as project director.

We quickly recruited a number of prospective collaborators, most of whom became contributors to this volume. Almost all of these people were in residence in Cambridge during academic year 1993–94, aided by support from the Center for International Affairs and the Center for Science and International Affairs at Harvard University, by the Thyssen Foundation, and, centrally, by the Rockefeller Brothers Fund. This is a genuinely collaborative project, since we met weekly or biweekly during the fall term and frequently during the next spring to thrash out the common framework around which this volume is constructed. Our files are full of "analytical frameworks," summaries of discussions, critical comments on previous drafts, and research designs, as well as elaborate charts summarizing how everything should fit together. The "scaffolding" for the edifice we were creating was extensive, if somewhat ramshackle at times, and many contributors played a major role in creating it. This volume is neither a mere set of papers separately written and put between the same covers, nor an elaboration of the editors' arguments, embellished and detailed by contributors. Bill Clark, Barbara Connolly, and David Fairman, in particular, took the lead at particular points to reorganize our collective thinking and point the way toward more coherent joint work. Other members of the working group in residence, all of whom made significant contributions, included authors Hildegard Bedarff, Thomas Bernauer, Elizabeth DeSombre, Tamar Gutner, Cord Jakobeit, Joanne Kauffman, and Martin List. In addition, Suzi Kerr played a major role in conceptualizing issues of contracting, as did Lisa Martin of Harvard's Government Department; and Miranda Schreurs brought the valuable perspective of someone who does research in Asia. Wendy Franz, our graduate student assistant, was a valuable contributor in her own right, as the appendix to this volume demonstrates. Vicki Norberg-Bohm and Lin Gan were also frequent participants in the discussions.

In addition to our regular luncheon discussions, we held two workshops at the Center for International Affairs on this project. At the first one, on 22 December 1993, we benefited from the participation of Michael Ross from Princeton University, who became one of our authors, and from Jeffrey Leonard, Konrad von Moltke, Jane Pratt, Theodore Panayotou, and Eugene Skolnikoff. The perspectives of our guests, who had worked extensively on policy issues, were particularly valuable. When draft papers had been prepared, we held a larger workshop on 6 and 7 October 1994. At that workshop we profited from the comments of Richard Ackermann, Abram Chayes, Raymond Clemencon, Edmundo de Alba, Judy Harper, Xue Mouuhong, Kenneth Oye, Rob Paarlberg, Ted Parson, Patti Petesch, Thomas Schelling, and Helen Sjoberg.

Several members of the project participated in the first open meeting of the Human Dimensions of Global Environmental Change Community at Duke University, 1–3 June 1995, where we conducted a panel at which Barbara Connolly, Beth DeSombre, Tamar Gutner, and Michael Ross presented papers. Robert Keohane presented a draft of the introduction at the Overseas Development Council in November 1994. We learned from the reactions of participants on both occasions, and at other gatherings at which portions of this book have been discussed. The incisive, informed, and detailed written comments of Ronald Mitchell were especially valuable on two occasions, in December 1993 and again in June 1995.

Throughout this project our work has been aided by the excellent staff of the Center for International Affairs, led by its director, Robert Putnam, and its executive officer, Anne Emerson. At the Center, Marissa Murtagh played a key role by facilitating the reproduction and distribution of various drafts, and by efficiently arranging for our workshops. Madeline Sunley, acquisitions editor at The MIT Press, was both persuasive and supportive; and we gained insight from the three anonymous reviews for the press as well as two others written for another university press.

Above all, however, we are grateful to the Rockefeller Brothers Fund and to its president, Colin Campbell, for having faith in us even when we could not explain very coherently what we were doing!

As editors, we shared responsibility. Marc Levy played a critical role in early stages of the project, first proposing that we study financial transfer mechanisms, recruiting authors, managing logistics, helping develop our initial analytical framework and research design, and commenting on chapter proposals and early drafts. Robert Keohane was active throughout in conceptualization and research design, commented several times on each chapter, and guided the project to completion during the last phase of our work. Both of us are profoundly grateful to all of our collaborators for their contribution to our joint deliberations, their ingenuity and care in scholarship, and for their cooperativeness and good humor through a long but rewarding process. We especially want to thank William C. Clark, whose intellectual insights were accompanied by a wit and warmth of spirit that make him a pleasure to know.

I

Analytical Framework and Context

1

Analyzing the Effectiveness of International Environmental Institutions

Robert O. Keohane

Imagine two maps of the world. One displays the relative severity of environmental problems—air pollution, soil and water degradation, desertification, destruction of habitat—and therefore of biodiversity. The other map shows the capabilities that governments have to cope with these problems: the material resources at their disposal, the level of education of their people, the competence and honesty of their governments. Juxtaposing these maps would graphically reveal that environmental problems are most serious in those parts of the world with the least capacity to deal autonomously with them. Almost all residents of rich countries in Europe, North America, and the Pacific rim have access to safe drinking water; a large proportion of people living in Africa, Asia, and Latin America do not. Except for some areas of the formerly socialist countries of Eastern Europe, few citizens of the rich countries find their health seriously endangered by air pollution; but such a situation is common in the big cities of the Third World. Temperate forests, although under pressure in a variety of countries, are not being destroyed at the same rate as tropical forests—not so much because of their inherent biological features but because of the greater wealth and more responsive political systems of the countries in which they are located.

It is futile to demand of poor countries that they give sufficient priority to environmental degradation to resolve these problems. Even those Third World governments genuinely dedicated to the welfare of their people and concerned about environmental degradation—and many are neither—face more severe trade-offs than governments of rich countries. In the long run, economic growth is likely to alleviate certain environmental problems, at the same time as it creates others. Certain measures of environmental health, such as the efficacy of sanitation facilities and air quality, tend to improve as per capita income rises beyond a certain point (*World Development Report 1992*: 39–41). Much could be done simultaneously to promote environmental protection and real economic growth and to avoid the illusions resulting when "growth" is based on deforestation, desertification, or soil degradation, which is not adequately computed in national accounts (Repetto 1989). But when resources are highly constrained, choices involve severe opportunity costs: resources devoted to environmental protection cannot be used to grow more food or provide more shelter for poor families. Hence the discrepancy between levels of environmental quality of rich and poor countries has structural roots: it will continue as long as large per capita gaps in income persist.

Yet industrial growth—conventionally measured economic development—hardly guarantees environmental protection, as the experience of the Soviet Union and its former satellite states, and recent efforts in the United States Congress to eviscerate laws protecting the environment, show. The degraded condition of the natural environment in much of Eastern Europe at the end of communist rule in 1989 demonstrates that other factors, particularly what we label in this volume as "concern," play an important role. Although the most prominent and enduring environmental asymmetries reflect differences in levels of development between rich and poor countries, differential levels of concern between and within middle-income or rich countries, as reflected in public awareness and mobilization on environmental issues, are also important. So are fluctuations in concern over time.

Often damage to the natural environment has minimal effects outside the country in which it occurs. When discrepancies in income or concern lead only to local damage, it would be naive to expect a massive international effort. Some public and private aid may be given to alleviate disaster, but countries will have to rely on themselves for sustained and effective remedial efforts. Many environmental problems, however, have international effects. Ozone-depleting chemicals, produced anywhere in the world, deplete the life-preserving stratospheric ozone layer for everyone. Climate change, to which carbon dioxide from energy production and destruction of tropical forests both contribute, can have global effects, even though these remain largely unpredictable. When a nuclear plant at Chernobyl melts down, people in other European countries suffer effects of the resulting radiation. Future generations may suffer, although in ways never known, by losses of biodiversity occurring in tropical forests now.

People in rich pluralistic societies worry about the effects of environmental degradation elsewhere, both due to concern about planet earth and for more self-interested reasons. Eventually, they recognize that preaching to poor countries will not solve the problem, and some of them organize or agitate for effective action. If effective action is to be taken, the rich will have to provide resources to the poor, so that the problems can be addressed where they matter, with capabilities that only the wealthy can provide. Hence, proposals are made for environmental aid, and when political pressure to act is sufficiently intense, these proposals are taken up by governments.

For aid to be provided, it is necessary to establish what we call *financial transfer institutions for the environment:* sets of rules, typically linked to one or more international organizations, established to govern a flow of funds from richer to poorer countries to achieve specific environmental purposes. This book is about how these institutions operate in the contemporary world. Under what conditions are they effective or ineffective in achieving their manifest purpose: strengthening environmental policies in recipient countries and ultimately improving the quality of the natural environment? The authors of this collective volume seek to understand obstacles to effectiveness, drawing on literatures from economics, international relations, and development assistance, as well as the growing literature on international environmental relations.

Effectiveness depends partly on the magnitude of resources provided. But resources do not constitute the whole story: equally important are the institutional arrangements governing their transfer. Indeed, many of the same dilemmas that have bedeviled development assistance can be expected to arise with respect to international financing of measures for environmental protection. In the broadest sense, therefore, this project seeks to discover to what extent and in what ways the lessons of traditional development assistance apply in the environmental realm. How are environmental issues similar, and how are they different? What international institutional arrangements tend to facilitate or hinder effective resource transfers for environmental purposes?

In seeking to answer these questions, we have tried to gather evidence on the most important international financial transfer institutions for the environment. The domains of four of our case studies are delineated by specific financial transfer institutions, involving rules and organizations: the chapters on the Global Environment Facility (GEF), the Montreal Protocol Fund, debt-for-nature swaps, and efforts to solve problems of chloride pollution in the Rhine River fall into that category. The other three case studies focus on specific issues—protection of tropical forests, East European nuclear power, and East European environmental protection in general—and explore the role played by financial transfer institutions and their effectiveness or ineffectiveness. We have not selected our cases because of either success or failure; indeed, they represent quite a comprehensive set of financial transfer institutions, with considerable variation among them in processes and outcomes.

The first lesson of the studies reported in this volume is that establishing effective financial transfer institutions is difficult: manifestly ineffective institutions have been more common than effective ones. None of the case studies in this book represents an unqualified success; indeed, the discussions below of East European environmental activity since 1989, the Global Environment Facility, efforts to stop destruction of tropical forests, and arrangements to reduce chloride pollution of the Rhine River show that establishing effective arrangements is remarkably difficult. Not only is financial and political commitment in both donor and recipient countries often lacking, but there may be better ways to deal with environmental problems, such as liability rules, issue-linkage, or supranational management, as Thomas Bernauer emphasizes in his analysis of the Rhine case. There are many paths to environmental failure: unwillingness of publics in rich countries to provide sufficient funding; indifference by governments of countries in which environmental damage is occurring; lack of administrative and political capacity in those countries to implement policies effectively; capture of problem definition by industry, or sheer corruption; struggles between North and South for control of allocation processes; organizational self-interest and resulting "turf struggles"

among international organizations; high bargaining costs among participants, each seeking to make gains at the expense of others. The list is
varied and daunting.

Indeed, the simplest message of this book is cautionary: the formulation and implementation of coherent and well-designed policies governing the transfer of funds from rich to poor countries, to protect the
natural environment, are subject to severe political constraints. Ineffectiveness of various sorts is endemic.

Yet there is variation in effectiveness. The Montreal Protocol Fund,
providing financial transfers to reduce production and use of ozone-
depleting chemicals, is administered by a coherent set of organizational
arrangements that bode well for long-term effectiveness, although results
have been slow in coming. Part of the experience of debt-for-nature swaps
is a story of creativity and remarkable impact for a very small investment
of resources. Although attempts to protect tropical forests through international action have been remarkably sporadic and weak in most countries, the Philippines during the 1980s provides an exception: a "window
of opportunity" opened for effective action, and such action made a big
difference. The Global Environmental Facility has been largely hamstrung
by political conflict, but after its restructuring it has a second chance to
prove itself. Even in Eastern Europe, where the first five years after the
collapse of the Soviet empire have shattered many environmentalists'
dreams, there are some areas of progress.

Variation is to the social scientist what money is to a banker or a hoop
and ball to a basketball player: the necessary condition for our activity.
Given that we observe variation within and across our cases, we can ask
questions about its causes and conditions. What accounts for the occasional successes, as compared to the frequent failures? Under what conditions do we observe effective action? How can policy changes affect these
conditions and these outcomes?

From a normative or policy standpoint, variation is also welcome. For-
tunately, policies involving financial transfers sometimes seem to promote
environmental protection. Even if failure seems more common, occasional triumphs may enable us to understand the conditions under which
success is possible, enabling us to suggest policies, and institutional arrangements, that have higher probabilities of success. Hence, we do not

despair, but hope that our analysis—even of situations that are largely ineffective—may contribute to greater effectiveness in the future.

Section 1 of this introduction describes the fundamental concepts, drawn from past work, that are employed in our political analysis: the concepts of concern, contracting, and capacity. Section 2 discusses the elusive notion of effectiveness, and section 3 presents the self-conscious analytical framework that we devised in a series of group discussions while research for this book was under way. This framework was used in preparing each of the empirical chapters, although in the interests of readability we have tried to "take down the scaffolding" so that the chapters are not filled with the jargon that we invented to structure our investigations in comparable ways. Section 4 reviews the major findings of this volume, and section 5 suggests some of our conclusions, elaborated in more detail in chapter 10, the conclusion to this volume, by Barbara Connolly.

1   Concern, Contracting, and Capacity: The "Three C's"

The design of a coherent analytical framework for the description, explanation, and evaluation of international financial transfer mechanisms for the environment has been a major objective of this study. Our efforts have drawn from the literatures of international relations, development assistance, and policy analysis. These literatures suggest three major themes that have shaped our thinking. First, international financial transfers for the environment will not be effective unless both their purposes and their institutional arrangements are consistent with the interests of the most powerful actors involved in the issue—often states and international organizations. Second, financial transfers will not be effective unless both their purposes and their institutional arrangements are consistent with the interests of those whose ultimate behavior they are intended to alter— often low-level state organizations involved in policy implementation, firms, and households. Third, "conditionality" will be a central tension of most efforts to design effective financial transfer mechanisms, given that the specific purposes that motivate funders to provide resources may not be shared by the recipients of those resources. Financial transfers for the environment will not be effective unless there exists sufficient agreement between funders and recipients that mutually beneficial procedures and practices can be followed. Environmental aid institutions will only work if they create incentives for cooperation; otherwise, they will merely become sites for repetitive and costly political struggle.

Our analytical framework is built around the conditions for effectiveness dubbed the "three C's" in earlier work at the Center for International Affairs: sufficient *concern*, solutions to *contracting problems*, and adequate *capacity* for designing and implementing specific measures. We do not conceptualize these conditions as explanatory variables, because they are less causal factors than reflections of social forces, economic resources, and established institutional arrangements. Rather, we view the three C's as *causal pathways*, identifying clusters of factors that must change for international environmental institutions to become effective: "any effective action of international institutions with respect to the global environment is likely to follow a path that increases concern or capacity, or improves the contractual environment" (Haas, Keohane, and Levy 1993: 21). Showing how concern, contracting, and capacity have changed, or failed to change, does not constitute a true explanation of outcomes, unless the conditions under which these pathways are activated have been specified. But the three C's do provide a convenient set of rubrics to facilitate the insightful description of complex events.

Concern

Concern refers to the interests in preserving the environment expressed by potential funders, recipients, and governments involved in a financial transfer. At least one party has to take the initiative, putting issues on the agenda, or nothing will happen. For serious action to be taken, the interests of powerful governments must to some extent be complementary. However, if the home state (in which the environmental degradation is taking place) has very high levels of concern about the issue, it will act on its own. Hence for financial transfers to occur, there has to be asymmetrical interest in action—higher in potential funders than in the home state—and the level of concern of the home state must be sufficient to make action feasible, but not so high as to induce it to go ahead on its own.

Whether financial aid for the environment, and the institutions within which it is provided, can affect levels of concern depends both on internal political conditions and on the tactics employed by funders. Given that aid can induce governments temporarily to act in environmentally benign ways, it may be difficult to determine whether more fundamental change is taking place. The Global Environment Facility and the Montreal Protocol Fund, for instance, support the incremental costs of specific actions to improve the quality of the global environment; but it is not yet clear whether they have systematically altered attitudes or governing coalitions in recipient countries in ways that generate local concern. When governments lack information on the costs and benefits of environmental protection, or when the governing coalition contains both advocates and opponents of effective action, timely international action—as documented in Michael Ross's discussions of forestry programs in the Philippines—may play a catalytic role in changing governmental organization, priorities, and personnel. As in the case of debt-for-nature swaps, financial transfers may strengthen nongovernmental organizations that are committed to environmental protection, thus creating institutions that will act as lobbyists for these causes. Nongovernmental actors, intergovernmental conferences, and the media are almost certainly as important as states in fostering increased concern among both potential funders and recipients (Levy 1993b).

The asymmetrical levels of concern inherent in international aid for the environment mean that relations between funders and recipients are not harmonious: their priorities differ. The resulting political tensions are endemic to financial transfer mechanisms and cannot be avoided by clever management. Programs that are well designed to increase the resources available to sympathetic organizations and individuals in recipient countries may indeed increase support for environmental protection, and therefore ameliorate these tensions; but the tensions will not go away.

Contracting

Any arrangements for international environmental aid entail solving difficult negotiating problems involving both distributional and informational issues. Conflicts over distribution are inherent in any financial transfer mechanism, because participants will seek insofar as possible to maximize the benefits they receive and shift costs by others. Information is also crucial: unless participants have at least some confidence in the credibility of their partners' commitments, agreements will not be reached. Yet even when agreements are reached, participants have incentives to hide or manipulate information, and more generally, to behave strategically. Hence, at the core of institution building with respect to financial transfers are contracting problems: how to draw up contracts (in the absence of a world judicial and police system that can enforce them) that are robust to attempts at manipulation and in which the parties can have confidence.

Many potential contracting problems exist. In some cases, recipients may face the danger that after making policy changes that are politically or economically costly, funders will renege on their financial commitments. Developing countries expressed these concerns in negotiations about the Global Environment Facility, discussed in chapter 3. More typically, however, the funders are more worried than recipients about contracting issues. Recipients may renege on their commitments: to fail to protect forests they have undertaken to protect or to divert funds to projects favored by influential local elites. In chapter 2, Fairman and Ross discuss the endemic problem of failure of recipients to fulfill commitments under policy-based lending, or strict conditionality, practices. Fears of reneging are evident in a number of our case studies, notably those of the Montreal Protocol Fund (chapter 4), debt-for-nature swaps (chapter 5), and nuclear safety in Eastern Europe (chapter 8). In all three cases, funders made large contributions in return for promises of future performance by recipients—promises that are difficult to enforce. Having received funds, recipients have incentives to revert to actions oriented toward their own priorities, rather than those of the funders: to continue producing ozone-depleting chemicals after an extended deadline has passed; to let nature preserves be invaded by loggers and poachers; to extend the lives of retrofitted nuclear plants rather than closing them down.

Providers of funds hope that recipients' concerns for their reputations will provide strong incentives for compliance with their commitments; but few experienced practitioners are willing to rely on vague concerns about reputation alone. Hence, institutions for environmental aid are always designed with contracting problems more or less explicitly in mind; and the efficacy of action depends to some extent on the quality of the institutional design. The financial transfer institutions discussed in this book contain features designed to control opportunism, including reporting and auditing requirements; disbursements of funds in tranches, with each successive disbursement subject to a decision by the funding organization; and the embedding of particular transfers within ongoing organizations such as the World Bank. Each of these measures can be interpreted as an institutional attempt to create a network of incentives to comply with the conditions originally agreed on by the parties.

Capacity

With the partial exception of the European Union (EU), international organizations are not authorized to enforce rules within the jurisdiction of sovereign states, nor do they usually have the ability to carry out large-scale projects on their own. These organizations must rely on the bureaucracies of national governments or on domestic nongovernmental organizations (as in the debt-for-nature swaps discussed by Cord Jakobeit). Hence, the capacity of recipient governments to implement policies designed to protect the natural environment and assure sustainability is crucial. Often, as the case studies show, the lack of such capacity is a critical source of policy ineffectiveness. Also important is the capacity of the donor institutions. In chapter 9, Barbara Connolly, Tamar Gutner, and Hildegard Bedarff show that in providing environmental aid to Eastern Europe, the World Bank, the European Bank for Reconstruction and Development (EBRD), the European Union, and national bureaucracies administering bilateral aid all had distinctive capabilities and operating styles. They sought to find problems that fit the solutions they had devised, which led to a process that was far from the ideal of rational bureaucratic problem solving. Likewise, a key problem of the Global Environment Facility has been its lack of strong internal lines of authority enabling it to develop its own capacity for decisive action.

Our use of "capacity," however, is broader than simply administrative capacity. We are also interested in the ability of nongovernmental organizations and domestic political institutions to translate concern about environmental effects into policy. For instance, in the chapter on debt-for-nature swaps, Cord Jakobeit shows that Costa Rica benefited from a disproportionate number of these financial transfers because it had a dense and relatively effective network of nongovernmental organizations that had the capacity not only to implement policy themselves but to put pressure on the government.

Capacity is often the condition that is most lacking with respect to international financial transfers to deal with environmental problems. As we have noted, these arrangements are typically initiated because of concern in rich countries that poorer countries are not adequately protecting the global or regional environment. These perceived policy failures always reflect, in part, lack of human, organizational, and financial resources in the poor countries; although they may also result from a lack of local concern about the environmental problems. So almost by definition, national capacity is deficient on issues with which financial transfer mechanisms are concerned. Developing countries may typically have environmental ministries, but many of them lack the technical competence or political clout to make much difference. Given that arrangements to transfer financial resources in support of environmental goals are relatively new, potential funders may also lack appropriate analytic capability, local knowledge, or long-term ties to recipient counterpart agencies.

"Capacity building"—defined in terms of the institutional capabilities and networks enabling countries to integrate development and environmental policies—is a key concept in Agenda 21, the program of action for the twenty-first century developed in connection with the UN Conference on Environment and Development (UNCED) in 1992 (Koch and Grubb 1993: 148). However, it is often used rhetorically as a mantra, ritually chanted without being analyzed, while on the ground, as David Fairman and Michael Ross argue in chapter 2, "many development practitioners have found the problems of 'institutional weakness' so intractable that they prefer to ignore the issue altogether." Capacity building is not a matter just of administration but of politics: as Fairman and Ross also point out, capacity-building efforts often fail due to a lack of commitment by the recipient country—that is, due to a lack of what we have called concern.

Each of the three C's is an analytical lens through which to view features of the environment (e.g., "structures"). Concern directs attention to preferences; contractual environment, to problems of information and enforcement that may prevent credible commitments, and therefore agreements, from being made; capacity, to the ability of organizations (governmental or nongovernmental) to implement decisions that they have taken. But as just noted, the C's interact. Perhaps most obviously, capacity is affected by concern. The ability of governments of wealthy countries to provide funding for financial transfers is limited by the degree of domestic support for such transfers and by characteristics of domestic political systems that facilitate or obstruct efforts by supporters to appropriate funds. Environmental capacity in recipient countries is also to some extent a reflection of concern: decisions on where to allocate scarce resources depend on domestic priorities. Conversely, however, development of national capacity, broadly defined, affects subsequent concern: for example, nongovernmental organizations are likely to promote environmental awareness in publics as well as to put pressure on governments. Clearly, the contractual environment is affected by capacity, because capacity affects the ability of governments and other organizations to make credible commitments. Hence, although it is convenient to distinguish among concern, contracting, and capacity for analytical reasons, they may be hard to disentangle in practice.

2   Effectiveness

We seek in our analysis to connect concern, contracting, and capacity with effectiveness. "Effectiveness" is a difficult concept that can have many meanings (Bernauer 1995). Ultimately, we are interested in the impact of financial transfer mechanisms on the solution of significant environmental problems. In this broad normative and analytical sense, the proof of effectiveness is to be seen in the improvement of the targeted aspect of the natural environment. The measures taken have to have changed the behavior of actors, and the policies and performance of institutions, in ways that contribute to management of the perceived problems; but the problem also has to have been perceived more or less correctly.

Hence, effectiveness depends on choosing a significant problem, defining its scope in a manageable way, proposing solutions, devising institutional arrangements to implement the solutions, and actually implementing the solutions that have been agreed to. Because it is unlikely that any policy solution will work perfectly the first time it is implemented, or that initially effective solutions will continue to work as conditions change, we are also interested in effectiveness as a dynamic concept. Over the long run, the initial impact of a given arrangement may matter less than whether it is effective in promoting continuous improvement in environmental conditions.

The previous discussion has emphasized that financial transfer mechanisms are not necessarily effective. Indeed, our research has turned up some activities (such as the Rhine arrangements discussed in chapter 7) that had very small positive effects on the environment and may even have diverted attention from more significant issues, and others that seem positively perverse (such as the "slippery slope" toward reinforcing rather than reducing reliance on dangerous nuclear power plants in Eastern Europe, discussed in chapter 8). Such consequences may have been unintended by at least some of the policy innovators involved in the financial transfer mechanism, but may have been intended by others. Agents with various interests behave strategically, often concealing their true objectives. Attaching a "green label" to a segment of international relations hardly guarantees good outcomes.

It may be useful to distinguish more systematically among types or levels of effectiveness, although we did not do so in the analytical framework that underlies each empirical chapter. Here and in the conclusion, we distinguish project effectiveness from aggregate effectiveness. *Project effectiveness* refers to how well, relative to costs, a single financial transfer or set of transfers contributes to solving a particular environmental problem or set of problems, given the way in which problems are defined and preexisting institutional arrangements. For example, did particular environmental projects in Eastern Europe that were supported by the Global Environment Facility, the Montreal Protocol Fund, or the World Bank achieve results commensurate with costs? Sources of *ineffectiveness* can usually be traced to issues of deficient concern or capacity, as Barbara Connolly shows in the conclusion to this volume.

Aggregate effectiveness refers to the extent to which environmental aid programs fulfill their potential ability to solve or alleviate environmental problems. Some sources of aggregate ineffectiveness are institutional, deriving from organizational inertia and failures to coordinate policies, as the conclusion also points out. For instance, the Global Environment

Facility has been plagued both with problems of inertia and lack of coordination. Another component of aggregate ineffectiveness refers to the priorities reflected in environmental aid: how well does the overall conception of the problem that underlies institutional responses and project-specific activity match environmental priorities? As Connolly and her colleagues point out in the two chapters on Eastern Europe, failure to adequately specify environmental problems produces unsatisfactory results even if institutions work and projects are selected with care. These distinctions among types of effectiveness were not incorporated explicitly into our analytical framework, but they are implicit in many of our chapters and are explicitly addressed in the conclusion.

3   A Framework for the Description and Analysis of Our Cases

The case studies in chapters 3–9 are written according to a common analytical format to ensure consistency and comparability across cases. We are interested in description, explanation, and evaluation as three "lenses" through which to view the activities of financial transfer institutions. Consequently, section 1 of each chapter describes what happened at each of the five phases indicated below; section 2 seeks to explain the observed patterns; and section 3 evaluates the success of the strategy that was followed, making an overall judgment about the extent to which the financial transfer mechanism contributed to improvements in environmental quality. Each chapter focuses on the three C's: the three causal pathways through which change can take place, concern, contracting, and capacity. Hence we strive for thematic as well as organizational unity in our analysis.

   In thinking about our cases during our joint deliberations, we began with a conventional distinction among three aspects of problem solving: (1) how the problem is defined; (2) what solutions are envisaged; and (3) mechanisms for implementing such solutions. With respect to solutions and mechanisms, we distinguished between the specification phase—how the solutions and mechanisms are characterized or "specified"—and the implementation phase—how the mechanisms envisaged are implemented, and what the consequences were for the implementation of the solutions. We summarized the resulting five phases of problem solving as follows:

1. collective problem definition
2. solution specification
3. mechanism specification
4. mechanism implementation
5. solution implementation.

Issues involving concern dominate phases 1 and 2, and remain important in phase 3, in which issues of contracting and capacity become significant, and in later phases, where capacity is a critical factor.

This structure of analysis should be visible in all of the chapters, although in later drafts we have sought to eliminate jargon and, as noted above, to dismantle some of the "scaffolding" of our work in the interests of readability. Nevertheless, we have insisted on such a systematic approach for two reasons: (1) to ensure that each chapter systematically considers the sequence of action relevant to the effectiveness of financial transfers, from explanatory and evaluative standpoints as well as descriptively, and (2) to facilitate a process of drawing out generalizations, across cases, about the conditions for success and failure of financial transfer mechanisms. The conclusion offers a systematic discussion of the generalizations that seem to be supported by the evidence in this volume.

4   The Chapters and Major Findings

As noted briefly above, conditionality—setting conditions on financial transfers and seeking to enforce those conditions on recipients—is an integral component of the financial transfer process, given that the funders' purpose is to promote environment-protecting policies in the recipient country that would not otherwise be pursued by its government. Accordingly, chapter 2 reviews the lessons for international environmental practice of experiences with conditionality in the arena of economic development. David Fairman and Michael Ross point out that attempts to rely principally on strict conditionality have generally failed. Recipient governments have stronger incentives to renege on their agreements than funding agencies have consistently to enforce them. Because the potential aid for environmental issues is so much lower than that for economic development, this experience does not augur well for environmental conditionality.

Fairman and Ross argue that a long-term commitment to policy dialogue, in which persuasion plays a larger role than coercion, is more promising than strict conditionality, but only in conjunction with a commitment by funders to develop strong local knowledge and long-term ties to recipients. Effective coordination among donor agencies, and efforts to improve recipients' administrative and political capacities to use aid, are also important. Unfortunately, such commitment and coordination are rarely observed for sustained periods of time, let alone over the five- to fifteen-year period required to institutionalize major policy reforms. For conditionality to have the desired results, they argue, it must be used in politically sophisticated ways that draw on deep local knowledge: funders should focus on areas where there is domestic support for their objectives and build on the efforts of committed local politicians, officials, and members of the public. In general, the level of recipient commitment and their capacity to run aid programs have more impact on the success of environmental aid than tactics involving carrots and sticks.

Each of the case studies discussed in chapters 3–9 is organized according to the framework outlined in section 3 above. Each study contains descriptive, analytical, and evaluative arguments. Part II (chapters 3–6) discusses issues that primarily involve transfers from rich developed states ("the North") to poorer developing countries ("the South"). The institutional arrangements devised to facilitate North-South transfers include the Global Environment Facility, discussed in chapter 3, and the Montreal Protocol Fund, analyzed in chapter 4. North-South exchanges were also the focus of debt-for-nature swaps (chapter 5) and attempts to use conditionality to protect tropical forests (chapter 6). Part III (chapters 7–9) focuses on regional efforts in Europe. Chapter 7 discusses the one intra-Western environmental financial transfer mechanism that we have been able to identify: an arrangement to cope with chloride pollution in the Rhine River. Environmental aid activities in Eastern Europe are discussed in chapters 8 and 9: nuclear cleanup (chapter 8) and the array of other environmental efforts in Eastern Europe since 1989 (chapter 9).

The Global Environment Facility, discussed in chapter 3, was established in 1990 as a pilot program with $1 billion for three years, and renewed for three additional years at triple the budget, after protracted and acrimonious negotiations. Evaluations of the GEF indicate that during its first three years it had little impact on developing countries' environmental policies: its project portfolio was unimpressive; and it did not effectively promote concern or capacity building. As David Fairman shows, the GEF was thoroughly politicized, becoming part of a North-South struggle over control of financial transfers for environmental purposes. Governments of poor countries feared that conditionality would undermine their autonomy, while governments of rich countries fretted that without tight controls funds would be wasted. Both developing and developed countries worried about setting adverse precedents that would affect major future allocations of resources. Conflicts among the implementing agencies—especially between the World Bank on the one hand and the UN Development Program and the UN Environmental Program (UNEP) on the other—worsened the situation.

These conflicts of interest and perspective made it impossible in negotiations for the GEF to devise clear institutional arrangements with coherent centers of authority; instead, the compromises reached were ambiguous, which in turn undermined both the GEF's political legitimacy and its operational effectiveness. That is, rather than helping to resolve underlying conflicts, the GEF's governance arrangements, during its first three years, were obstacles to conflict resolution. However, just as "the prospect of a hanging concentrates the mind," the prospect of the GEF's collapse led both developed and developing countries, at the United Nations Conference on Environment and Development (UNCED) in 1992, to compromise on key issues of governance. Subsequently, new institutional arrangements were constructed that may well increase the authority of the GEF secretariat over the operating agencies, increase the organization's legitimacy, and therefore improve its performance. Whether the "new GEF" will be more effective than the old one, however, remains to be seen.

The Montreal Protocol Fund, established in 1990 as part of the London agreements to strengthen the Montreal Protocol on Substances That Deplete the Ozone Layer, has had a more encouraging, but hardly problem-free, history. Before the developing countries were brought into negotiations, developed countries were already taking steps, individually and collectively, to reduce production of ozone-depleting substances. Hence they were keen to ensure that developing countries would not undermine their efforts by sharply increasing their own production of these substances. The combination of high rich-country demand for agreement and strong bargaining power for developing countries led to productive negotiations to create the Fund. The subsequent evolution of the Fund involved the creation of a decision-making structure in which scientific criteria play a remarkably large role. The secretariat for the Fund has even managed to influence the programs put forward by the implementating agencies (IAs)—three UN agencies and the World Bank. Hence in the case of the Fund a coherent organizational structure was put into operation. On the other hand, planning and implementation of projects has been very slow, and the ten-year grace period for developing countries has not only slowed down reduction of production of ozone depleting substances, but raises contractual questions for the future. In chapter 4, Elizabeth DeSombre and Joanne Kauffman analyze the conditions that enabled the Montreal Protocol Fund to begin its operations much more smoothly than the GEF, which was created in the same year; but they also carefully consider the continuing uncertainties about its ultimate success.

In chapter 5 Cord Jakobeit describes how nongovernmental organizations (NGOs) innovated debt-for-nature swaps in the 1980s, seeking to take advantage of the debt crisis by buying debt at low secondary market prices and donating it to local nongovernmental organizations for the creation of nature reserves, thus protecting tropical forests from destruction. This idea "caught on" because NGOs, governments of recipient countries, and banks all benefited; the absence of preexisting intergovernmental organizations operating in the area meant that there was no central source of resistance. Contractual issues and capacity issues could have posed difficulties to effective implementation, but through experience and organizational learning, many of these problems were overcome. These "first generation" nongovernmental debt-for-nature swaps, however, were limited in scale. The organizations controlling large resources—governments, multilateral development banks, or commercial banks—were unwilling to invest substantial resources in these schemes.

In the second phase of debt-for-nature swaps, relaxation of the debt crisis made these swaps less attractive for NGOs, because less leverage was provided for their contributions. The magnitude of resources devoted to such arrangements increased dramatically, however, when governments such as Germany and the United States became involved. Whether environmental goals were actually fostered, however, depended'on the conditions imposed (or not imposed) on debt forgiveness. In the end, debt-for-nature swaps only made a small impact on the issues of deforestation and loss of biodiversity for which they were initially designed, but they did show that nongovernmental organizations could foster institutional innovation. They also revealed the absence of sufficient capacity in poor countries and demonstrated the limits of actions not supported by major governments and international financial institutions.

NGO interest in debt-for-nature swaps was at a peak while intergovernmental efforts to protect tropical forests were floundering. Such international agreements to protect tropical forests, including the Tropical Forestry Action Plan (TFAP) and the International Tropical Timber Agreement (ITTA), have failed to affect tropical forestry practices, largely due to low concern on the part of the governments of the tropical countries. In chapter 6, Michael Ross discusses efforts by the World Bank and other international financial institutions to remedy this situation by making aid conditional on more sustainable logging policies. On the whole, such attempts have also been ineffective: when the logging industry is part of the government coalition, as in Indonesia, funders cannot persuade governments to implement genuine reform. In the Philippines in the late 1980s, however, conditionality led to a remarkably far-reaching series of reforms, because the World Bank was able to work with activists in the new Aquino government, who sought conditionality in order to increase their leverage vis-à-vis the logging industry. The implication of Ross's study, consistent with the more general findings about conditionality in chapter 2, is that effective international financial transfers to protect tropical forests depend on active forces for reform within the countries concerned. "Buying reform" through conditional aid does not work.

In chapter 7 Thomas Bernauer discusses one of the most sharply focused instances of international financial transfers, arrangements to reduce pollution of the Rhine River by chlorides, first negotiated in the

1970s, and in effect since the beginning of 1987. These agreements included payments both to an upstream state, France, to compensate it for reducing chloride emissions from a state-owned potash mine, and to a downstream state, the Netherlands, for a project to reduce chloride pollution within its territory. Conditions might have seemed conducive to success: the issue was narrowly defined, only four states, all of them rich (France, Germany, the Netherlands, and Switzerland), participated, and results were easily monitored. Yet bargaining was protracted and the effect on the environment was minimal. Moreover, the riparian countries focused too heavily on a problem (chlorides) that was of minor importance compared to other Rhine pollution issues, such as heavy metals and pesticides. The results of these arrangements can be characterized as "wrong problem, wrong solution."

Chapter 8 discusses the reactions of governments, the nuclear industry, the European Community, and nongovernmental organizations to the opportunity, after 1989, to scrap or retrofit dangerous Soviet-built nuclear power plants operating in Eastern Europe. Barbara Connolly and Martin List describe how, despite great apparent concern in both the East and West about nuclear safety, steps actually taken to improve the safety of East European power plants have been remarkably modest: reliance on still-dangerous nuclear power plants in the area is not being phased out, and in some respects is even being reinforced. The nuclear industry has defined the problem as one that is susceptible to technical fixes rather than as a more fundamental problem of inefficient and wasteful patterns of energy use. This problem definition suggested a programmatic emphasis on profitable supply-side actions rather than on efforts to reduce energy demand.

Having defined the problem as a technical one of ensuring that nuclear power in Eastern Europe is safe, the West has found it difficult to achieve this objective. East European governments have consistently put power generation priorities over environmental ones. Competition between suppliers, higher priority for energy and economic development than for nuclear safety, and inherent contractual problems have made it impossible to impose effective conditionality. Consistent with other chapters in this volume, chapter 8 points to the difficulties in imposing conditionality in the absence of sympathetic partners with whom coalitions can be formed.

Some efforts to reward East European countries whose governments are concerned about safety have been tried; and programs to train local personnel in safety procedures, thus increasing capacity, seem to have been valuable; but these successes are overshadowed by the overall pattern of failure. Indeed, Connolly and List argue that Western efforts may not only be ineffective but may lead to a "slippery slope": reducing safety risks in the short term may lead East European governments to prolong the operational lifetimes of dangerous reactors, especially when opportunities arise to sell electric power to the West. In general, Connolly and List argue, the history of nuclear safety politics in Eastern Europe since 1989 is a history of missed opportunities.

In chapter 9 Barbara Connolly, Tamar Gutner, and Hildegard Bedarff analyze environmental aid efforts in Eastern Europe since 1989, under the auspices of the World Bank, the European Bank for Reconstruction and Development, the European Union (EU), and bilateral aid programs. Since 1989, economic issues have gradually eclipsed the early concern about environmental problems in Eastern Europe. The resources available from the rich democracies have been less than anticipated, and East European governments have assigned these issues a low priority. Furthermore, the funds that have been used have not been allocated in anything like an efficient manner, partly because the behavior of each of the four sets of financial transfer institutions has been driven more by institutional interest and practice than by functional need.

Actual practice in Eastern Europe resembles the famous "garbage can model" of organizational choice (Cohen, March, and Olsen 1972). Some of the most dangerous environmental situations have been neglected. Funders determine priorities, and the solutions they have devised—their standard operating procedures—determine which problems are identified as important. For example, the World Bank emphasizes loans for large power plants in preference to support for local air pollution projects; the European Union emphasizes institutional strengthening of government ministries; bilateral programs select problems that provide them with export opportunities, or result in environmental benefits for the donor. Behind these patterns lie the shadows of economic interests in donor countries and the organizational interests of the programs that they establish. One result is frequent reliance on consultants from donor countries rather than local experts, whose employment would enhance national capacity. Another is policy fragmentation, because the funding organizations have different priorities and compete with one another. Against this background of fragmentation, however, there are signs of some progress. An "Environment for Europe" process, which has been established to define environmental priorities and facilitate coordination, shows signs of improving the effectiveness of environmental assistance to Eastern Europe. Connolly, Gutner, and Bedarff emphasize the difficulty, but also the necessity, of combating the organizational inertia that has bedeviled East European environmental efforts.

5   Conclusion

Social science is often characterized as the study of choices subject to constraints. In world politics, the most basic constraints reflect fundamental social forces such as nationalism, sovereignty, and the distribution of economic and military resources; they also reflect the capabilities of international institutions, prevailing levels of scientific knowledge, and the attitudes of publics. When constraints on action are strong, we may observe nothing: no action is taken, because none is possible. But insofar as constraints are relaxed, there may be opportunities for choice—by individuals in civil society, nongovernmental organizations, bureaucrats, or political leaders. Initiatives may be taken to achieve objectives that formerly seemed out of reach. If constraints remain significant, however, these initiatives are likely often to be deflected, distorted, or stymied: even if policy entrepreneurs understand the constraints they face, risks must be run; and unexpected shocks may worsen the odds.

This book studies risky initiatives that have been taken under severe constraints. The initiatives themselves reveal the constraints: we perceive inconsistencies between objectives and results; observe the political processes that took place; and make inferences about the social forces that accounted for the gaps. Given that ineffectiveness is common, the constraints are fairly clear, and most of our systematic findings have to do with these constraints. Our distinction among concern, contracting, and capacity is designed to help us understand these constraints by associating them with different causal pathways that must operate for effective action to be taken.

At the most basic level, *self-interest* is the key constraint on concern: on the willingness of rich countries to fund financial transfers; of recipient governments to take effective action against privileged families or groups that benefit from environmental exploitation; and of international organizations to cooperate with one another. Because international financial transfers involve the provision of funds by one set of countries in order to alter the operational priorities of others, conflicts of interest are endemic. Characteristic patterns of North-South politics afflict the Montreal Protocol Fund, the GEF, forestry issues, and government attempts to use debt-for-nature swaps. Conflicts of interests do not just come from governments: as the studies of tropical forest protection efforts and East European environmental aid show, industries with economic stakes in environmental issues are likely to be influential players. So are international organizations, which have their own interests at stake, as the struggles over control of the GEF and lack of coordination in Eastern Europe reveal.

Financial transfer institutions can only work well when the interests of powerful actors intersect. But even if such complementarity exists, lack of organizational or national capacity can thwart effective action. Shortages of national capacity pervade the process. As we noted earlier, these shortages are usually conditions for the development of financial transfer mechanisms, so their prevalence should hardly be surprising. The resulting shortcomings of international institutions, and of the organizations that operate within them, should not necessarily, however, be seen as signs of ineffectiveness, because we must ask, "ineffective compared to what?" The criticisms of institutional performance in each chapter must be tempered by recognition, in each case, of some elements of accomplishment and often prospects of reform. In any case, we need to remember that even if effectiveness has often been low, doing nothing could have been worse.

It is often difficult for funders to distinguish problems that arise from low recipient government concern for the environment from problems that arise from an absence of capacity. On the issues of debt-for-nature swaps, forestry protection, and East European environmental aid ineffectiveness within recipient countries reflects a difficult-to-disentangle mixture of low priority for the environment and inability to implement desired actions. Contractual problems are often derivative from these issues of concern and capacity. If the fundamental issue is a lack of concern, aid to enhance capacity may have little effect and conditionality may be essential—although, as Fairman and Ross indicate in chapter 2, under these circumstances such conditionality is unlikely to be very effective. By contrast, insofar as a lack of capacity is the key constraint and underlying concern is relatively high, providing aid to generate capacity could lead to greater efforts by the country involved and obviate the need for strict conditionality.

The generalizations of this book concern constraints—on concern, contracting, and capacity. But some of the most interesting stories have to do with leadership and the exploitation of "windows of opportunity." Social scientists find it more difficult to discuss leadership than to focus on constraints, because leadership depends on idiosyncratic and conjunctural factors that defy generalization. But for better or worse, we cannot omit human initiative and personality from our stories. Who could understand the New Deal without knowing about Franklin D. Roosevelt, or the turn to the right in American politics during the 1980s without Ronald Reagan? On a much smaller scale, the same is true for several of our case studies: where effective action took place despite severe constraints, it reflected, in part, the initiatives of particular individuals. Thomas Lovejoy first proposed debt-for-nature swaps and pursued the issue tenaciously for several years. Mostafa Tolba, executive director of the United Nations Environmental Program, seized the opportunities provided by concern over stratospheric ozone depletion to establish the Montreal Protocol and an associated fund for financial transfers. Human rights lawyer Fulgencio Factoran played a key role in reforming Filipino logging policies.

Social scientists have difficulty explaining why certain individuals run personal risks to improve the quality of the natural environment, while others, in similar positions, are more passive. But we can study whether individual action has catalytic effects. Lovejoy's debt-for-nature initiative was consistent with the interests of a variety of disparate actors, including governments in debtor and creditor countries, banks, and environmentalists. Hence it sparked real action. When the debt crisis passed, however, debt-for-nature swaps also lost their appeal; and attempts by governments to use the same concept met with mixed results. Factoran's activism benefited from his control over financial flows from the Asian Development Bank (ADB) and from conditions attached to World Bank and United States Agency for International Development (USAID) loans; but whether enlightened forestry policies continue depends on whether institutionalized capacity is built. Tolba's initiatives were effective largely because they became institutionalized in the Montreal Protocol and its Fund, which continue to exercise an impact on policy even during periods of reduced environmental concern.

These successes temper the bleakness of our generalizations about constraints. International politics is resistant to reform; and much activity within international environmental institutions is doomed to frustration. Yet opportunities arise: when scientists make environmentally frightening discoveries, when new governments come into office, or when interests temporarily converge as a result of an unexpected crisis. Reformers need not merely understand the constraints on action, but also be ready to seize opportunities for change. Students of international institutions, like ourselves, should not be satisfied to show how political constraints make effective action difficult, but should also point where and how it is possible for dedicated people and well-structured organizations to make a difference.

Note

The author is grateful to William C. Clark for work on an earlier draft, and to members of the environmental transfers working group for comments, especially to Thomas Bernauer, Barbara Connolly, David Fairman, Wendy Franz, Tamar Gutner, Cord Jakobeit, and Michael Ross for written comments on earlier drafts. I am also grateful to Ronald Mitchell and the anonymous reviewers for their comments and suggestions.

2

Old Fads, New Lessons: Learning from Economic Development Assistance

David Fairman and Michael Ross

Aid funders are routinely criticized for following ephemeral "development fads"—pouring money into programs for "the poorest of the poor," "women in development," or "sustainable development," then dropping these programs when they go out of style. Often the criticisms are unfair: each of these "fads" has had a useful and enduring impact on the way funders approach development assistance. But all new themes in development assistance share at least one fault: the belief that their new approach will effortlessly circumvent the political and institutional problems that plagued earlier approaches.

The recent enthusiasm for sustainable development has affected development assistance in two ways. It has spawned a new set of institutions, projects, and programs devoted to environmental protection, many of which are analyzed in this volume. And it has made conventional aid programs more environmentally sensitive, often by promoting the use of environmental impact statements and curtailing funds for environmentally harmful projects. These changes have reshaped the landscape of development assistance, resurrecting old problems and creating new ones.

Many "green aid" programs face common predicaments. They are often characterized by conflicts between funders who want to ameliorate regional or global environmental problems and recipients who are more concerned with local issues—a tension adeptly described in this volume's studies on efforts to preserve tropical rain forests, promote nuclear safety in Eastern Europe, and protect the global commons through the Montreal Protocol Multilateral Fund and the Global Environment Facility. On each of these issues, funders have turned to conditionality and policy dialogues to influence the policies of aid recipients. As these studies show—and as we argue in this chapter—conditionality and policy dialogues are often difficult for funders to use and may only be effective in exceptional settings.

Because aid recipients often have little experience in environmental management, funders have also concentrated on capacity-building programs and institutional reform. The chapters here on debt-for-nature swaps, logging in tropical forests, nuclear safety, and environmental aid for Eastern Europe all show funders grappling with—and sometimes surmounting—the perplexing obstacles to effective capacity building, which remains the Achilles' heel of many assistance programs.

Finally, the profusion of green aid projects has heightened the problems of aid coordination, as new funds and funders have entered an already poorly organized field. The dilemmas of coordination are especially evident in this volume's two studies on aid to Eastern Europe, where funders outnumber recipients and poor coordination has sapped the effectiveness of many programs.

The purpose of this chapter is to draw some cautionary lessons about international aid programs from the fifty-year history of development assistance—lessons that are salient for environmental aid. We believe these lessons help explain many of the institutional failures described in later chapters. We also believe they point toward solutions. After reviewing the past experience of funders with conditionality, policy dialogues, capacity building, and coordination, we show how the studies in this volume illustrate, emphasize, and sometimes amend these earlier lessons. We conclude with some broad recommendations about the design of environmental aid programs.

1 Conditionality

At first glance, conditionality is an attractive tool for promoting environmental reform, because it appears to require little effort on the part of funders and can target the policies of recipient states that perpetuate environmental degradation. It also appears suited to a world where new money for international assistance is scarce: it is typically used with loans, not grants; it can be "piggybacked" on to some existing aid programs, giving funders the promise of more political leverage for the same amount of money; it appears to overcome the complicated shortcomings of project-oriented environmental assistance programs; and it offers aid agencies in wealthy states a politically defensible rationale for their programs. For all these reasons green conditionality has been on the rise since the late 1980s.

Conditionality can be defined as the exchange of financial assistance for policy reforms.[1] There is a large literature on the most common use of conditionality, the structural adjustment and sectoral adjustment loans offered by the World Bank and the International Monetary Fund (IMF) to developing countries in exchange for a variety of economic reforms. There is a much smaller literature on—and much less real-world experience with—the use of conditionality for the pursuit of noneconomic goals. Both literatures, however, suggest that conditionality is a frustrating and elusive policy tool that fails more often than it succeeds.

Conditionality can take two forms. The more common type, which has been used in the case of structural adjustment, is "policy-based lending." In this approach, funders supply loans or grants to recipients who pledge to implement an accepted set of policy changes. The second type is "allocative conditionality," in which funders distribute varying amounts of aid to countries based on their adherence to specified standards, but (in theory) do not attempt to bargain with them to bring about reforms. This section reviews the findings of recent scholarship on the effectiveness of each type of conditionality and attempts to apply them to environmental conditionality.

Policy-Based Lending

The effectiveness of policy-based lending has been closely studied and yields six broad lessons on the limits and uses of conditionality. First, international funders have only a limited ability to bargain with recipients over policy reforms; and once these bargains have been struck, funders have only a limited ability to enforce them. The World Bank, for example, has virtually no flexibility in the size of the loans it can offer to a recipient: the amount is set far in advance and is known to both parties. Hence the Bank is unable to "buy" more reforms from potential recipients by offering them more money (Mosley, Harrigan, and Toye 1991: 115). More important, most funders are under great institutional pressure to disburse their funds, regardless of the compliance of recipients with the prearranged conditions. The larger the country and the more important its status as a borrower, the less able lenders are to refrain from offering them loans while still spending their budgets and remaining solvent.

In addition, the World Bank has periodically been coerced into making loans to states that are poor credit risks but strategic allies of the Bank's major benefactor, the United States. For these reasons, the most politically important borrowers are generally given the "softest" conditions, and lenders virtually never cancel a loan once disbursement has begun, no matter how poorly the recipient is complying with the conditions (Mosley, Harrigan, and Toye 1991: 48, 71–72, 165).

Moreover, if a borrower has fulfilled at least 50 percent of the conditions in its previous loan (and if it is an important borrower or has a close relationship with the Bank, fewer than 50 percent), the Bank is almost certain to offer it additional loans once the initial loan has been disbursed. There is now so much evidence of the Bank's laxity that borrowers no longer feel obliged to fulfill more than half the conditions they initially agree to. The Bank's eagerness to disburse its funds has one additional consequence: it gives Bank officials an incentive to overestimate what a recipient can accomplish with a conditional loan. Given that their counterparts in national ministries of finance share this incentive, they tend to collaborate in setting unrealistic targets for recipient countries, which they cannot meet under the best of circumstances (Mosley, Harrigan, and Toye 1991: 109, 170–171).

Second, because international lenders have a limited capacity to monitor their agreements, there is a trade-off between the number and complexity of the conditions in the agreement and the ability of lenders to enforce them. Agreements with many intricate conditions are difficult to enforce for three reasons: they overburden the recipient government (particularly when funders are unable to prioritize their conditions); they overburden the funder, which has a finite ability to monitor compliance; and because many of the conditions have interactive effects, they make it difficult to distinguish between governments that are committed to reform and those that are covertly violating the agreement (Mosley, Harrigan, and Toye 1991: 44–45).

The third lesson is that compliance with a conditionality agreement has little to do with the recipient's type of government and much to do with the characteristics of the issues at hand. The compliance of all types of governments is likely to be greater on issues with the following attributes:

1. they are not politically sensitive (Kahler 1992: 97);
2. they can be put in place by a small number of officials, and do not require the cooperation of large numbers of bureaucrats or agencies;
3. they can be easily monitored;
4. they can be accomplished quickly and in a single step (Nelson 1989);
5. they do not require extensive institutional change to implement;
6. they "break bottlenecks," opening the way for further actions by the government or private actors;
7. there is a strong technical consensus on their utility (Nelson and Eglinton 1993: 69).

Fourth—and perhaps most surprisingly—there is little evidence that policy-based loans actually induce recipients to embark on difficult reforms. Borrower governments have usually been either firmly opposed or strongly committed to the type of reforms favored by the funder, quite apart from the influence of conditionality. Even when there are internal battles between pro- and antireform factions, the outcome is usually clear and has little to do with any enticements offered by international lenders. Governments that wish to obtain the loans and then renege on the conditions have been able to do so, sometimes by taking advantage of the limited enforcement capacity of the funder, and sometimes by engaging in "false compliance" by fulfilling the letter of their obligations but nullifying their impact with countervailing measures. The ease of cheating, plus the common eagerness of funders to disburse loans, gives those governments lacking a strong commitment to reform few incentives to take painful measures.

Fifth and related, conditionality ironically can play an important role in supporting governments already committed to reform. Imposing conditionality on a reform-minded government can aid reformers in five ways: first, it may give public officials a more defensible rationale for taking politically unpopular measures; second, it may raise the political authority of pro-reform technocrats who work closely with the funder, or are responsible for carrying out the policy changes; third, it can provide the recipient with the resources to buy out opponents of reform or build a new, proreform constituency; fourth, it enables the recipient to use the technical skills of the funder to implement complex reform programs; and finally, the use of cross-conditionality—making funds for one sector dependent on reforms in another—can build reformist alliances within the government and strengthen the hand of key players (Mosley, Harrigan, and Toye 1991: 119–120, 160).

Still, even though policy-based lending may help fortify like-minded technocrats, the prior commitment of the political elite to reform remains crucial: in both Zaire and the Philippines, autocrats temporarily empowered reform-minded technocrats while negotiating policy-based loans, then rescinded their authority once the loans had been approved (Kahler 1992: 128).

The sixth and final lesson is that there are techniques that can and should be employed to force recipients to reveal their support or opposition to reform before a policy-based loan has been disbursed. Having belatedly realized the importance (and the opacity) of recipient commitment, funders now frequently insist that prospective borrowers make large "down payments" by enacting painful reforms in advance of any disbursement, to demonstrate their commitment to reform and reduce the chance of defection or false compliance (Kahler 1992: 117; Mosley, Harrigan, and Toye 1991: 128–129). The form of conditionality discussed below—allocative conditionality—attempts to take this last lesson one step further.

Allocative Conditionality
Allocative conditionality has recently drawn a good deal of attention in the foreign aid community. Funders who follow this approach make no attempt to bargain with recipients, as they would when engaging in policy-based lending. Instead, they try to allocate their funds according to the recipients' prior commitment to policies or values that the funder wishes to promote. Since 1989, most bilateral funders have announced that the allocation of their development assistance will be based, in part, on the recipients' poverty-reduction programs, human rights standards, reductions in military spending, and progress toward "good governance" and democratic reforms. The European Bank for Reconstruction and De-

velopment goes even further, offering its services only to states that uphold human rights and the rule of law and have competitive elections.
Scholarship on the effectiveness of these policies is still preliminary (Ball
1992; Lewis 1993; Nelson with Eglinton 1992; Nelson and Eglinton
1993).

Nelson and Eglinton (1993) examine the impact of recent decisions by
funders to reduce or suspend their aid to countries that egregiously violated human rights or had been "backsliding" on democratic reforms.
The findings are mixed, but suggest that an aid cutoff was more likely to
catalyze policy changes if the recipient had a strong, organized opposition
that supported reforms (e.g., Guatemala) rather than a weak or divided
opposition (e.g., Kenya and Malawi); or if the country was governed by
an autocrat who demonstrated concern for the public welfare (e.g., Peru's
Alberto Fujimori) rather than simply his own survival (e.g., Haiti's
Raoul Cedras).

There is not enough data on the use of allocative conditionality to reward good behavior to draw any conclusions about its efficacy; the foreign aid budgets of most Organization of Economic Development and
Cooperation (OECD) countries have been too tightly constrained to permit much experimentation. Nonetheless, by announcing explicit standards for allocative conditionality, bilateral funders may have some
impact on the expectations of current and potential aid recipients, who
conceivably may alter their behavior in anticipation of the judgment of
funders—an impact that would, however, be difficult to measure.

Nelson and Eglinton find that allocative conditionality has thus far
"been implemented only partially and inconsistently." Funders have penalized a handful of extreme violators (Zaire, Myanmar, Haiti) and rewarded a handful of countries undertaking democratic reforms (Zambia,
Poland, Russia), but have failed to systematically reallocate aid among
recipients (Nelson and Eglinton 1993: 50). They also suggest that allocative conditionality will necessarily remain a crude and partially implemented foreign policy tool, because bilateral funders have neither the
administrative capacity to carefully gauge the progress of recipients toward a complex set of goals, nor the desire to forsake their prior commitments and interests in allocating foreign aid. In general, however,
allocative conditionality faces far fewer implementation and compliance

problems than policy-based lending, because under allocative condition-ality, the funder rewards recipients for their prior behavior rather than bargaining with them over their future conduct.

Lessons for Environmental Assistance

The structural and sectoral adjustment loans offered by funders to debt-ridden countries involve large sums of money—often hundreds of millions, or even billions of dollars—in a single package. Environmental aid programs are usually much smaller. But the lessons drawn from these cases may be useful for at least three reasons. First, the institutions that make loans and grants for environmental protection face many of the same constraints (and are often the same institutions) as those involved in other types of conditionality. Second, structural adjustment condition-ality provides an opportunity to study these institutions under a set of extreme conditions, in which funders have maximum leverage over recipients. By extrapolating from this magnified setting—in which the stakes are high for both lenders and recipients—it may be possible to gain insights into the limits of conditionality under more mundane circumstances. Finally, we may yet see high stakes "green conditionality" that more closely resembles structural adjustment conditionality, if global environmental problems grow more severe.

Although green reforms may be less distasteful to government leaders than structural adjustment, a decade of experience with conditionality suggests that environmental aid packages will fail to persuade reluctant governments to implement policies they oppose—particularly when those policies are politically costly, cumbersome to implement, or difficult to monitor. On the other hand, when properly structured, conditional environmental transfers may help a "green" political elite overcome substantial technical and political obstacles to reform.

Conditionality has already frustrated proponents of green aid. Two case studies in this volume illuminate the problems that afflict even the most highly motivated funders when they try to use conditionality. The chapter on aid programs to promote nuclear safety in Eastern Europe, authored by Connolly and List, illustrates the special difficulties faced by bilateral funders who use policy-based lending: domestic commercial pressures (in this case, from the nuclear industry) can undermine the funders' resolve; coordination problems among funders—some who favor conditionality, some who oppose it—can dilute its impact; and the relatively modest quantities of the aid typically offered by bilateral funders may be too small to persuade recipients to alter their policies. The study also shows how conditions that take many years to implement—such as shutting down aging nuclear reactors—can be difficult to enforce, because the aid must be disbursed long before the recipient takes decisive action. Unable to make policy-based lending work, some of these funders now follow a much simpler strategy of allocative conditionality, directing their aid toward those states whose nuclear policies they approve of.

Michael Ross's study of multilateral attempts to promote sustainable logging in the tropics takes a detailed look at how the domestic politics of aid recipients can influence the success or failure of policy-based lending programs. His study compares the World Bank's failed efforts to use conditionality in Indonesia with its success (and the success of other funders) in the Philippines. He finds that in Indonesia, the logging industry was in the hands of government allies, and was protected by the government against reform. As in the Eastern Europe case, the effort to impose conditionality on Indonesia was also undermined by the availability of alternative sources of nonconditional funding. In the Philippines, however, the logging industry had been discredited by the fall of the Marcos government in 1986 and received no special favors from the Aquino government. As a result, the funders were able to form an alliance with a handful of government reformers. Together they used the conditional aid package to push through a series of dramatic policy changes.

2   Policy Dialogue

When funders wish to promote policy reform, conditionality is their most coercive tool. Their most common tool, however, is the "policy dialogue," a term that denotes funder attempts to persuade or coerce recipients into policy reform without using explicit conditionality. Several factors determine its effectiveness. For recipients, a political commitment to reform and strong analytic and negotiating skills increase the chances that funder initiatives will be well conceived and politically sustainable. For the funders, their depth of country experience, their analytic capacities, and their ability to make long-term resource commitments all increase the chances that their policy advice will be useful.

Policy dialogue assumes both shared interest in policy change and conflict over its form, pace, and magnitude. The degree of conflict depends on the initial disparity between funder and recipient viewpoints, their respective analytic and negotiation skills, and the leverage each can exert to secure the other's agreement on contentious issues (Jay and Michalopoulos 1989: 91). Analysts and practitioners of policy dialogue generally agree that funder leverage is neither necessary nor sufficient to induce recipient policy reform (see e.g., Mosley, Harrigan, and Toye 1991, chapters 4 and 5). If a recipient's policies reflect fundamental political goals and constraints, policy dialogue can only affect them at the margin. The use of leverage may even backfire if it alienates the recipient government or causes it to lose credibility in the eyes of powerful domestic constituencies (Berg 1991: 219–220). Funder insistence on ill-considered policy reforms has often led to disappointment and backlash.

When a recipient's analytic and negotiating skills are weak, funders sometimes seek to promote policy change by giving them technical and capacity-building support. During the 1980s, for example, the World Bank tried to strengthen the analytic skills of governments receiving structural and sectoral adjustment loans, with mixed results. Paul (1990: 53–54) notes that the World Bank did less to build capacity for policy analysis than to build operations-related technical capacity. Bank staff found it difficult to determine the appropriate organizational location, staffing needs, and audience for policy analysis. It was also difficult to measure the contribution of investments in analytic capacity to the goal of sectoral adjustment. Paul argues for earlier, more intensive, and longer-term support for the development of sector-level policy analysis capability, both within government ministries and in nongovernment settings.

Studies of policy dialogue emphasize the need for funders to base their policy advice on a sophisticated understanding of local economic and political conditions, and the need for funders to make credible, long-term aid commitments in support of policy reforms. The need for sophisticated funder analysis seems obvious, yet in practice funders frequently advocate similar remedies for economic reform in very different economic and political contexts (e.g., a standard package of devaluation, price decontrol, and deflation). Self-serving funder interests in opening markets, ensuring debt service, and employing expatriate consultants may explain some of the failures of funder analysis. The ideological underpinnings and standard operating procedures of key multilateral funders may be another source of bias in policy advice. Bilateral funders often follow the lead of the World Bank and IMF in policy dialogue and hence tend to adopt their biases, whether or not they are appropriate (Cassen 1986: 74–85).

Analyses of policy dialogue often focus on the need for credible recipient commitment to policy reform, but long-term commitments by funders are equally important. A long-term approach can have political, economic, and institutional benefits. Politically, it may take several years to strengthen and stabilize economic and political constituencies with an interest in reform, and to provide adequate compensation to groups adversely affected by reform. Economically, long-term assistance can provide a buffer against external shocks and slower-than-anticipated supply and demand responses to policy reforms. Institutionally, it may take several years of financial and technical support to build the capacity necessary to implement reforms, particularly when the reforms require increased technical sophistication. Recipients recognize the importance of these long-term commitments and are more likely to be swayed in the policy dialogue when they believe them to be forthcoming.

Implications for Environmental Policy Dialogue

The major lessons of funder and recipient experience with economic policy dialogue apply *a fortiori* to environmental policy dialogue. The need to build constituencies for environmental reform, the need to strengthen both funder and recipient understanding of complex environmental systems, and the need for an enduring political commitment by the recipient and economic commitment by the funder are all critical.

Although funders have increasingly embraced the rhetoric of sustainable development, it is not clear how much their words represent a real change in beliefs and values, and how much they are a symbolic response to political pressure from developed country environmental NGOs. In some cases, however, networks of environmental NGOs from developed countries and pro-poor NGOs from developing states have first pressured funders and recipients to reform their environment and development policies, then subsequently held them to their word by monitoring implementation. When NGOs can make alliances with actors who stand to gain from a change in development policies—including government agencies, elected politicians, and businesses—they may become a powerful political force for promoting sustainable development (Rich 1990; WRI, 1992: 226–232).

In this volume, Cord Jakobeit's study of debt-for-nature swaps illustrates the importance of political commitment and organized reform constituencies for environmental aid. Northern NGOs tried to use debt-for-nature agreements to open a broader policy dialogue in several Latin American countries, but they succeeded best in Costa Rica, whose government had already demonstrated a commitment to nature conservation and whose domestic NGO network acted as an organized reform constituency. Similarly, the chapter on Eastern European nuclear safety shows the importance of the recipient's technical capacity in fostering a fruitful policy dialogue. Where aid has improved the ability of recipients to analyze their own domestic nuclear risks, a stronger policy dialogue has resulted, contributing to a tighter regulatory framework for nuclear plant operation. On the other hand, David Fairman's chapter on the Global Environment Facility shows how the absence of a long-term financial and political commitment by both funders and recipients can undermine the implementation of even well-endowed environmental aid programs.

Despite their shared interest in reducing transboundary environmental impacts and risks, funders and recipients will continue to have mixed motives for engaging in environmental policy dialogue. As the cases of the Global Environment Facility, environmental aid for Eastern Europe, debt-for-nature swaps, and tropical forest logging reform illustrate, funders and recipients have embraced the rhetoric of sustainable development as much for its symbolic political benefits as for its policy impact.

Based on our observations in this volume and on other studies of environmental aid (Rich 1990; WRI 1992: 226–232), it appears that advocates for environmental protection *can* use policy dialogue to push funders and recipients from rhetoric toward concrete policy reforms, but they must combine concern-raising political mobilization, capacity building, and contracting institutions that commit funders and recipients to a long-term process of analysis and action.

These political, technical, and institutional requirements, daunting on their own terms, may still not be sufficient to ensure the success of environmental policy dialogue. Greater integration of environmental and economic policy issues in the policy dialogue and a more serious search for complementarity in economic and environmental policies are prerequisites for success in countries that depend heavily on natural resources for subsistence and growth. Without integration and a search for "win-win" strategies, a narrowly focused environmental policy dialogue is likely to be undercut by economic and social policies working at cross purposes with environmental goals (World Bank 1992a, chapter 3).

3   Capacity Building

Perhaps nothing in the field of development is as popular to promote and as difficult to accomplish as capacity building. For decades, funders have tried to foster what we now refer to as "capacity building" under the rubrics of institution building, institutional development, public sector management, and public administration. Environmentalists have a special interest in these initiatives: calls for capacity building are sprinkled throughout Agenda 21, the plan for sustainable development produced by the 1992 UN Conference on Environment and Development; UNDP has embarked on a major effort to promote environmental capacity building, called "Capacity 21"; and capacity building has become a ubiquitous feature of bilateral environmental aid programs. The great majority of what has been written on capacity building is normative and concerned with practical management questions; relatively little evaluative work has been done. Still, there are at least six lessons to be gleaned from the handful of careful studies.

First, capacity building has been the Achilles' heel of a great many development projects. Most aid projects can be broken down into a "hard investments" component (equipment and physical infrastructure) and an "institutional development" component (which includes the promotion of planning, accounting, maintenance, staff training, and other forms of technical assistance). Assessments by the World Bank have found that the institutional components of programs fail about twice as often as the physical components, and that the most common obstacles to the successful implementation of projects are "managerial" or "institutional" problems (Israel 1987, 3). A review by USAID of its support for institutions found the results "about half positive and half negative." The Canadian development agency discovered that two-thirds of the recipient country institutions it had assisted were unprepared for self-reliance at the end of the project (Cassen 1986: 111). In fact, many development practitioners have found the problems of "institutional weakness" so intractable that they prefer to ignore the issue altogether and instead worry about more tangible problems.

Second, ex post evaluations of World Bank programs have shown that the success of institutional development programs varies significantly by sector, subsector, and activity. The most successful programs have been in industry, telecommunications, some utilities, and nonrural finance; the least successful in agriculture, education, and railways. Capacity-building efforts on technical and financial issues have done relatively well; initiatives promoting interagency coordination, personnel management and training, maintenance of facilities and infrastructure, and sectorwide reforms have done relatively poorly (Uphoff 1986: 196; Israel 1987: 19–20; Cassen 1986: 195–198). Good overall explanations for these patterns are lacking, although it seems clear that setting up effective and efficient institutions is much more difficult in rural than in urban areas (Rondinelli 1986).

Third, in many aid projects there is a conflict between the short-term interest of funders in efficiently completing hard investments and the long-term goal of fostering institutional development. Funders are usually anxious to identify problems in recipient countries and to supply the organizational skills and technologies (often originating in the funder's own country) to solve them. This may be the most efficient way to approach a discrete "development" problem in low-income countries. But it also may have adverse consequences for any capacity-building efforts, because it tends to deprive local officials of the training and experience they will ultimately need to maintain the project. The problem is exacerbated by the desire of funders to show their own taxpayers the tangible results of foreign aid programs as quickly as possible, and their interest in hiring their own citizens and firms to perform much of the work.

Fourth and related, the institutional components of development projects frequently fall apart once funders leave, because the recipient country is unable to afford the recurrent costs of staffing and maintaining a project. Governments of low-income countries typically find it easier to acquire funds for hard investments from international funders and financiers than to raise the operating costs to maintain the institutions surrounding the project once the funder leaves—commonly because the recipient must operate from a weak and politically volatile tax base, and/ or has poor macroeconomic policies. The scarcity of funds to maintain funder-initiated institutions means, on a practical level, that new bureaucracies cannot fund their operating costs; and that roads, vehicles, and other infrastructure must operate well below capacity, and over time may have to be abandoned (a ubiquitous phenomenon in Africa). Capacity-building efforts, if they are to be sustained, must include provisions for long-term recurrent financing (Duncan 1985: 136–137; Mosley, Harrigan, and Toye 1991: 30–31).

Fifth, even when funders provide both training and recurrent cost finance, capacity-building efforts may fail if the recipient lacks a durable and bureaucratically widespread commitment to the success of the new institution. Funders and recipients commonly have differing views on the goals or structure of the targeted institutions. If the recipient fails to give the institution sufficient economic, political, and bureaucratic support once the funder leaves—or fails to sustain that support over a period of decades—the institution may well not survive (Israel 1987: 40–41).

Finally, developing country NGOs are playing an increasingly important role in augmenting the capacities of their governments. The number and resources of southern NGOs vary widely: they are thought to be strongest in Asia and Latin America, weak in Africa, and virtually nonexistent in the Middle East, North Africa, and northeastern India. Although the figures are unreliable, the number of NGOs in developing states appears to have tripled between 1977 and 1992. In some states, NGOs have formed alliances with governments and have taken on quasi-governmental functions that governments themselves may be ill-equipped to perform, such as delivering services in rural areas and mobilizing resources at a local level. Southern NGOs appear to be increasingly successful in attracting financial support directly from the North (often from other NGOs), thus serving as conduits for capacity-building efforts that bypass the state completely (Livernash 1992; Dichter 1988).

Implications for Environmental Aid

Perhaps more than any previous form of development assistance, environmental aid transfers are aimed at capacity building; hence the lessons from previous efforts to foster "institution building," "institutional development," and "public sector management" are especially salient. Many environmental assistance projects treat capacity building far too casually, assuming that a short-term barrage of technical training programs will bring about sturdy environmental institutions in low- and middle-income countries.

This volume richly illustrates both the importance of capacity-building programs and the many ways they can fail. Three of the chapters—on debt-for-nature swaps, logging reforms in the tropics, and environmental aid for Eastern Europe—offer evidence that recipients with higher technical capacities are more likely to attract foreign aid. This is often a poor way for funders to parcel out their resources. States with lower technical capacities may need more aid, not less. In the case of Eastern Europe, funder bias toward high-capacity recipients led to an inefficient distribution of resources. In the case of tropical forest protection, a different problem emerged: Indonesia's remarkable ability to attract aid from a wide range of donors made it difficult for a single large funder—the World Bank—to put conditions on its loans.

In their chapter on Eastern Europe, Connolly, Gutner, and Bedarff show how capacity-building efforts can be thwarted by the political interests of funders, who feel compelled to hire experts from their own countries, instead of developing the skills of local peoples. But the study of nuclear safety in Eastern Europe, authored by Connolly and List, also shows how important capacity-building programs can be. Although funders were unable to use conditionality to force aid recipients to close their nuclear plants, their capacity-building programs helped local experts better understand the risks they faced and how to address them.

Jakobeit's study of debt-for-nature swaps illustrates the many ways that NGOs can augment the capacities of both funders and recipients. NGOs pioneered the idea of debt-for-nature swaps, and NGOs later persuaded Western governments to fund them as part of their bilateral aid programs. NGOs also played a critical role in recipient states, taking on many quasi-governmental functions. But the study also finds many drawbacks to using NGOs as capacity-enhancing institutions. They are, unfortunately, not a panacea for states with low administrative or technical capacities.

Finally, Ross's chapter on logging reform in the tropics explains the many ways funders can enhance the *political* capacities of a recipient, particularly when the recipient government favors environmental reforms but is too weak to prevail over its domestic opponents. In these settings, funders can take the blame for unpopular measures, provide money to buy out opponents to reform, and wield their influence to build pro-environment alliances—a formula that helped produce a surprisingly strong package of forestry reforms in the Philippines.

4   Coordination

In many developing countries, twenty-five to thirty bilateral and multilateral aid agencies may be operating at any given time. Most recipient governments have no central agency or process to integrate aid programs into government planning and budgeting. Some have no centralized mechanism for coordinating aid agency activities (Cassen 1986: 219). The arrival of new environmental funds and funders has made the problems of coordination more complex.

Even though some decentralization and competition may make the allocation of aid more efficient, most observers agree that the costs of uncoordinated aid programs outweigh their benefits. Many aid projects and programs are duplicative; others work at cross-purposes. Underlying the coordination problems are substantial conflicts of interest among funder agencies, among recipient agencies, and between funders and recipients. Both funders and recipients have attempted to overcome coordination problems, occasionally with success. In general, coordination has been more effective when recipients have a centralized agency or process for integrating aid with their overall development strategies, and when funders organize national roundtables or consultative groups to coordinate their programs (Danida 1991: 31).

Cassen (1986: 220–224) identifies three major problems caused by co-ordination failures. First, projects proliferate, causing duplication of effort, conflicts among project and program goals, and deadweight loss from incompatibilities. In some cases, aid "fads" lead funders to over-commit resources to a particular sector, so that the marginal value of aid approaches zero. In other cases, funders may form alliances with rival ministries pursuing conflicting objectives at the sectoral or regional level. In still others, funders may provide incompatible equipment for a project or sector.[2] Second, proliferation spurs the growth of ad hoc, project-related administrative structures, which lead to competition for scarce managerial talent in recipient agencies. Third, recipients lacking central-ized oversight of their aid programs may underestimate the longer-term impact of project operation and maintenance costs on their budgets and managerial talent. In times of budgetary stringency, many recipients cut operating budgets and staffing to inefficiently low levels, resulting in a large number of underperforming projects.

Funders and recipients have been well aware of the costs of uncoordi-nated aid efforts for at least twenty-five years (Pearson Commission 1969: 227–228; OECD 1985: chapter 8). The continuing difficulty in achieving coordination reflects several conflicting funder and recipient interests (Cassen 1986: 230–232).

Funders' motives for giving aid include short-term commercial and po-litical objectives (particularly export promotion) and longer-term devel-opmental and political goals. Coordinating aid—either through multidonor meetings or by channeling funds through multilateral agen-cies—both eliminates projects and programs that could be used to boost funder exports and reduces the political leverage of individual funders.

Funders may also genuinely disagree with each other on development goals for the recipient or on the best way to allocate aid. When funders believe that further discussion will not change their position, and that key recipient decision makers share their views, they may reasonably resist coordination. As Pressman and Wildavsky (1984: 133–135) point out, "coordination" may serve the interests of some participants at the ex-pense of others, particularly when the party initiating coordination seeks to set the agenda and terms of reference to favor its own views.

Finally, funders may feel that the administrative costs of coordination outweigh its substantive benefits. Organizing and attending meetings at the country level may be logistically difficult for understaffed field offices. Following up planning meetings by coordinating projects and programs demands even greater staff resources.

Although recipients seem to have much to gain from coordination, they may also have much to lose. Even when the aggregate benefits of coordination outweigh the aggregate costs, the benefits may be diffuse, while the costs to influential segments of the government may be concentrated and sharply felt (cf. Wilson 1980). Coordination may also provide funders with an opportunity to "gang up" on the recipient. If funders reach an agreement about the recipient's policy shortcomings, they may apply collective pressure to change them. In contrast, when funders do not coordinate their approach, the recipient may be able to further its interests by playing funders off one another.

Over the past decade, funders have developed a more sophisticated understanding of the need for coordination and the need to expand the scope of issues coordinated. From a primary emphasis on pledging, coordination has evolved through a phase of "problem-centered" approaches to drought, famine, and balance of payments crises, to current concern with the interaction of economic, social, and environmental projects and policies.

Most of the successes in aid coordination are due to strong leadership by the aid recipient.[3] On the funder side, coordination has been more successful when it has taken place under the auspices of bilateral and multilateral agencies with relatively impartial views of the development process, strong analytic and administrative capacity, and substantial country-level staffing and experience (OECD 1992a: 26–28).

On the other hand, the 1992 annual report of the OECD Development Assistance Committee (DAC) highlights continuing major problems in aid coordination: tying aid to purchases in funding countries, funder competition for projects, project proliferation, ad hoc administration without capacity building, recipient administrative overload, inadequate integration of projects into policy frameworks, and failure to follow up planning meetings with ongoing operational coordination (OECD 1992b: 48).

Finally, although funders have an increasingly sophisticated understanding of the interaction of aid programs with local, national, and global economics, they have only just begun to analyze the interaction of economics with demographic and environmental systems; they are further still from reaching a political consensus on how to approach trade-offs among competing objectives. As the DAC report notes,

A large number of interlinked economic and policy issues affect development cooperation: population, regional security, environment, migration, drugs, public health, monetary policy, fiscal policy, taxation, trade, export credit, and agriculture. Each of these issues is associated with domestic interest groups as well as one or several government ministries whose primary concerns and responsibilities are unlikely to be directly connected with development. Obtaining consensus among these interests and government bodies to improve coherence in individual policy cases or on a broad front poses a challenge for political leadership. (OECD 1992b: 46–47)

In short, although both funders and recipients recognize the need for improved coordination, conflicting interests remain powerful barriers despite a broad consensus on economic goals.

Implications for Environmental Aid Coordination
Environmental policy goals will remain subsidiary to economic policy goals for most developing countries for years to come. Still, over the next two decades environmental projects will likely constitute a growing fraction of development assistance. Hence, funders and recipients should try to coordinate environmental projects and programs with nonenvironmental ones, identify the trade-offs between environmental and nonenvironmental goals, maximize the number of win-win strategies, and minimize the environmental impact of nonenvironmental projects and programs.

Accomplishing these tasks may be more difficult for environmental aid than it has been for other types of development assistance. The base of consensual knowledge is smaller, the issues are lower on both funder and recipient agendas, and funders are likely to underinvest in institutions for coordination. In this volume, Connolly and List show how a common aid problem—funder competition for export markets—undermined the coordination of nuclear safety aid for Eastern Europe. Connolly, Gutner, and Bedarff show that the problems of coordinating environmental aid programs for Eastern Europe stemmed from both incomplete information about environmental damage and the roller-coaster rise and fall of concern about environmental problems in the region. The initial alarm among both funders and recipients about environmental problems in the region spurred a proliferation of multi- and bilateral aid efforts, but concern declined equally rapidly as the trade-offs between economic and environmental goals became clearer. What remained was a new set of aid programs and staff interested in their own institutional survival, without sufficient high-level political commitment to eliminate program overlaps and conflicts.

In addition, the achievement of environmental goals is so heavily dependent on the broader economic and social policy framework that both funders and recipients will find it difficult to justify providing substantial amounts of environmental aid if it is not coordinated with other policies, particularly economic policies. Funders and academic analysts are paying greater attention than ever before to the environmental consequences of macroeconomic and trade policies (Reed 1992). Nevertheless, both the technical complexity of coordination problems and the greater weight virtually all governments give to economic considerations make it unlikely that environmental concerns will drive economic policy reform in the near future.

5   Conclusion

Our review of major themes in the development assistance literature suggests four general lessons for the designers of environment aid programs. First, the level of recipient political commitment has a dramatic impact on aid effectiveness, yet attempts by outside funders to create this commitment where it does not already exist have generally failed. For complex tasks like capacity building, the recipient's commitment must be especially durable and bureaucratically widespread. Determining a recipient's "true" level of commitment in advance of a program is difficult, because recipients can easily feign commitment and funders have a limited ability to monitor compliance. The best solution for large funders may be to insist that prospective recipients demonstrate their commitment to environmental programs by implementing vital (and ideally, hard-to-reverse) portions of any aid-related agreements before funds are disbursed. The best solution for smaller funders, who have less leverage, may be to use allocative conditionality, granting aid to countries that already exemplify those values the funder seeks to promote.

Second, regardless of their size, environmental funders may be able to change the policies of recipients in some settings. Funders with detailed knowledge of local conditions and strong ties to their recipient counterparts may become influential, even if they do not contribute a preponderant share of the recipient's aid. On the other hand, "major" funders with substantial aid resources and the ability to use added incentives and sanctions may fail to influence environmental policy over the long term if they lack local knowledge and contacts. Funders who specialize in a small number of countries and sectors and target their interventions at a subset of institutional actors (e.g., ministers, sectoral agencies, local governments) may also have a disproportionate influence in their areas of specialization. But specialization calls for coordination, which often fails. The major multilateral agencies must do more to facilitate country-level coordination and to identify the trade-offs between environmental and nonenvironmental goals.

Third, the indigenous capacity of recipients to analyze, implement, and administer aid programs is a key variable in the long-term success of these programs; but these capacities are often lacking in the environmental sector. Moreover, efforts by outside funders to foster these capacities have often failed. More research is needed on the determinants of effective capacity building. As the studies in this volume demonstrate, many environmental assistance programs have not yet taken seriously the painful experiences of development programs with institution-building efforts.

Fourth, there are large gaps between what aid analysts and practitioners say they have learned and what they continue to do in practice. In the areas of policy dialogue, conditionality, capacity building, and interagency coordination, best practice and standard operating procedures continue to diverge. It is not enough for analysts to review evidence, modify theories, propose recommendations, and have them nominally adopted by policymaking bodies. The literatures on international organization and implementation suggest that well-specified targets and timetables, clear lines of accountability and responsibility, and periodic review of both practice and principles are essential for institutional learning (Haas 1990; Mazmanian and Sabatier 1983).

Given the multitude of problems with development assistance, we cannot be optimistic that the same institutions will do a better job at environmental assistance. But green aid is a new realm for funders and recipients; it has also led to the formation of new institutions, including the Montreal Protocol Fund and the Global Environment Facility. Environmental aid may be dismissed as a fad. But as the studies in this volume show, it can also spur new efforts to solve old problems, and open the door to innovation.

Notes

The authors are indebted to Robert Keohane, Henry Lee, and members of the Financial Transfers Project for their comments on an earlier draft of this chapter, and to the Harvard University Financial Transfers Project for its support. Michael Ross would also like to thank the Center for Energy and Environmental Studies at Princeton University, the Social Science Research Council, and the Institute for the Study of World Politics for their support; David Fairman would also like to thank the U.S. Department of Education, Jacob K. Javits Fellowship Program, for its support.

1. All loans and grants have conditions that limit their use and (for loans) are meant to ensure their repayment. The term "conditionality" here denotes the exchange of financial assistance for broader policy changes, beyond those strictly needed to define a program or ensure repayment.

2. For example, donors have provided the Bangladesh railway system with a variety of wagon cars that cannot be connected and several types of locomotive requiring substantially different mechanical skills and maintenance schedules (Cassen 1986: 221).

3. India, Papua New Guinea, and Indonesia are often cited as countries that developed strong central finance or aid administration agencies and pushed for donor coordination (Cassen 1986: 236).

II

Promoting Environmental Protection in Developing Countries

3

The Global Environment Facility: Haunted by the Shadow of the Future

David Fairman

Introduction

The Global Environment Facility (GEF) was established in November 1990 as a three-year, $1 billion pilot program to finance developing country action on four global environmental problems: global warming, biodiversity loss, pollution of international waters, and depletion of the ozone layer.[1] Throughout its pilot phase, the GEF's operations were hampered by political, organizational, and administrative conflicts among its major stakeholders—developed and developing country governments, implementing agencies, and interested nongovernmental organizations.

In June 1992, at the UN Conference on Environment and Development (UNCED), these stakeholders reached agreement in principle to "restructure" the GEF and to use it as the "interim" financial transfer mechanism for the climate change and biological diversity conventions. In March 1994, after eighteen months of laborious negotiations, the GEF's participating governments and implementing agencies established the restructured GEF as a permanent facility.

Like most of the financial transfer mechanisms examined in this book, the GEF has made only a modest contribution to alleviating the environmental problems it was designed to address. Yet unlike many of them, it has received high-level political attention from developed and developing country governments and major international organizations, sizeable financial commitments from developed country funders, and a great deal of policy, program, and project-level input from a wide range of NGOs and academic experts. Why, given the seemingly high level of international interest in achieving the GEF's stated goals, substantial financial and technical capacity, and painstaking policy development to resolve potential contracting problems between donors and recipients, did the GEF engender such high levels of conflict during its pilot phase? What changes in the period after UNCED allowed the GEF's stakeholders to resolve many of these conflicts? And finally, what are the prospects for the GEF to make a substantial contribution to solving the global environmental problems it has been mandated to address?

This chapter argues that the GEF's key stakeholders initially agreed only at the most general level on the need for a new financial mechanism to facilitate developing country action on global environmental problems. Their interests diverged when it came to deciding which problems the GEF should address, allocating policymaking authority among funder and recipient governments, and allocating operational authority among governments, implementing agencies, and nongovernmental organizations.

The World Bank, tasked by its funders to turn a vague proposal into a concrete institutional design, put together a "package deal" combining a mandate and governance and administrative structures acceptable to all major stakeholders. This package sidestepped potentially irreconcilable differences by leaving a number of major institutional questions unresolved. The GEF's designers hoped that the GEF's pilot phase operations and the parallel negotiation on climate change, biodiversity, and other UNCED-related issues would build a global consensus on the GEF's mandate, structure, and operations.

By creating a single financial mechanism with an open-ended mandate to address a range of global environmental problems, the GEF's stakeholders gained immediate political benefits but raised the political stakes associated with the GEF's pilot phase. Believing that the institutional decisions they made about the GEF would affect not only financial transfers for current global environmental problems, but also North-South diplomacy, the stature and authority of major international organizations, and norms for the participation of nongovernmental agents in international environmental governance for years to come, the GEF's stakeholders invested the GEF's pilot phase with meaning and significance far greater than it would otherwise have justified.

As the GEF became more and more closely linked to the climate change, biodiversity, and UNCED financial negotiations, its stakeholders did not use it systematically to "learn by doing" in the new field of global environmental management. Instead, they tried to use the GEF to influence the financial decision-making rules of current and future environmental conventions, the allocation of funds and authority among the GEF's implementing agencies, and the norms for participation by nongovernmental actors in intergovernmental negotiations and organizations.

By the time the convention and UNCED negotiations entered their endgames in the spring of 1992, the divergent interests of GEF stakeholders on these issues threatened to overwhelm their shared interest in addressing global environmental problems. At UNCED, developing countries made a credible threat to abandon the GEF unless it were restructured to give them greater decision-making power and authorized to fund action on the "sustainable development" problems identified in Agenda 21. Developed countries made an equally credible threat to cease funding the GEF unless (1) they retained substantial authority over GEF policy and financial decisions; (2) the GEF's mandate remained focused on global commons problems; and (3) the World Bank retained a leading role in managing the GEF.

On the brink of deadlock at the Earth Summit, both developed and developing country governments blinked. Unwilling to risk the political fallout from a failure to reach agreement at UNCED, and recognizing that developed and developing country blocs each held effective veto power over the GEF's future, the GEF's stakeholders agreed in principle to restructure it to meet the concerns of both blocs, and to make it the "interim" financial mechanism for the climate change and biodiversity conventions.

Having decided to institutionalize the GEF as the leading financial mechanism for global environmental problems, the stakeholders continued to bargain hard on the issues that had nearly destroyed the GEF during the pilot phase. After UNCED, however, the shadow of the future spurred consensus building as well as conflict. Bargaining with the knowledge that there was no other politically viable alternative for managing large-scale financial transfers, most stakeholders were more willing to search for creative compromises than they had been during the run-up to UNCED.

The restructuring agreement appears to have resolved many of the disputes that plagued the GEF during its pilot phase, and to have increased the likelihood that the GEF will be designated as the permanent financial mechanism for the climate change and biodiversity conventions. It is too soon to tell whether these institutional changes will increase the GEF's ability to leverage its limited budget and staff in ways that produce substantial global environmental benefits. Reform of environmentally and economically unsound sectoral policies in developing countries may offer the greatest potential leverage, but to catalyze policy reform the GEF will have to develop a more flexible interpretation of its global environmental mandate and a greater institutional capacity to build on existing structural and sectoral adjustment programs.

Section 1 describes the GEF's development, pilot phase operations, and restructuring negotiations. Section 2 analyzes the interests and capacities of the GEF's stakeholders to show how a limited convergence of interests and an overarching, shared concern about the GEF's precedent-setting role facilitated the GEF's creation, hampered its operations, and forced its restructuring. Section 3 evaluates the GEF's effectiveness in meeting its stated goals and the potential for the restructured GEF to leverage its human and financial capacity to manage global environmental problems.

1   The GEF's Design, Pilot Phase, and Restructuring

During the late 1980s, a growing number of international environmental problems—ozone depletion, climate change, loss of biological diversity, deforestation—received international, diplomatic, and organizational attention. Scientific and technical experts, agency managers, and nongovernmental advocates generally agreed that developed and developing countries would have to work together to manage these "global commons" problems. They also agreed that most developing country governments had little information about, interest in, or capacity to take action on these problems. Several intergovernmental and nongovernmental organizations argued that developed countries should stimulate developing country participation by subsidizing capacity-building programs and environmental investments (WCED 1987: 338–339; WRI 1989).

As developed country environmental groups intensified demands for action on global environmental problems, and as work on several international environmental treaties accelerated, developed country governments began to explore options for financing developing country action. In September 1989, developed country governments asked the World Bank, the leading multilateral aid agency, to design an international environmental aid facility.[2] They gave the Bank a great deal of discretion in defining the facility's mandate, strategy, and operational structure.

Designing the GEF's Mandate and Strategy
From the fall of 1989 through early 1990, Bank management and staff drafted a proposal for a "global environment facility." This proposal was circulated to developed country governments in the spring of 1990 and to developing country governments in the summer of 1990. A revised version (World Bank 1990a) was approved as the basis for the GEF's operations in November 1990.[3]

The Bank proposed that GEF address four global environmental problems: climate change, biodiversity loss, ozone depletion, and pollution of international waters. To help developing countries deal with these problems, the facility would support "programs and activities for which benefits would accrue to the world at large while the country undertaking the measures would bear the cost, and which would not otherwise be supported by existing development assistance or environment programs" (World Bank 1990a: 2).[4]

The facility would provide grants rather than loans, because the activities it funded would not generate net financial or economic benefits to the countries undertaking them. Grant money from the GEF would in effect compensate developing countries for undertaking activities that were costly to them but that generated benefits for the global environment.

The facility's basic operational strategy was to provide

modest incremental resources to finance programs and projects affecting the global environment . . . in a manner designed to test how best to strengthen local analytical, regulatory and monitoring capacity and to test means for sharing existing and emerging technologies. (World Bank 1990a: 4)

In line with this strategy, the GEF would support projects with significant innovation and demonstration potential. Cost-effectiveness (the amount of global benefit achieved per dollar spent) was also an important goal, but the Bank acknowledged that it would need expert advice to develop quantitative indicators of projects' global benefits (World Bank 1990a: 7).

Developing countries would not have to prepare country strategies as a framework for individual GEF project investments. On the other hand, without requiring extensive preparation of a national framework, or explicitly imposing policy conditionality, the GEF would give preference to countries that had demonstrated both commitment to and capacity for sound environmental management.

The GEF's Institutional Structure

During 1990, the Bank, potential donor and recipient country representatives, and representatives of UNDP and UNEP defined four core institutional features of the GEF: its status as a pilot program, governance rules, administrative structure, and relationship to nongovernmental actors.

First, the GEF would be a pilot program, with potential links to the Montreal Protocol on ozone depletion, the climate, and biodiversity conventions, and UNCED-related programs. From the beginning of 1990, the GEF's designers presented the facility as an experiment, "complementary" to the convention and UNCED processes, but not necessarily destined to become a permanent financial institution in support of UNCED or the conventions. Initially, it would have a three-year operating life. It could become a permanent facility, however, if it were chosen as the financial mechanism for the ozone, climate, or biodiversity conventions or for UNCED-related programs.[5]

Second, the GEF would have informal governance rules. Participation by interested developed and developing country states would be voluntary: there would be no membership requirements beyond a *de minimis* financial contribution,[6] no reporting obligations beyond those required for project proposals, and no formal decision-making rules. During the pilot phase, representatives of the GEF's participating governments ("participants") would meet twice a year at meetings convened by the GEF's chairman, a senior World Bank manager.[7] At these meetings, participants would review and approve the GEF's work program by consensus, monitor and review its operations, and consider its relationship to the ongoing convention negotiations. Participants would not, however, review and approve individual projects. Final project approval decisions would be left to the managements of the three implementing agencies.

Third, the GEF's operations would be administered jointly by three international organizations with potentially complementary administrative and technical capacities. The World Bank would serve as the GEF's lead administrator, manage the GEF's finances, and handle its investment projects. UNDP would handle technical assistance projects and coordinate GEF activities with national environmental programs in recipient countries. UNEP would provide liaison between the GEF and the UNCED and convention processes; it would also organize and support an independent scientific and technical advisory panel (STAP). STAP would refine the GEF's proposed overall strategy and the strategies proposed for each problem area, and would propose criteria for project selection (World Bank 1990: 6,7).

Fourth, NGOs, including scientific and technical organizations, grassroots development agencies, and international advocacy groups, would have some opportunities to participate in GEF policy, program, and operational activities. There was general agreement among developed and developing country representatives whom the World Bank's design team consulted that NGOs could provide useful advice on project design and might be eligible for GEF funds for specific projects. Government representatives were divided, however, on the question of whether NGOs should be represented in discussions of GEF policy and strategy. The GEF's foundation documents indicated that NGOs could assist in project design and implementation, particularly in the biodiversity area, where NGOs' ability to represent the concerns of local people might be critical to project success. The implementing agencies would also consult with NGOs "concerning the Facility's programs and the progress in implementation," using "mechanisms already established by [each agency] for . . . cooperation with NGOs" (World Bank 1990a: 12). NGOs would not, however, be allowed to observe meetings of the GEF's participating governments.

The GEF's Pilot Phase

Each aspect of the GEF's design—its mandate and strategy, governance and administrative arrangements, and opportunities for NGO participation—proved problematic during the GEF's pilot phase. The mandate and strategy proved difficult to operationalize, a problem exacerbated by the GEF's extremely compressed timeline for project development.[8] The

GEF's governance arrangements and relationship to the conventions became increasingly controversial as UNCED approached. Its implementing agencies failed to work out a mutually acceptable division of responsibility for GEF administration. As the GEF began to design and implement projects, NGOs increasingly criticized it for offering only limited opportunities for local participation in the project cycle, and for excluding NGOs from its semiannual participants' meetings.

Implementing the Mandate   Most of the GEF's participants were eager to demonstrate the viability and financial benefits of the GEF prior to UNCED. They encouraged the implementing agencies to accelerate the process of identifying and proposing GEF projects (UNEP, UNDP, and the World Bank 1993: 15). The rapid pace of portfolio development meant that most of the GEF's projects were designed before the facility's strategies and project selection criteria were well established. Consequently, there was no way to ensure that these projects met the facility's innovation, demonstration, and cost-effectiveness goals.

One major casualty of rapid institutional development was the scientific and technical advisory panel, the body that was supposed to refine and operationalize the GEF's criteria in each of its problem areas. STAP did not convene until the spring of 1991, and did not complete its criteria development work until after the third participants' meeting, when the majority of pilot phase projects and funding had already been committed.

The incremental cost criterion was another casualty of haste. Operationally, the GEF was supposed to pay only for projects (or project modifications) that developing countries had insufficient economic incentives to undertake themselves, but that provided significant global benefits. In practice, however, the implementing agencies were unable to develop a coherent methodology for calculating incremental costs. The wide variety of problem and project types, and the pressure to produce project proposals quickly, led to great variation in the use of the incremental cost criterion. In the vast majority of project proposals, the agencies used the criterion as a heuristic justification, not as a rigorous tool for decision analysis (UNEP, UNDP, and the World Bank 1993: 42–44; see also King 1994: 6–9).

As a consequence of the hasty and ad hoc project development process, the quality of GEF projects varied greatly within and across problem areas.[9] Most of the GEF's biodiversity projects followed traditional protected area strategies. The GEF's most successful biodiversity projects were those that built on existing initiatives, or that funded innovative ideas (e.g., Bhutan Trust Fund [GEF 1991a: 31–36; Danish 1994]) previously developed by NGOs and national governments (Bowles and Prickett 1993).

The GEF's global warming portfolio integrated international expert consensus and "state of the art" strategies more successfully than its biodiversity portfolio. The implementing agencies were far less successful, however, in developing an overarching strategic framework for their climate change work (UNEP, UNDP, and the World Bank 1993: 64–73). Consequently, the GEF's global warming portfolio was a mixed bag of projects that scored high on one or more strategic criteria, and a few whose main purpose seemed to be to ensure that the GEF global warming portfolio included projects from every region.

The GEF's international waters portfolio was particularly deficient in strategic thinking (UNEP, UNDP, and the World Bank 1993: 74–84). Many projects were marginal modifications to World Bank port and harbor projects, adding grant elements for reception and treatment of ship wastes. Few sought to identify the major inland and coastal sources of marine pollution and to set priorities for pollution reduction. Nor were most international waters projects well linked to existing regional initiatives such as UNEP's regional seas programs.

North-South Divisions on the GEF's Structure and Operations   Arguments about the relationship between the GEF and the conventions and about the GEF's governance structure led to a widening North-South split in GEF participants' meetings. Operationally, the GEF's tightly focused mandate, complex approval procedures, and limited outreach to recipient government agencies also undermined developing country support for the facility.

At their initial meetings, GEF donor and recipient country delegations focused on the development of the GEF's project portfolio, and the tone was generally cordial. As the convention negotiations progressed, however, fundamental differences between developed and developing countries on the relationship among the GEF, the conventions, and UN-CED began to surface (Jordan 1994a, 1994b). The GEF's developed country participants (particularly the United States, France, and the United Kingdom) became increasingly anxious to establish links between the GEF and the convention processes. They argued that the GEF should be used to manage financial transfers for the conventions for several reasons: to avoid a proliferation of convention-specific funds; to use the comparative advantage of the World Bank, UNDP, and UNEP in managing environmental investment and capacity-building programs; and to avoid disputes over decision rules (particularly one-nation, one-vote versus contribution-weighted voting) by continuing to use the GEF's informal governance structure. At the same time, developing countries began to fear that developed countries were using the GEF as a way to preempt discussion of alternative institutional arrangements and voting rules more favorable to their interests. They also feared that developed countries would use their support for the GEF as an excuse to avoid providing additional aid for the implementation of Agenda 21.

Operationally, developing country government agencies became frustrated with the GEF's focus on global problems, the relatively small amounts of funding available, the extensive project preparation and approval procedures, the limited opportunities for consultation with the GEF's implementing agencies, and the facility's extensive information disclosure requirements. It became apparent that GEF implementing agency staff, rather than developing country agencies enthusiastic about the GEF's mission, were doing most of the work to generate GEF project proposals. A survey of developing country government officials and NGOs undertaken during the independent evaluation of the GEF revealed a widespread lack of knowledge about the GEF and general dissatisfaction with the implementing agency–driven style of its project development process (UNEP, UNDP, and the World Bank 1993, chapter 7).[10]

Implementing Agency Turf Battles   Disagreements among the implementing agencies had begun during the design phase, when both UNDP's and UNEP's senior managers sought a greater role in GEF administration than initially proposed by the World Bank's design team (UNEP, UNDP, and the World Bank 1993, chapter 9). During the pilot phase, UNDP and UNEP asked for the right to review and comment on World Bank investment project proposals at an early stage, but were rebuffed by the Bank. UNDP and the Bank also fought over shares of the GEF's budget (GEF 1992a: 5; Thalwitz 1991; Tolba 1991).

As disagreements among the implementing agencies persisted, the interagency implementation committee, which had begun the pilot phase with meetings almost every month and intensive interaction over project and program design, became a forum for brief, formal reviews of each agency's project proposals prior to submission to the participants' meetings. The agencies ceased attempting to coordinate their GEF project activities at the country level, or to coordinate their approaches to each of the GEF's problem areas.

Limited Integration of NGOs in GEF Policy and Operations   During the pilot phase, the GEF partially satisfied some of the demands of international and developing country environmental NGOs for participation in its policy and project work. Some NGOs responded positively to these opportunities for dialogue and pursued a policy of "constructive engagement" with the GEF. Others viewed the GEF's efforts to involve NGOs as token gestures, and they became increasingly critical of the GEF's mandate, structure, and operations.

GEF implementing agencies arranged for "NGO consultations" prior to the second and subsequent participants' meetings, and transmitted NGO statements from these consultations to the participants. The implementing agencies also held a series of consultations with NGOs on problem-area strategies and regional programs. At the project level, they consulted with and hired NGOs to help them design and implement specific projects, particularly in the biodiversity area. Several major international NGOs (including the World Conservation Union [IUCN] and the World Wildlife Fund [WWF]) began to work as executing agencies on GEF projects, and UNDP worked with NGOs in several developing countries on the design and implementation of its small grants program.[11] Activist NGOs (most notably the Environmental Defense Fund, Natural Resources Defense Council, Greenpeace, and the Rainforest Action Network) were not satisfied with the GEF's attempts to involve them in its operations. As UNCED approached, they became more vocal in their criticism of the facility. They denounced the GEF for failing to involve international NGOs in the facility's design, limiting the involvement of local NGOs and communities in project design, excluding NGOs from participants' meetings, and withholding information on GEF projects from public scrutiny. They argued that the GEF's selection of problems reflected Northern government priorities, and that its strategies ignored the need to reform aid, trade, and debt regimes (GEF 1992b, annex 4; GEF 1992d, annex 4; GEF 1993a, annex 7; GEF 1993e, annex 7).

The Restructuring Process and Outcomes

In the spring of 1992, roughly a third of the way through the GEF pilot phase, negotiations on financing for the climate change and biodiversity conventions and for sustainable development activities under Agenda 21 (the action plan for sustainable development drafted as the centerpiece of UNCED) entered their endgames. Developed and developing country governments disagreed sharply on whether and how the GEF should be linked to the conventions. Developed countries stated that unless the GEF were used as the financial mechanism for the conventions, they would not be willing to finance developing countries' convention-related activities. Developing countries continued to argue for either convention-specific funds under the direct control of their respective parties, or a new "green fund" for sustainable development in which developed and developing countries would have equal decision-making authority.

During intensive, last-minute negotiations at UNCED, developed countries won agreement to designate the GEF as the "interim financial mechanism" for the two conventions. In exchange, developing countries won agreement that the GEF's governance structure would become more "democratic" and "transparent." Lewis Preston, president of the World Bank, also promised to seek a $5 billion "green increment" for the next International Development Association (IDA) replenishment, dedicated to supporting implementation of Agenda 21 in developing countries (Johnson 1992: 449).

Following UNCED, the GEF's participants quickly reached agreement on the GEF's relationship to the conventions, and on the outlines of a governance mechanism with an assembly, an executive council, and a "functionally independent secretariat." By the spring of 1993, negotiations had narrowed down to three core issues: the independence of the restructured facility's chairman and secretariat from the World Bank and other implementing agencies; the potential for broadening the GEF's scope to include nonglobal environmental problems identified in Agenda 21; and the balance of representation and decision-making authority between developed and developing countries (see bracketed text in GEF 1993c).[12]

The GEF's relationship with the conventions, one of the most contentious issues during the pilot phase, was relatively easy to resolve after the parties to the conventions agreed to make the GEF their interim financial mechanism. GEF participants agreed that the parties to the climate change and biodiversity conventions would establish the GEF eligibility criteria and program priorities for their respective problem areas. The decisions of the convention parties would thus replace the ad hoc combination of participant, STAP, and implementing agency views that had "guided" the GEF's pilot phase work program. The convention parties would also determine eligibility for GEF funding, presumably limiting funding to countries that had ratified the conventions. The convention parties would not, however, review the work program or make decisions on individual project proposals. These responsibilities would remain with the GEF's participants, reconstituted as a "participants assembly."[13]

The participants assembly (PA) would oversee the GEF's operations in essentially the same way it had during the pilot phase, with two important differences. First, not all countries would have individual representatives. Instead, multicountry constituency groups would send representatives to an executive body, the GEF council. The council would achieve "balanced and equitable representation" of developed and developing countries. It would normally make decisions by consensus. The constituency structure, as well as council voting and decision rules in the absence of consensus, would be negotiated by the participants. Second, the PA and the council would act as the watchdog for the parties to the conventions, ensuring that the GEF's convention-related activities followed their program goals and priorities. A more controversial issue was the division of authority between the GEF secretariat and the council, and the independence of the GEF secretariat from the World Bank and the other implementing agencies. Developed countries pushed hard for a "functionally independent" secretariat supported administratively by the World Bank, but accountable only to the GEF's participants. They also wanted the GEF's chairperson (to be called the "GEF CEO") to chair meetings of the participants. Developing countries pushed equally hard for the creation of a legally separate secretariat and for election of the chairperson from and by the participants. The compromise was a functionally independent secretariat and a CEO-participant co-chair arrangement, with the participant co-chair elected at each semiannual meeting of the GEF's council.

To improve interagency coordination, the GEF chairperson would become an appointed CEO. The GEF CEO would organize meetings of the PA and the council, and would serve as the primary liaison among GEF participants, convention parties, and implementing agencies. The CEO would oversee the work of the implementation committee, mediating interagency conflicts like those that had undermined the GEF's program and project work during the pilot phase. This arrangement would provide the CEO with more independence from the World Bank and more resources than the GEF chair and administrator had commanded during the pilot phase.

After reaching agreement on a broad framework for GEF governance and administration, the participants focused on the issues of the GEF's scope, constituency representation, and decision making in the GEF council. These issues dominated the negotiations from the summer of 1993 until the final agreement was reached in March 1994.

On the question of the GEF's scope, developing countries sought to broaden the mandate to include land degradation, particularly desertification, drought, and deforestation (G-77 1993). Developed countries maintained a united front: they would only fund developing country action on a limited set of global commons problems, using the incremental cost concept as the basis for determining project eligibility. The final agreement was a compromise that allowed funding of the incremental costs of land degradation activities "as they relate to the four focal areas."

The allocation of voting power among developed and developing country constituencies became the final sticking point in the restructuring negotiations.[14] Negotiations nearly collapsed in December 1993 over the linked issues of how many seats each bloc would receive and how large a majority would be necessary if and when votes were taken. The final agreement (GEF 1994a) established a GEF council with fourteen developed country members, sixteen developing country members, and two members from countries in Central and Eastern Europe and the former Soviet Union. Each of these three blocs would decide how to select its representatives. When there was no consensus on an issue and a decision was necessary, approval of a proposal would require a 60 percent majority of donation-weighted votes and a 60 percent majority of countries represented. The CEO and a council member elected at each semiannual meeting of the GEF's council would co-chair council meetings. In a separate agreement, donors agreed to provide roughly $2 billion for the next three years of the GEF's operation, double the amount of the pilot phase.

2 Analysis

This section argues that the GEF's design, pilot phase operations, and restructuring resulted from a limited and shifting convergence of interests among its stakeholders. In 1990, most developed countries were eager to demonstrate their commitment to act on global commons problems, and were willing to experiment with funding developing countries' efforts to assist in global environmental management. Most developing countries were fundamentally less concerned about global commons problems, had far less capacity to analyze or address them at the national level, and were mainly interested in using environmental concern in developed countries as a way to increase aid for their domestic economic and environmental priorities.

These differences in concern and capacity raised potentially serious contracting problems for both blocs: developed countries feared deliberate and/or involuntary redirection of GEF funds by developing countries; developing countries feared de facto redirection of aid flows by developed countries. The GEF's design tried to bridge this gap in two ways: by framing the mandate at a politically acceptable level of generality, and by establishing an institutional structure with diffuse enough lines of authority that each stakeholder would have some power to interpret and renegotiate the mandate during pilot phase operations.

Unfortunately, as the GEF's pilot phase continued, the open-ended and ambiguous quality of its mandate and structure actually intensified conflict. The GEF's lack of institutional clarity gave its stakeholders—most importantly developed and developing country blocs, but also the implementing agencies and interested NGOs—strong incentives to "bargain on the rule" in order to set precedents within the GEF and in the parallel convention and UNCED negotiations.

At UNCED, developed and developing country blocs brought this process to a head by testing each other's willingness to do away with the GEF. Ultimately, each bloc calculated that it would require less time, political, human, and financial resources to restructure the GEF than to start a new round of financial mechanism negotiations. During the subsequent restructuring process developed and developing countries, implementing agencies, and the NGO community continued bargaining on constitutional issues and decision rules. They now recognized, however, that holding to their most extreme positions could only lead to another impasse; their positions were also informed at the margin by a formal process of institutional evaluation and learning. As a result, the restructured GEF incorporates a number of creative compromises on both mandate and structure; as a package, the restructured GEF may satisfy the core interests of its stakeholders far more effectively than it did during the pilot phase.

Defining the GEF's Mandate: The Logic of Incremental Cost
The GEF's mandate resulted from a limited convergence of developed country, developing country, and World Bank interests in linking global commons problems and incremental cost finance. As noted above, during the late 1980s a combination of increased scientific concern, NGO advocacy, and media coverage of global environmental problems increased public concern in developed countries and stimulated developed country political leaders to "do something." On the other hand, not all developed country political leaders experienced the same type or intensity of political pressure to act, and some (particularly the governments of the United States and the United Kingdom) were pressured not to act. Nevertheless, in the fall of 1989 most developed country leaders felt that it would be politically expedient to make at least a symbolic gesture demonstrating their commitment to protect the global environment.[15]

Developing country governments agreed in principle on the need for action on global environmental problems, but expressed a number of important reservations. They warned aid donors against attempting to impose "environmental conditionality" in the allocation of aid funds (NAC 1989; Sjoberg 1994: 26). They also insisted that any new environmental aid funds should be additional to current aid flows, should compensate developing countries for action on highly concessional terms, and should facilitate developing country access to advanced industrial technologies. They also wanted to leave open the possibility that the GEF's mandate could expand to include some of their higher-priority domestic environmental problems, such as desertification, declining agricultural productivity, and fresh water availability.

The World Bank was also ambivalent about the need for a new global environmental finance mechanism. On one side of the Bank's internal debate were members of the Bank's new environment department, who were eager to improve the Bank's treatment of environmental issues and to establish their department as a strong voice in the Bank's strategic decision-making process. On the other side were senior managers in the Bank's finance vice presidency, who viewed the proposal for a new program as a potential threat to their core mission—generating donor contributions for the periodic replenishments of IDA, the Bank's concessional lending arm. The decisive factor resolving this debate may have been senior management's concern that if the Bank failed to respond aggressively to the September 1989 request for a detailed proposal, developed countries would give the global environmental funding mandate to another international organization or bilateral consortium.

The GEF's focus on global commons problems and its incremental cost finance strategy thus satisfied a number of core interests of its most important stakeholders, without resolving several important underlying disagreements. It allowed developed country finance and aid ministry representatives to demonstrate to green constituencies that they were taking action, while also assuring aid and environment skeptics that the new facility would address a need not met by existing programs and that costs would be contained by applying the incremental cost criterion. It allowed developing country representatives to argue that the GEF was an additional, concessional, and unconditional aid window, and that it could be expanded to address more-localized "sustainable development" problems. Finally, it allowed the World Bank to distinguish the GEF from its other lending windows, bolster its credentials as an environmentally friendly aid agency, and establish its leadership in an area of increasing interest to its major donors. It did not, however, resolve the questions of whether or when the GEF's mandate would actually be expanded beyond the initial suite of four global commons problems, nor did it clarify how incremental costs would be measured or used for project selection and funding decisions.

Institutional Structure: Deferring Conflict by Diffusing Authority
As noted above, the key elements in the GEF's institutional structure were its status as a pilot program, informal governance rules, World Bank–led interagency administrative structure, and the significant but limited role of NGOs in its policy and project work. These elements preserved ambiguity on many issues where initial agreement was not possible, creating the preconditions for serious political and operational problems during the pilot phase.

The pilot program designation, as Sjoberg (1994: 29–30) points out, allowed the facility to begin operation without resolving disagreements among its government and implementing agency stakeholders on the scope of the problems it should address, on whether and how it should be linked to the conventions and UNCED, or on whether and how work on global environmental problems should be integrated with the ongoing work of the World Bank and the other implementing agencies.

The GEF's informal, consensus-based governance arrangements used neither Bretton Woods (contribution-weighted votes) nor UN (one-nation, one-vote) rules, thus avoiding immediate conflict over the allocation of decision-making authority. Initially, a majority of government representatives at GEF participants' meetings came from finance and aid ministries. Many served as World Bank executive directors or assistants, and most were comfortable with the relatively informal and consensus-oriented style of World Bank directors' meetings. The informal governance arrangements began to break down, however, as the GEF became more closely linked to the UN-based convention and UNCED negotiations on finance.

The World Bank–led interagency administrative arrangements also occupied a narrow area of common ground among competing institutional interests. Government finance and aid ministry representatives preferred that the World Bank play a leading role in GEF administration, both because they themselves would then play a leading role in its governance, and because they believed that no other multilateral aid agency could manage a $1 billion trust fund with financial prudence. Foreign and environment ministry representatives preferred a balance between the Bank and the UN agencies, based on a similar combination of self-interest and belief that global environmental partnership between developed and developing countries should be reflected in a partnership among Bretton Woods and UN agencies. Environmental NGOs, who exercised indirect influence through their government delegations, opposed giving the Bank a leading role because of its poor environmental track record (see Le Prestre 1989; Rich 1985, 1990). They recognized, however, that some Bank involvement was a prerequisite for donor funding.

The implementing agencies had strong institutional incentives to compete for control over the facility's administration. World Bank managers recognized that some type of interagency arrangement was a prerequisite for funding, and acknowledged the comparative advantages of UNEP in linking the GEF to the conventions and UNDP in providing country-level support for GEF operations. But also they wanted to maintain control of major GEF program, project, and budget decisions. The heads of the other two agencies had each hoped for a leading role in new environmental initiatives, and negotiated hard to expand their institutional influence on the GEF's development and operation and the share of GEF resources their agencies would receive. The initial interagency agreement attached to the GEF's foundation document (World Bank 1991a, annex C) satisfied no one, but met the initial minimum requirements of each actor.

The Pilot Phase: Political Solutions Became Operational Problems
In theory, each of the GEF's main design features could have helped solve the contracting problems that the stakeholders' divergent interests created. In practice, however, these features actually exacerbated underlying conflicts.

Implementing the Mandate: Disjointed Incrementalism   During the pilot phase, the growing conflict over financing the conventions and Agenda 21 undermined the initial agreement on the GEF's mandate and strategy.

As it became clearer that developed countries were not prepared to provide substantial additional resources for implementing Agenda 21, developing countries became more strident in demanding that either some of the GEF's resources be spent on sustainable development problems, or that the GEF be replaced by a new "green fund" that would address both global commons and sustainable development problems (First Ministerial Conference 1991).

The GEF's participants and their counterparts in the UNCED and convention negotiations might have resolved this dispute by recognizing the large area of overlap between the GEF's nominally global problems and many of the problems identified in Agenda 21, and authorizing the GEF to fund programs to address both domestic and global environmental problems. To do so, however, would have undermined the "incremental cost" rationale for the GEF's existence, potentially threatening the facility's political viability in developed countries. Ironically, while the political dispute about the GEF's incremental cost mandate was raging, it was becoming increasingly clear that the incremental cost criterion itself would have to be applied "pragmatically," because of the conceptual and technical difficulty of distinguishing global from local benefits.

Institutional Structure: Ambiguity Undermined Authority   As noted above, the aid and finance ministry staff who created the GEF were initially insulated from North-South diplomatic debates on the governance of international environmental institutions. Many developing country diplomats involved in the convention negotiations, however, interpreted the GEF's "informal" governance structure as a deliberate attempt by developed countries to place control of financial decisions within the World Bank, an institution that they perceived as favoring the interests of developed countries.

As the convention and UNCED negotiations progressed, developed countries might have increased the GEF's legitimacy with developing countries by agreeing to change the GEF's governance arrangements, for example, by accepting a governance structure that would give developing countries greater formal authority in GEF decision making, along the lines of the MPMF executive committee. Instead, they refused to make concessions that would have increased developing countries' ability to determine the allocation of GEF resources. Developed country representatives at the GEF and in the convention negotiations did signal their willingness to renegotiate the GEF's governance structure in the spring of 1992 (see GEF 1992c). By then, however, the gulf between developed and developing country positions had become so wide that the GEF participants' vaguely worded proposal for a "balanced and democratic" GEF governance structure had little impact. Instead of facilitating agreement on financial issues in the convention and UNCED negotiations, as developed countries had hoped, the GEF's governance arrangements became an obstacle to agreement.

The implementing agencies' failure to achieve effective coordination resulted from an analogous conflict of institutional interests in the absence of clear procedures for resolving disagreements. The GEF's strategy of organizational specialization based on comparative advantage presupposed that the agencies and the participants agreed not just on the principle of comparative advantage, but also on the specific areas in which each agency actually possessed a comparative advantage. In several areas—investment project identification and preparation, provision of technical assistance, and refinement and application of technical criteria—there was no such agreement. The GEF's foundation documents and procedural guidelines allowed scope for conflicting interpretations of agency roles in each of these areas, and the GEF's design did not provide any mechanism to resolve the agencies' disputes.

These disputes might have been resolved by the GEF chairman, but because the chairman was a World Bank official, UNDP and UNEP managers did not accept his authority to arbitrate interagency disputes. The agencies might also have brought their disputes to the participants, but they usually did not, perhaps because they feared revealing that they were having trouble meeting the GEF's coordination goals. The participants might have tried to identify or resolve interagency conflicts, but their prior agreement to delegate the GEF's administration and their own disagreements on the relative strengths of the agencies prevented them from doing so. As a result, the agencies (particularly the World Bank and UNDP; UNEP's participation in GEF operational decision making was minimal) reached a de facto agreement to disagree: they established duplicative, uncoordinated procedures for project identification and preparation at the country level, and they limited their efforts to review and critique each other's projects.

Restructuring the GEF: Power and Learning

Political interests and resources and a process of institutional learning combined in the GEF restructuring process to produce changes in the GEF's governance and administrative structure. Politically, in the final days of negotiation on the biodiversity and climate conventions, the developing countries made a credible threat to reject the GEF as the financial mechanism for conventions unless the GEF was restructured to give them a greater share of decision-making power. Developed countries made an equally credible threat not to provide more than token funding for the conventions unless they used the GEF as their financial mechanism. Developing countries wanted additional financial resources. Developed country political leaders did not want a summit of heads of state to fall flat because of what might appear to be "stinginess" on their part (Johnson 1992: 443–450; Ricupero 1993). Agreement was almost as inevitable as its eleventh-hour quality.

After UNCED ratified the GEF's status as the major source of public finance for global environmental problems, both developed and developing country governments continued to have a strong interest in maximizing their control over GEF decision making. At the same time, each bloc recognized that the other could and would desert the facility unless it were assured an equal share of decision-making authority.

A process of self-evaluation and institutional learning also tempered the bargaining during the restructuring process. By commissioning an independent evaluation, the GEF's participants both responded to NGO criticism and also created the potential for an "objective" analysis of the GEF to become a focal point in the restructuring negotiations.

In practice, the evaluation did not have a major impact on the mandate or governance negotiations, but it did significantly affect the negotiations on administrative arrangements. By the end of the pilot phase it had become very clear that the implementing agencies were not able to coordinate their operations effectively, and that participants were not prepared to "micromanage" the interagency programming process. The independent evaluation of the GEF (UNEP, UNDP, and the World Bank 1993: 9

and chapter 9) highlighted these problems and called for the creation of a functionally independent secretariat to manage the program development process. Participants endorsed this recommendation and the implementing agencies accepted it, although not without some resistance.[16]

Although hard North-South bargaining dominated the restructuring process, formal evaluation and informal collective learning appear to be having a greater effect as the restructured GEF begins to clarify its operational strategies. At the first meeting of the GEF council, the evaluation's recommendations were cited explicitly and extensively as the basis for rethinking the GEF's mandate and strategies (see GEF 1994b). The council also agreed to slow the pace of project development so that it does not outrun political consensus on the GEF's mandate or organizational capacity for implementation. The evaluation has also influenced the development of the restructured GEF's operational strategy; the draft operational strategy paper (GEF 1995b) sets programmatic priorities, notes potential trade-offs, and delimits the GEF's scope far more clearly than any previous GEF policy document. Whether the GEF's stakeholders will be able to translate this document into a set of cost-effective investments acceptable to all major stakeholders remains to be seen.

3 Evaluation

During its four-and-a-half-year operating life, the GEF has not been notably successful in meeting its stated goals. According to several independent evaluators (Bowles and Prickett 1993; GEF STAP 1994a, 1994b; UNEP, UNDP, and the World Bank 1993; Wells 1994; Wolf and Reed 1994), it has done little to increase developing country governments' interest in or capacity to manage global environmental problems. Its project portfolio, although still at an early stage of development, has not received high marks for innovation or demonstration value, or for cost-effectiveness. Rather than improving the "contractual environment" (cf. Robert Keohane's introduction to this volume) by facilitating agreement in the UNCED and convention processes, the GEF became a major obstacle to agreement.

The GEF's shortcomings are significant. Its direct contribution to the solution of the problems it is supposed to address will remain small, given the gap between the resources governments are prepared to commit and the human and physical capital investments needed in each of its problem areas. Nevertheless, the GEF has given its stakeholders an opportunity to experiment and learn (though mostly the hard way) about the political, bureaucratic, and technical challenges of managing global environmental problems.

Institutional Design

Institutionally, GEF stakeholders have learned that the problems the facility is designed to address are so closely tied to intergovernmental negotiations and institutions that policy decisions cannot be delegated to a set of implementing agencies. Instead, they have given the parties to the climate and biodiversity conventions the authority to set the GEF's policies and strategic priorities in these areas. They have also recognized the need to create a neutral intermediary body between governments and implementing agencies, in the form of a "functionally independent" CEO and secretariat. Finally, the GEF's governmental stakeholders have decided that the benefits of NGO participation in the GEF's policy deliberations outweigh the drawbacks; in early 1995, the council authorized NGOs to attend meetings of the council as observers.

On the other hand, several important institutional questions remain unresolved. First, there is not yet a clear dividing line between convention "policymaking" and GEF "implementation." The literatures on international organizations (e.g., Cox and Jacobson 1973) and policy implementation (e.g., Mazmanian and Sabatier 1983) and the GEF's operational experience suggest that there will always be a grey area of implementing agency discretion, even with tripartite oversight of the facility by the convention parties, the GEF council, and the CEO. In fact, the tripartite oversight structure itself may reduce the implementing agencies' accountability by generating multiple, potentially conflicting interpretations of policy directives.

Second, the CEO's authority to conciliate, mediate, and arbitrate conflicts among the implementing agencies will not be clear until it is tested and supported by the GEF council. In July 1994, the GEF's council members approved the nomination of the current GEF chairman, Mohamed El-Ashry, as the GEF's first CEO. El-Ashry was "the devil they knew"—

a forceful personality who had not endeared himself to participants, implementing agencies, or NGOs, but whose competence and commitment to the facility were widely acknowledged. Despite the fact that he was nominated jointly by the three implementing agencies, and the fact that the World Bank's relationships with the other implementing agencies have improved in the wake of leadership changes at both UNDP and UNEP, El-Ashry will have to work hard to establish the CEO's legitimacy and authority as an honest broker among the GEF council, the implementing agencies, the convention parties, and interested NGOs.

Third, the extent of NGO and local community representation in the GEF's policy, program, and project activities has not been fully clarified. The GEF secretariat's proposal for mixed government-NGO "national contact groups" as counterparts to the GEF implementing agencies in developing countries was rejected by a majority of developing country participants as an infringement of sovereignty (USCAN 1995). Procedures for information disclosure and local participation in project planning and implementation are still being debated.

Strategy

Political and organizational rather than technical rationality drove the development of the GEF's mandate and strategy. The independent evaluation of the GEF found conceptual and operational problems with several elements of the strategy (UNEP, UNDP, and the World Bank 1993, chapter 4). It faulted the incremental cost and additionality criteria as inadequate and sometimes counterproductive tools for priority setting. It praised the GEF's commitment to innovation and national policy integration, but found that the GEF's implementing agencies generally failed to achieve either goal during the pilot phase. It also noted that participants' desire for "regional balance" (i.e., for spreading the GEF's investments across a large number of countries around the world) directly contradicted their desire to maximize the cost-effectiveness of GEF projects.

The findings of the official independent evaluation are consistent with the findings of several unofficial evaluations of the GEF (Bowles and Prickett 1993; Reed 1993; Wells 1994). All of these evaluations called for the GEF to slow the pace of project development while it reviewed and refined its generic and problem-specific strategies. The independent evaluation also called on the GEF to establish systematic procedures for monitoring and evaluating its work, rather than relying solely on periodic outside evaluation (UNEP, UNDP, and the World Bank 1993: 5,6). When the results of the independent evaluation were presented to the GEF's participants at their December 1993 meeting, they endorsed the idea of a slowdown and reassessment.

In response, the GEF's new secretariat and the core GEF staff at the three implementing agencies have developed a "two-track" strategy for the GEF's second phase (GEF 1994b). The first track involves a comprehensive review of the GEF's strategies in each problem area, with the active participation of the parties to the climate and biodiversity conventions, to be completed by the end of 1995. The goal of this review is to develop a new set of policy and strategy guidelines for GEF operations. The second track authorizes funding for some high-priority projects while the long-term guidelines are being drafted. Top priority during this interim period goes to helping developing countries meet their reporting and planning obligations under the climate and biodiversity conventions. The two-track plan was criticized by some NGOs for allowing the GEF to commit a substantial amount of funds ($300–$400 million) before developing its long-term strategy (Horta 1994).

Concern about rushing into the next phase notwithstanding, the two-track approach appears to be providing the GEF with a useful "breathing space." Since the restructuring agreement was concluded in March 1994, the GEF secretariat and the council have been working with the implementing agencies, convention secretariats, and interested NGOs to develop an overall operational strategy and problem-specific strategies for climate change, biodiversity, ozone depletion, international waters, and land degradation.

The draft operational strategy document (GEF 1995b) reiterates the GEF's commitment to incremental cost financing, shifts the GEF's operational focus from individual projects to "work programs" in each problem area, gives examples of the kind of global work programs the GEF may develop (e.g., supporting the commercialization of renewable energy technologies, protecting classes of critically endangered ecosystems), and prioritizes long-term program investments, with enabling activities, short-term project investments, and targeted research as other possible areas for funding. The current strategic planning process does not aim to resolve the trade-offs among competing strategic goals (e.g., innovation versus regional balance versus cost-effectiveness; additionality and incremental cost versus integration into national programs), but it is likely to make the trade-offs more explicit, and to increase the political legitimacy of the GEF's future operations by building consensus on which types of projects should have the highest priority for GEF funding.

Operations

From the outset, the GEF's stakeholders recognized and debated the merits of two different approaches to GEF operations: one that sought to integrate global issues into the operations of existing international organizations and national agencies, and another that aimed at differentiating global issues by creating new organizations, programs, and procedures focused solely on them (cf. Kjorven 1992). The GEF was supposed to achieve some of the merits of each approach by adding new units and a coordinating structure to existing international agencies, using their links to national environment and development agencies and programs to identify and address global concerns, and providing incremental cost finance to catalyze and compensate national action where necessary.

Operationally, however, the combination of time pressure, staff and financial constraints, ill-defined strategies for either integration or differentiation, and constant suspicion from NGOs that "integration" would amount to nothing more than "green window dressing," led the GEF's implementing agencies to "high-differentiation, low-commitment" responses. Operational differentiation made it clear that global concerns were being addressed, but the small amount of resources devoted to GEF operations made it very difficult for GEF staff to change their agencies' mainstream programming priorities or project selection criteria, much less affect the national priorities of recipient governments.

The independent evaluation highlighted these integration problems; but in this area, the GEF's ability to learn from its experience is highly constrained by continuing ideological and interest-based conflict at the international, national, and organizational levels. Internationally, developing countries remain suspicious that "integration" is the Trojan horse by which developed countries will impose green conditionality on economic development aid. Nationally, many economic policy makers in developing countries remain indifferent to national environmental concerns and are openly dubious about the need to consider the global impacts of their activities (see e.g., WRI 1994, chapter 14). Organizationally, environmental ministries and agencies remain weak actors in national economic decision making, and there are few career positions or rewards for environmental staff in public sectoral agencies or private firms (see e.g., Dick and Bailey 1994).

If the GEF were to pursue a strategy of integration, it would have to begin with the three implementing agencies themselves, particularly the World Bank, which has the most leverage over sectoral agencies and policies in developing countries. The Bank's global environment coordination division (ENVGC), the unit that manages the Bank's GEF operations, is aggressively pursuing a strategy of "mainstreaming" global environmental concerns in the Bank's lending activities (see World Bank ENVGC 1994). According to ENVGC staff, the unit has had some success in informing Bank operations staff about the GEF, and in encouraging staff to consider global issues in their policy and project work. On the other hand, the Bank continues to be internally divided between staff who see the global environment as a significant element of the Bank's overall commitment to supporting sustainable development and staff who see the GEF and the ENVGC unit as largely symbolic gestures to appease donor environmental constituencies. Similarly, UNDP seems strongly committed to its Capacity 21 initiative, which is designed to support the professional and institutional development needs identified in Agenda 21. Its institutional commitment to global environmental concerns is less clear (Williams and Petesch 1993: 38–45). UNEP's staff have developed a strategy paper for increasing UNEP's participation and effectiveness in the GEF's next phase (UNEP 1994), but staffing and resource constraints may limit its ability to play a major role in the GEF.

In short, the goal of integrating global concerns into the activities of existing environmental and economic agencies faces serious structural obstacles at many levels, both within and beyond the GEF. The next phase of the GEF's operations should provide an opportunity to experiment with more intensive and sustained country programming exercises than the facility undertook in its pilot phase, involving not only core GEF staff, but also operations staff and national counterpart agencies. The results of these exercises will give GEF stakeholders a better sense of how serious the obstacles to integration really are, and how they may be overcome.

Conclusion

The GEF's design, operations, and restructuring cannot be neatly explained by any one theory of international institution building. Throughout its brief history, the GEF has experienced structural conflict between developed and developing countries (Krasner 1985), bureaucratic politics within and across international organizations (Allison 1971; Cox and Jacobson 1973), political entrepreneurship by "issue networks" of environmental NGOs and their allies in governments and international organizations (Sikkink 1993), and institutional learning among GEF stakeholders (Haas 1990).

Nevertheless, the analytic framework employed in this volume, focusing on problems of concern, capacity, and contracting, may shed some additional light on the GEF's clouded path. As in the cases of the Montreal Protocol Multilateral Fund, nuclear safety aid, and environmental aid to Eastern Europe, the major actors involved in the GEF's design have been interested not only in increasing aid recipients' concern and capacity to address environmental problems, but also in maximizing their political and organizational control of the contracting arrangements. What makes the GEF distinctive are its "one-stop shopping" mandate to finance action on an indefinite set of global commons problems, the highly uncertain magnitude and distribution of costs and benefits associated with these financial transfers, and the complex and forward-looking institutional linkages among national governments, multilateral negotiations, international agencies, and nongovernmental organizations that are required to implement the transfers.

Compared to most other financial transfer mechanisms analyzed in this volume, the GEF's mandate has commanded only very qualified support from its stakeholders, while the details of its governance and administrative structures have been extremely important to them. Consequently, the GEF's stakeholders have placed a great deal of emphasis on attempting to improve the contractual environment in which financial transfers for current and future global environmental problems might take place. They have spent far less time considering how the GEF might best leverage its limited resources to integrate global environmental concerns into developing countries' economic and environmental activities.

Ironically, the decline in international concern about global environmental problems since UNCED may actually have facilitated the GEF's restructuring and the development of its operational strategy. It now appears that for the next decade at least, developed countries will not substantially increase the amount of official development assistance channeled to global environmental problems, nor does it appear likely that any major new international institution will be created to promote joint management of the global commons. The stakes associated with the GEF's governance rules and policy priorities have become far clearer than they were before UNCED, and for most of the GEF's stakeholders, lower than anticipated. With the reduction in high-level concern about the "implications" of the GEF's mandate and structure, each stakeholder group is adopting a less anxious, less ambitious, and more pragmatic approach to the GEF's operations.

For the GEF to come into existence at all, it had to claim the small area of common ground among its diverse stakeholders, but this area later proved too cramped to allow the GEF to function effectively. During its operating life, the need for a strategy that would allow the facility to address problems that were both of pressing concern to developing countries and directly relevant to its global mandate became abundantly clear. The restructuring agreement and the current strategy-setting process may allow the GEF to achieve greater integration of national and global priorities. If so, the GEF's future contributions to protecting the global environment may greatly overshadow its past shortcomings.

Notes

Acknowledgments: The author wishes to thank the editors and members of the Harvard CFIA/CSIA Environmental Transfers group, participants in its workshops and in the Harvard-MIT Joint Research Seminar on International Environmental Affairs, three MIT Press reviewers, and several government, implementing agency, and NGO representatives involved in the design and operation of the GEF for their comments on drafts of this chapter. He also wishes to thank the Harvard Center for International Affairs, the U.S. Department of Education, and the Program on Negotiation at Harvard Law School for financial and logistical support. In addition to the published sources cited in this chapter, the author is indebted to numerous members of GEF participant governments, implementing agencies, the Scientific and Technical Advisory Panel, and interested NGOs for providing information in not-for-attribution interviews, and providing access to documents that were not part of the public record. All errors of fact and interpretation that remain are the responsibility of the author.

1. The GEF has been administered by the World Bank, in close collaboration with the UNDP and UNEP. Its participating governments (participants) include most of the OECD states and a growing number of developing countries (initially 7, currently more than 140). As of the summer of 1995, it had authorized $736 million in grants for 115 projects in 63 countries. The GEF funds three types of projects: investments, technical assistance, and applied research. The GEF also administers a small grants program for NGOs in developing countries; to date, this program has funded over 200 small NGO projects in 23 countries. Most of the GEF's projects are still at an early stage in their development; of the $742 million authorized, less than $100 million has been disbursed (GEF 1995a).

2. French and German finance ministry representatives made the request at a meeting of the development committee (the advisory and planning committee of the board of governors of the World Bank and the International Monetary Fund). Several other developed country representatives supported the request. France also promised to commit $100 million if other donors contributed enough to create a $1–1.5 billion fund.

3. Sjoberg (1994) provides an excellent account of the GEF's design process, with particular attention to actors and debates within the World Bank.

4. In consultation with potential donors, the GEF's designers also established a rough division of GEF funds among its four problem areas. They proposed that 40 percent to 50 percent of the GEF's resources be used for greenhouse gas reduction projects, 30 percent to 40 percent for biodiversity protection, 10 percent to 20 percent for international waters, and roughly 5 percent for ozone-depleting substance (ODS) reduction projects (GEF STAP 1992: 3).

5. Although the GEF's designers originally expected the facility to serve as the financial transfer mechanism for the Montreal Protocol on ozone depletion, the parties to the protocol decided in the summer of 1990 to create a separate mechanism, the Montreal Protocol Multilateral Fund (MPMF). (See DeSombre and Kauffman in this volume). Subsequently, the GEF's mandate to address ozone depletion was limited to a small number of Eastern European and former Soviet bloc countries which were ineligible for support through the MPMF (GEF 1991b: 8).

6. Initially, both developed and developing countries were required to contribute at least SDR 4 million to become GEF participants. This requirement was waived after UNCED. The GEF did restrict eligibility for project funding to countries with GNP per capita of $4,000 or less, the same threshold used by the World Bank for its International Bank for Reconstruction and Development (IBRD) loans.

7. The chairman at the first participants' meeting was Wilfried Thalwitz, then the Bank's vice president for policy, research, and external affairs. He was replaced during the spring of 1991 by Mohamed El-Ashry, who continued to serve as the GEF chairman until the end of the pilot phase. El-Ashry was an Egyptian-born environmentalist who had served as vice president of the World Resources Institute. He became the second director of the Bank's environment department in 1990, and continued to chair the GEF and head the environment department throughout the pilot phase. He has subsequently been appointed the chief executive officer of the restructured GEF.

8. In their three pre-UNCED meetings (May and December 1991, April 1992), participants committed $584 million for 72 projects (GEF 1992a: 9). In the three meetings that followed UNCED (December 1992, May and December 1993), the participants committed an additional $143 million for 41 additional projects (calculated from GEF 1992a, 9 and GEF 1993e, 6). Thus, the GEF's project portfolio was "front-loaded": 60 percent of its projects and 80 percent of its funds were committed in the first half of its three-year operating life.

9. Because its work on ozone depletion was largely preempted by the MPMF, the GEF funded only one ozone depletion project during the pilot phase. The following discussion focuses on climate change, biodiversity loss, and international waters, the areas in which the GEF committed substantial resources and for which independent evaluations are available.

10. The GEF's participating governments commissioned an independent evaluation of the GEF in December 1992, at the urging of environmental NGOs. Its results were presented to the participants in draft form in September 1993, and a final report (UNEP, UNDP, and the World Bank 1993) was presented in December 1993. See Fairman (1994) for a review of the independent evaluation.

11. This program established national NGO committees to review and select small (less than $50,000) proposals submitted by national and local NGOs. NGO committee activities are supervised by UNDP's resident representative (GEF 1991b: 17–18).

12. Parallel to the restructuring negotiations, GEF donors discussed the replenishment of the GEF's resources. At UNCED, donors had indicated that they would be willing to contribute up to $3 billion to a three-year "GEF II" (GEF 1993b). The replenishment negotiations continued through the spring of 1994, not primarily because of donor disagreements on burden sharing, but rather because donors wanted to make it clear that the replenishment amount would be proportional to their influence in governing the restructured GEF.

13. Although it was relatively easy for the GEF negotiators to reach agreement on the GEF's relationship with the conventions, it is not yet clear whether the restructured GEF will satisfy the concerns of developing country parties to the climate change and biodiversity conventions. As of the summer of 1995, the first Conference of the Parties to the Framework Convention on Climate Change (FCCC) has endorsed the continuing use of the GEF as the interim financial mechanism, but will not make a final decision until 1999. It seems likely, however, that

agreement will be reached and that the GEF will become the primary financial mechanism for the FCCC (Victor and Salt 1994). There is more uncertainty about the GEF's relationship to the Convention on Biological Diversity. Although the first Conference of the Parties in late 1994 agreed to continue using the GEF as its interim financial mechanism, developing country representatives continued to complain that they do not have adequate representation in GEF decision making on biodiversity issues, and that the GEF has too much autonomy from the Conference of the Parties (Earth Negotiations Bulletin 1994).

14. McNeil (1994) documents proposals and negotiations on voting mechanisms during the GEF's restructuring.

15. The United States position on the GEF was particularly complex. On the right, Bush administration and congressional conservatives who distrusted multilateral environmental initiatives and institutions did not want to signal any U.S. commitment to sign the proposed climate or biodiversity conventions, let alone finance them. On the left, NGOs and their congressional allies wanted to use the political window of opportunity provided by the GEF to leverage change in the World Bank's economic development lending practices. They also supported multilateral action on global environmental problems, but did not want the Bank to dominate implementation of the UNCED and convention agreements. In the bureaucracy, Treasury officials did not wish to see the new initiative evolve into a "free-standing" facility outside the multilateral bank system and therefore outside their control. Occupying the narrow area of common ground among these contending factions, the U.S. GEF delegation evolved a position strongly supportive of the GEF's global commons and incremental costs mandate, but initially noncommittal on the question of whether the GEF should be used as the financial mechanism for the conventions.

In contrast to the United States, most European governments were less constrained by domestic skepticism about global environmental problems or hostility to multilateral institutions, and more interested in capturing support from domestic environmental constituencies. Most wanted the proposed facility to evolve into some sort of permanent financial mechanism related to the ozone, climate, and biodiversity conventions, and perhaps to a sustainable development program created through UNCED. On the other hand, most European donor agency representatives agreed that they could only gain political and financial support for a new environmental aid program if it were clearly distinguished from existing aid flows. Most of them indicated that political support would be strongest for an initiative specifically focused on global environmental problems (Sjoberg 1994: 24).

16. The implementing agencies endorsed the independent evaluation's recommendations, but argued that cooperation and coordination had improved during the pilot phase (GEF 1993d: 6). They underscored their responsibility and authority for the GEF's work program and resisted the independent evaluation's recommendation (UNEP, UNDP, and the World Bank 1993: 10) that the GEF's project identification process be modified so that other agencies could submit project proposals directly to the new GEF secretariat.

4

The Montreal Protocol Multilateral Fund: Partial Success Story

Elizabeth R. DeSombre and Joanne Kauffman

The Montreal Protocol on Substances That Deplete the Ozone Layer was a landmark agreement in international cooperation. The goal of the treaty and subsequent strengthening amendments was to prevent depletion of stratospheric ozone—the earth's thin veil of protection against the potentially catastrophic effects of ultraviolet (UV) radiation from sunlight. Initially the treaty required that parties reduce their use of regulated ozone-depleting substances by 50 percent (relative to 1986 levels) by 1999; later amendments required phaseout of chlorofluorocarbons (CFCs) and some other ozone-depleting substances (ODS) completely by 1996, halons by 1994, and other substances in gradual steps by 2030. Developing countries initially had little incentive to participate in a global agreement that would preclude their use of a class of inexpensive industrial chemicals that had contributed significantly to development of the industrial world, despite provisions in the treaty to delay their obligations for ten years. Yet, they ultimately did. In a hard fought negotiation, developing countries won the commitment of developed countries to assist them financially and technically to meet their obligations under the Montreal Protocol.

The Montreal Protocol Multilateral Fund was established in 1990 by the parties to the protocol to assist developing countries with the costs of phasing out substances that deplete the ozone layer. This mechanism is surprising in several key ways. The strong North-South conflict evident in the negotiating stage, with both sides concerned about setting important precedents, could have blocked meaningful agreement. The Fund's use of existing organizations (whose ability to implement environmentally friendly financial policies was already in question) could have led to bureaucratic inertia, duplicated effort, and inefficiency. Most important, the financing could have been used simply as a bribe to buy the acquiescence of reluctant developing states, without assessing the merit of the uses to which the funding would be put.

The Fund has functioned more smoothly than might have been expected, however, avoiding some of the problems encountered by the similarly constituted Global Environment Facility (see Fairman, chapter 3). The Montreal Protocol process created a well-elaborated, finite, contractual mechanism for channeling developed country funding to projects specifically aimed at phasing out ozone-depleting substance use in developing countries. Each side had a strong bargaining position: the developed countries had the funding and technology the developing countries wanted, and the developing countries had a credible threat to undermine the process if their needs were not met. Compromises were therefore inevitable if a funding mechanism were to be created. The fact that all influential participants, North and South, environmentalist and industry, gained something from the creation of a funding mechanism made the necessary compromises possible. And characteristics of the environmental problem itself, combined with a negotiating process that kept the mechanism issue- and time-bounded, made the creation of the Fund easier and less risky than it might otherwise have been. Working within the framework of the Montreal Protocol aided agreement on the Fund since the parties (especially those from developed countries) had already agreed to take certain steps toward phasing out ODS that provided incentives to push for universal participation. The scientific process started under the Montreal Protocol both reduced uncertainty about the need and the possibility for phaseout, and gave industries an incentive to agree to regulation and push for funding so that developing countries could provide markets for their substitutes.

The Montreal Protocol process and the fund it created have not, however, solved all potential contractual problems. Contradictions in design, generated by the multistage process by which the Protocol and then the Fund were created, threaten to undermine some of the success brought about by having an otherwise well-constructed funding mechanism. In particular, the fact that developing countries are allowed to increase their consumption and even production of ODS for an additional ten years before meeting Montreal Protocol regulations could result in a situation in which developing country producer nations are able to demand additional compensation for eliminating an increased ODS capacity at a time when the developed countries will already have phased out their ODS industries. Compensating countries for phaseout activities when they are not yet legally required to phase out ODS may not be economically efficient; although in some cases countries may phase out earlier than required, in others they will retrofit old plants while building new ones that make or use ODS. Slowness in implementation, a side effect of the way the Fund is structured, means that its ultimate success is hard to predict, given that much of the work that it has undertaken has yet to be done. And industry support of the Protocol and the Fund, which helped both come into being, is a mixed blessing: it influences the types of solutions that are considered, leading to technical and chemical solutions in cases where those might not be the best options. In short, the successes in the design of the Fund may not result in as positive (or timely) an outcome for the ozone layer as originally foreseen by some.

In this chapter we argue that the Multilateral Fund is contributing to the worldwide implementation of the Montreal Protocol, and provides important lessons for all parties concerned about how to address environmental problems with implications for both developed and developing countries. At the same time, our assessment is tempered with caution. Contradictions between the design of the mechanism and its goal to facilitate an earlier phaseout of ODS in developing countries than that required by the Protocol itself raise the specter of potential contractual problems in the future. We first describe the negotiations that led to the creation of a relatively well-designed financial transfer mechanism for a global environmental goal. We analyze the factors that enabled such a mechanism to be created and that influence the way that it operates. In particular, the structure of the issue area and the relative power and interests of the actors involved allowed both developed and developing countries to influence the form of the fund and allowed for productive trade-offs between symbolic and material politics. We then evaluate the successes and failures of the Fund and derive lessons from the Fund's experiences for financial transfers for environmental protection in general. The oversight that has evolved from the structure of the Fund is generally positive (although cumbersome), whereas the contracting problems arising from the multistage process in which the Fund was created threaten to undermine its effectiveness.

The Montreal Protocol Fund is in some ways the "easiest" case examined in this book. The Fund involves a potential win-win situation for all parties concerned. If the Fund works as planned, donor countries gain the full participation of developing countries in ODS phaseout (thereby avoiding the negation of developed country sacrifices to save the ozone layer), the industries that developed substitutes for ODS (largely and importantly the same companies whose chemicals were regulated by the Montreal Protocol) gain new markets for their products, recipient countries gain access to new technology at minimal cost and receive assistance in building administrative capacity for dealing with environmental issues, and the world gains the long-term advantage of an improved ozone layer. The cost of phasing out the global use of ozone-depleting substances is low compared to the costs of addressing other environmental problems examined in this volume, consensus was high on both the causes of and solutions to the ozone depletion problem, and the countries most responsible for causing the problem were also the most concerned about solving it and the most capable of financing an effort to encourage participation by those less responsible and less concerned. The successes and failures of this mechanism thus have added significance for overall conclusions: if problems thwart the overall effectiveness in this simple case, it will be considerably more difficult to design financial transfer mechanisms to address more complicated environmental problems.

1   Description

Diminution of stratospheric ozone, a thin veil of gas that protects the Earth, its crops, and inhabitants from the harmful effects of ultraviolet (UV) lightwaves from the sun, is a global problem of potentially catastrophic proportions. The predicted effects of a significant increase in exposure to ultraviolet rays include large increases in skin cancer and eye diseases, damage to the food chain, a weakening of the immune system, and strains on food production (UNEP 1995, Report of the Economic Options Committee: 1–2). Although potentially devastating to the health, environment, and economy of every country on the face of the earth, the ozone depletion problem was most salient to the governments of northern industrialized countries. At the time of international agreement to regulate these substances, their industries produced nearly all of the ODS sold and consumed more than 80 percent of them. Moreover, concern about the potential effects of increased ultraviolet rays was much higher in their populations than it was in developing countries.

CFCs, the chemicals first seen to threaten the ozone layer, came onto the market in the 1930s. Because they were nontoxic, inert, and provided a safe and effective substitute for flammable chemicals that had been used in refrigeration prior to their development, CFCs were hailed as miracle chemicals (Cagin and Dray 1993). Used in such products as aerosol sprays, refrigeration and mobile air conditioning, packaging foams, and solvent cleaners for electronic components, they contributed significantly to advances in food preservation and distribution and to northern hemisphere industrial development over the last sixty years. In 1974 two U.S. scientists posited that CFCs released in production, use, and disposal would, because of their long life, eventually make their way to the atmosphere where in reaction with sunlight they would decompose and release chlorine. Over a long period of time this chlorine could degrade the stratospheric ozone layer (Molina and Rowland 1974). A few countries, led by the United States (home to DuPont, the world's largest CFC producer), took unilateral regulatory action to curb CFC consumption in the wake of the Molina/Rowland announcement. But it was clear from early on that to be effective any regulatory curbs would have to be international.

What made the problem of ozone depletion potentially difficult to address? First, although the scientific evidence for ozone depletion was well developed, the predicted environmental damage was not yet visible. The effects of ozone depletion can take place decades after the chemicals that cause the damage enter the atmosphere. Thus there is a need for regulation before the environmental harm has been felt. Second, ozone depletion is truly an international issue. The depletion of the ozone layer is not contained over the countries from which the ODS are released, and the likely growth in the use of these substances by developing countries meant that most countries of the world had to be involved in the control of these substances for the environmental damage to be prevented or reversed. Third, ozone depletion played into the existing North-South conflicts about environment and development. It fit the model of a growing number of environmental problems in which Northern development caused environmental damage and regulation of the activities in question threatened to prevent the South from using the same means of development. Finally, the economic consequences of regulation were potentially high for a number of the actors involved. For the developed country industries, losing their ability to market their ODS products could have involved significant economic disruption, and development of alternatives would not be economically feasible absent assured markets for these substances. For developing countries, the costs of developing without this safe inexpensive technology could be high. It is in the context of these problems that the negotiations for the Montreal Protocol and later the Multilateral Fund must be examined.

The Montreal Protocol and North-South Divisions

The challenge to negotiators in favor of global restrictions on ODS use was to demonstrate to developing countries that participation in CFC regulations was essential to their future well-being and that their development goals would not be retarded by such participation. At the time of the signing of the Montreal Protocol more than half of the developing country demand for CFCs and ODS-containing products was met by imports. The developing country fear was that a worldwide shift to more expensive substitutes would increase the price and decrease the availability of CFCs. Developing countries would thus pay a higher premium than the industrialized countries for comparable development that required CFCs or their substitutes. In other words, they would pay the price for environmental damage from industrial development in the North. Deeply suspicious of what was perceived as "environmental imperialism" to stunt their development, many developing countries refrained from formal participation in the Montreal Protocol negotiations (Tolba 1994).

By the late 1980s developing country markets for CFCs were expanding rapidly. Developed countries recognized that rising demand in the developing world could negate their efforts to reduce ozone depletion unless such demand were met through the use of ODS substitute technologies. The key countries to involve in any regulatory regime were likely to be India and China, with an expected joint consumption of around one-third of the world total consumption of CFCs (or their substitutes) by 2008 (FOE 1990; Greenpeace 1993). Moreover, although developing country demand was being met primarily by imports, at least eight developing countries had the potential or actual capacity to produce CFCs immediately. Without their participation, the long-term viability of international agreement to protect the world from stratospheric ozone depletion was in jeopardy.

The global environmental problem of ozone depletion thus led to two related political problems: how to guarantee participation of reluctant countries and how to ensure that such participation would not in any way compromise economic development in developing countries. The negotiations in Montreal in September 1987 began the process that brought these two problems together.

Several significant events that preceded the Montreal meeting strengthened support to act on the side of the precautionary principle—that is, before negative effects become manifest. First was the unexpected discovery in 1985 by British scientist Joe Farman of the Antarctic ozone hole, a vast thinning of stratospheric ozone over the South Pole. Although scientists could not at that point explain the source of the thinning, its unexpected appearance rekindled intense interest in the possibility of anthropogenic causes. The image of a hole in the sky raised concern about the Earth's vulnerability and galvanized public opinion in many industrialized countries, especially Germany, where the Green Party was rapidly gaining stature in national politics. Governments thus came under increasing domestic political pressure to take decisive action (Kauffman 1994). In addition, in contrast to the strongly held position of producers that CFCs were irreplaceable for many uses, DuPont announced in the fall of 1986 that the development of chemical substitutes for CFCs was possible given the right market incentives. In other words, the company would invest the necessary capital to produce more-expensive CFC substitutes given the push that regulation would give other industries to use them (see Oye and Maxwell 1995).

The combined effect of these forces led twenty-seven countries and the European Community to sign the Montreal Protocol in September 1987.

In it, parties agreed to a 50 percent CFC phaseout by the turn of the century and a ten-year delay in implementation for developing countries during which time they could continue to increase their consumption and even production of ODS to meet basic domestic needs. The time lag was coupled with trade sanctions against countries that stayed outside of the agreement. Parties to the Protocol are first required to ban trade with nonparties in controlled substances, then trade in products containing controlled substances, and finally to consider restricting trade in products produced with (but not containing) controlled substances. (Montreal Protocol, Article 4). These sanctions threatened export-based developing country economies with the prospect of finding themselves with outmoded technologies and undesired products to offer international markets in the future. Article 10 of the Protocol acknowledged the special needs of developing countries for financial and technical assistance to meet their obligations under the terms of the Protocol, without specifying a solution. Countries eligible for such assistance were designated in Article 5 of the Protocol as developing countries that consume less than 0.3 kilograms/capita ODS per year. The Protocol also allowed for revisions in light of new scientific evidence.

Within three months of the signing of the Montreal Protocol, a scientific study definitively linked CFCs with ozone layer thinning, and another revealed more rapid and widespread ozone depletion than had previously been predicted (Watson 1988). The flexibility of the protocol, combined with further scientific understanding of the seriousness of the problem and its cause, provided both a stimulus and a window of opportunity for developing countries to bargain for compensation for agreeing to regulate their production and use of ODS. Initially only one developing country CFC producer, Mexico, signed the Protocol. Others withheld support pending clarification of the vague promise of "assistance" in Article 10 of the new Protocol.

Specifying the Financial Transfer Mechanism

The developing countries feared that assistance to meet their obligations under the provisions of the Montreal Protocol might dilute existing development aid and insisted that ozone assistance be "additional to other financial transfers" (UNEP/OzL.Pro WGII(2)/7). For their part, the advanced industrialized countries that would pay the compensation were concerned about the precedent such funding would set for other global environmental initiatives. At the very least, they wanted to ensure accountability for oversight of any funds earmarked for ODS reductions, through such organizations as the World Bank. In contrast, developing countries, suspicious of the Bretton Woods financial institutions and particularly the World Bank for not respecting the autonomy of developing countries, called for the creation of new institutions to administer environmental aid. This position was supported by environmental NGOs, similarly critical of the World Bank for its historical lack of interest in environmental concerns.

A "synthesis report" of the Technical and Economic Options Committees confirmed that developing countries would, indeed, require assistance if they were to meet their obligations in an equitable way. Although the Parties ratified the synthesis report at their first meeting in Helsinki in 1989 in a formal declaration acknowledging the need for assistance, the intensity of the debate over the form such assistance should take did not diminish. Disagreements were aired in the numerous forums organized by UNEP Director Mostafa Tolba to try to reach consensus and a decision in the debate that followed the two proposals. While UNEP and some of the Nordic countries favored creation of a new global environment fund, the major aid donors—the United States, United Kingdom, Japan, and Germany—wanted such assistance to pass through existing aid agencies (in particular the World Bank) that they controlled.

Developing countries argued forcefully for additional aid. Malaysia, an early (1989) signatory to the Protocol, characterized it as "inequitable," claiming that developing countries had been unaware of the full socioeconomic implications of the treaty and had thus been duped at Montreal. Delegates from both China and India expressed outrage at the idea that developing countries would have to forego the CFC-containing technologies or pay more for substitutes and thereby enrich the very chemical industries that had created the ozone problem in the first place (FOE 1990: 13). India and China refused to ratify the Protocol until issues of assistance were resolved (Benedick, chapter 13).

Tough negotiations in working parties called by Tolba between 1988 and 1990 finally resulted in a compromise to which the Parties to the

Protocol could agree at their second meeting in London in June 1990. During this period case studies were carried out in several developing countries to predict what the price of assistance would be. Results indicated that a safety net to cover expected shortfalls or "incremental costs" associated with transition from regulated substances to alternatives and substitutes would require somewhere between U.S. $1.5 billion and $5 billion over a ten- to eighteen-year period (UNEP/OzL.Pro WG.II(2)/7:6). These figures helped reassure potential donor countries that the stakes of setting up an interim fund were small enough and time-bounded enough to warrant experimentation. Moreover, the largest donors from CFC-producing states could view contributions as an investment in future markets for their producers. Because their industries would be those with sufficient capital and other research and development (R & D) resources to develop new technologies, they could expect a substantial return on investment as Article 5 countries began the transition to non-ODS technologies.

Just when international agreement was emerging, conflict within the United States threatened to undermine such a funding program (Weisskopf 1990). As potentially the largest donor to any compensation mechanism, U.S. support was crucial. Yet, in the months leading up to the second meeting of the Parties to the Montreal Protocol in London in late June 1990, U.S. delegates were instructed by the White House to oppose plans for a separate $100 million fund to help developing nations reduce the use of ODS ("U.S. instructions to meeting in Geneva" 1990). This blocking action by the United States had the potential to derail the compensation effort and concomitantly the fragile alliance between developed and developing countries. Under pressure from U.S.-based NGOs (including, importantly, U.S. industry), other parties to the Protocol, and developing countries, the Bush administration ultimately reversed itself and supported a fund. With U.S. opposition appeased, the Fund could go forward as negotiated, and it was established at the London meeting in a new Article 10 to the Protocol (Dumanoski 1990; White House 1990).

The "Interim Fund" established by the London amendments to the Montreal Protocol represented a victory for developing countries in the existence of a funding mechanism in general, and a compromise between developed and developing countries in terms of the structure of the fund that was established. Most important was the representation and voting structure of the Fund. Developed and developing countries are given equal representation on the fourteen-member executive committee (Ex-Com), which is responsible for administration and implementation of the Fund. Members of the ExCom are selected by parties to represent them (individually and on a regional basis) for a three-year period. The chair and vice-chair, who serve for one-year terms, are chosen from among the fourteen members of the ExCom. One of the two officers represents the developed countries and the other the developing countries on an alternating basis. In accordance with procedures followed for the Montreal Protocol negotiations, decisions of the ExCom are taken by consensus. If consensus is not possible, decisions require a two-thirds majority vote that must represent a majority among both Article 5 and non–Article 5 countries. Although the specified voting procedures appear complicated, in practice no actual votes have ever been necessary for decisions on either the Montreal Protocol or within the ExCom.

The committee is assisted by a secretariat based in Montreal, Canada, which offered to host the administrative body and assume the related administrative costs. The separation of the secretariat from the UNEP headquarters in Nairobi annually saves the Fund a 13 percent administrative support charge that it would otherwise be required to pay, and emphasizes the independence of the secretariat from the agencies it oversees. The decisions of the ExCom are executed by four implementing agencies—three UN agencies and the World Bank. The terms of reference of the Fund suggest a dominant role in implementation for the World Bank and a straightforward preinvestment supporting role for the UN agencies with tasks designated to correspond with their particular expertise. UNEP, precluded by its charter from undertaking investment projects, acts as treasurer of the Fund. As such it receives contributions from donors and issues approved funds to the other implementing agencies. Its Paris-based Industry and Environment Programme Activity Centre (IE/PAC) also is charged with promoting the objectives of the Montreal Protocol in Article 5 countries and acting as a clearing house for information on substitute technologies and regulatory approaches to controlling ODS. The UNDP is charged with preinvestment feasibility studies, technology demonstrations, pilot projects, and the training associated with these projects. The UN Industrial Development Organization (UNIDO), which did not become an implementing agency for the Fund until 1993, was brought on board because of its existing environment program for industry including some ODS reduction projects. The World Bank is charged with managing the bulk of the Fund's investment projects—that is, money to firms through Article 5 governments for the transition to non-ODS technologies (UNEP/OzL.Pro.2/3). In addition to the four implementing agencies, the Fund rules also allow ongoing bilateral programs to be counted as part of a country's contribution to the Fund (up to 20 percent of its assessed contribution).

Funds are contributed by parties to the Protocol whose annual consumption of controlled substances exceeds 0.3 kilograms per capita. In a compromise that provided a symbolic victory to the developed countries and a material victory to the developing countries, contributions to the Fund are based on the UN scale of assessments. In effect this calculation means that the three largest contributors are those that contributed most significantly to the ozone depletion problem in the first place: the countries of the European Community (which pays 34.67 percent of the budget of the fund), the United States (25 percent), and Japan (13.39 percent). This system avoids the symbolism of creating a "polluter pays" system while nonetheless ensuring that the largest "polluters" pay the most.

Recipient countries are asked to submit country programs to the ExCom with information on the source, application, and use of controlled substances, as well as the policies governing these substances. Ideally, the programs include a description of government and industry activities undertaken in response to the Protocol's requirements and an action plan, timetable, and budget for ODS removal. Proposals for projects are evaluated on the basis of cost-effectiveness in light of the beneficiary's economic and industrial strategies, the accuracy of accounting to avoid double-counting, the savings or benefits derived from the proposed project, and the proposed time frame (UNEP/OzL.Pro.3/8). Investment projects costing more than $500,000 are reviewed and approved individually by the ExCom, while those under this figure are considered within the annual work programs of the implementing agencies. Once the ExCom approves the country program, along with any projects over $500,000, the agencies may proceed with full preparation, appraisal, and, subject to acceptance of their work programs by the ExCom, implementation.

The reward for the specification of the Fund was the immediate acceptance of the Protocol provisions by China, followed by Brazil and India in 1992, and ultimately by most developing countries. Initially capitalized at $160 million for the three-year period 1991–1993, the Fund was increased to $240 million when India and China joined. By 1994, ninety-three of the one hundred and forty signatories of the Montreal Protocol were Article 5 countries, and funding for the second three-year period of operation was increased to $510 million (with $40 million carried over from the preceding period) for the period 1994–1996. The Fund is expected to remain in operation until 2010, when the developing countries are required to have phased out ODS under the ten-year grace period.

The Operation of the Fund

The unique features of the Multilateral Fund, including equal voting power, an independent secretariat, and a mix of implementing agencies between the Bretton Woods and UN systems helped to build confidence in both donor and recipient countries despite a troubled first year of contentious bargaining over policy guidelines, incremental costs, and administrative procedures. In the first year of the Fund's operation, not a single investment project was launched. In 1992, when the Fund was reviewed midway through its first three-year funding cycle, arguments were advanced by three major donors, France, Japan, and the United Kingdom, to fold the Fund into the World Bank–controlled Global Environment Fund to consolidate all environmental commitments and to maintain donor control. But others, including the U.S. delegation (buttressed in part by congressional hearings at which U.S. industry expressed support for the Fund) argued for its continuation as an independent body and prevailed (U.S. House of Representatives 1992). Most donors by then had learned that one of their previous fears, lack of control over a process in a new organization, had not been realized.

2   Analysis

The creation of a funding mechanism along with a global regulatory agreement came largely because of the conjunction of interests of those most concerned, responsible, and capable of dealing with the problem.

For the ozone depletion problem, those most concerned initially were those most responsible for the environmental problem, and also those with the most money. Without the strong concern on the part of the donor countries the money for the fund would not have been provided. The fact that the amount of funding required was small, compared with the costs anticipated for addressing other, more complicated, environmental issues also helped make creation of the Fund possible. And the influence and unity of the developing countries in calling for a funding mechanism made creation of a multilateral fund necessary if universal participation in the Montreal Protocol process were to be achieved.

The Impact of High Levels of Concern

Scientific research helped to intensify concern in developed countries. As further evidence emerged in the course of the Montreal Protocol process connecting CFCs and other chemicals to ozone depletion and showing the seriousness of the ozone depletion problem, the level of concern likewise increased. In addition, this scientific evidence contributed to the belief that the problem had been correctly identified and that the control measures taken could stop and ultimately reverse ozone depletion—an added incentive for states to agree to control measures. The evidence supported the notion as well that the phaseout of these chemicals could happen within a finite and relatively short time period.

Economic incentives also contributed to "concern" broadly defined. Developed country industries that produced the regulated chemicals were largely the ones that also produced the substitutes and therefore gained financially from the Protocol provisions requiring the world to change over to their substitutes. They therefore contributed to the call for increasingly stringent regulations, or at least did not stand in the way of the regulatory process. Developing country industries, fearing they would have to change over to non-ODS technology anyway, wanted assistance to do so.

The Montreal Protocol itself addressed a number of the inherent contractual difficulties between developed and developing countries. One of these was the fear on the part of the developed countries about entering into open-ended funding commitments that might be used to extract indefinite reparations for environmental problems caused during developed country industrialization. The fact that the scientific process was increasingly well understood and that the ozone depletion problem itself seemed bounded contributed to the ability to create a finite mechanism for dealing both with phaseout of ODS and with compensation of developing countries for the costs of phaseout. Another was the fear by developing countries that environmental negotiations might be used to thwart their development. Measures within the Protocol for continuing scientific evaluation and acceptance of the principle of compensation for developing countries helped ease this fear. Thus, in addition to defining the problem and providing a framework for solution, the Montreal Protocol also provides legitimacy to the objectives and establishment of a financial transfer mechanism for phasing out ODS.

If the scientific evidence of the need to phase out ozone-depleting substances was growing more convincing, and the technology-forcing effects of the Protocol would serve to encourage major industries to phase out ODS, why was there a fund at all? The Montreal Protocol's ten-year grace period for developing countries and trade sanctions with nonparties would certainly have contributed to encouraging developing countries to join the Montreal Protocol, especially those with strong interest in developing export-based economies. But the developing countries knew their participation was essential for phasing out ODS, and thus had a lot of leverage. This influence led the push for a fund in addition to the other methods for encouraging participation. The costs of phasing out or forgoing use of ODS, although difficult to predict in the face of evolving substitute markets, are substantial for developing countries for whom these substances would have been an important part of industrial development. China and India in particular refused to join the Montreal Protocol if they were not compensated, and the fact that these countries produced ODS for growing domestic markets meant that trade sanctions would not wipe out their access to or profit from ODS.

In short, the development of a funding mechanism to address the problems of ozone depletion benefited from a variety of factors that did not exist in the otherwise similar Global Environment Facility. While both funding mechanisms arose initially from high levels of concern on the part of developed countries, in the case of ozone there was strong scientific evidence of a global environmental threat, its causes and solutions were understood, and it was widely believed the problem could be solved in a finite period of time. At the same time, developing countries knew their participation in the Montreal Protocol process was necessary, which gave them a lot of leverage to hold out for as much compensation as they thought they could get. These parameters help explain the shape of the ultimate agreement and funding mechanism that emerged to address the environmental problem of ozone depletion.

The Form of the Fund

The design of the Multilateral Fund was due largely to four factors: the structure of the issue area, the high level of concern on the part of the potential donor countries, the bargaining power of developing countries given their pivotal role in determining the future of world use of ozone-depleting substances, and successful negotiations over structure that involved the productive trade-off between symbolic and material politics and designated clear parameters for the fund.

The environmental problem was collectively defined as that of stratospheric ozone depletion from anthropogenic substances. As a true commons problem, ODS do not obey geographic boundaries and deplete the ozone layer globally (or in areas unrelated to where they enter the atmosphere). Although developing countries did not use or produce large amounts of ODS at the beginning of the negotiations over ozone policy, the potential for greater developing country use and production of these substances if unregulated would soon eclipse any advantage gained by an agreement among the developed countries to limit ODS use.

The trade sanctions under the Montreal Protocol would serve to encourage non-ODS-producing developing countries to join the agreement as long as all producing states were members. Only states producing ODS for a domestic market could continue ODS use in the face of trade sanctions. And if all ODS producers were members of the agreement, those without domestic ODS production capabilities would not be able to buy ODS unless they joined the agreement, thus restricting access to ODS to those states that had agreed to abide by Protocol regulations. The main developing countries that would need to be brought into the agreement for it to work, therefore, were those capable of producing ODS. But if gaining the participation of the developing countries with ODS-production capabilities were the main concern, it might have been more efficient only to "buy off" these countries, restricting the funding mechanism to China, India, and perhaps Brazil, Mexico, Venezuela, and Argentina.

The option of a fund (or bilateral assistance) only for the states with ODS-production capabilities was never formally considered by the parties to the Protocol, however. The lack of attention to this option was probably due in part to the negotiation of the agreement within the United Nations framework in which universal participation is the norm, and to the solidarity of developing countries around the symbolic idea of compensation for regulations taken for environmental degradation done by the developed world. Although economically efficient, it would have been politically impossible to "buy off" only the countries for whom the other provisions in the Montreal Protocol would have been insufficient to gain participation. Also, as CFC production does not require sophisticated technology or large capital investments, the absence of a global agreement could have created an incentive to relocate the regulated industries, rather than contributing to the desired phaseout.

Bargaining over the design of the mechanism was tough and intense. The eventual structure of the Fund resulted from a variety of compromises over difficult issues. The interests at stake were partly financial (as some studies showed high costs of phaseout), but also largely symbolic. Both the donors and recipients were concerned about the precedents to be set by this mechanism. Donors (especially the United States) were concerned about establishing a precedent for compensating developing countries for forgone development opportunities resulting from developed countries' destruction of the environment, given the number of environmental issues that were sure to follow a similar pattern. Additionally, the uncertainty and disagreement about the costs of phaseout meant that the mechanism was specified before it was clear how much would have to be contributed; donors wanted to have control over how this amount was ascertained and allocated. Recipients were determined not to sign on to any obligations to preclude development options until they had some assurance that they would receive funding to help them meet those obligations, and that they would have some control over the funding mechanism. These conflicts were reflected in the issues of whether the member states would create a new organization or use one or more existing ones, and what form representation and voting would take within the structure that was agreed upon.

The symbolism of creating the mechanism was as important as other details. The developing countries, for whom the existence of the Fund itself was an important symbolic victory, also gained the symbolism of having equal representation and equal voting power over how the donations of the developed states were to be distributed. At the same time, the voting structure does not actually impinge on the ability of the donor countries as a group to control the allocation of the funding. The use of consensus as an initial approach to decision making in essence gives each donor country a veto, given that funding is not allocated unless all Ex-Com members agree. The ExCom has thus far been able to operate by consensus, but even if the voting structure were invoked money could not be allocated unless a majority of donor countries agreed. This voting structure provided a procedural safety net so that both sets of states could enter the agreement knowing they could control egregious breaches of their understanding of the funding process. Once the funding process was operating they discovered that the concerns they had about funding decisions have not been cause for problems within the Fund.

The uncertainty over the costs of ODS phaseout that concerned the donor countries was balanced by creating this mechanism only for the Montreal Protocol and giving it a finite time frame within which to operate. Donors, particularly the United States, were concerned that an ozone funding mechanism not create a precedent for other environmental issues or be used to finance global warming efforts, likely to be much larger. Limiting the scope and time of the Fund assuaged some of these fears (Benedick, chapter 12).

On each of the contentious issues, therefore, the developed or developing countries gained either a material or symbolic victory. The creation of a new organization was important to developing countries concerned about the dominance of the World Bank by donors; the executive committee and secretariat oversee the World Bank and other implementing agencies, but the World Bank nevertheless is responsible for disbursing the largest percentage of funding. The donor countries allowed a voting structure to pass that was symbolically central to the developing countries, but in effect did not give up their control; the use of consensus as a general procedure means that funding is not allocated without the agreement of the donors. The willingness of each side to trade off material and symbolic victories allowed all parties to negotiate an agreement they could all live with.

The Operation of the Fund

Most states have joined the Montreal Protocol, and projects to phase out ODS in developing countries are progressing, albeit slowly. The fact that the Fund functions successfully to identify projects of merit seems to come from the complicated oversight structure put into place in the ExCom and the secretariat, and the incentive compatibility of the process in general. The Fund has not operated without difficulties, however. In its first four years of operation, the Fund succeeded in implementing only a handful of approved projects that phased out a little over 5,000 tons of ODS by the middle of 1995. Many problems can be attributed to the complexity of setting up a new organization and working within a changing scientific and regulatory environment; others reflect structural problems, political conflicts, and different expectations of what the Fund is meant to achieve. The important issues that have arisen during the course of the Fund's operation include the collection of payments from donor countries, the way to figure incremental costs, the roles of the implementing agencies, and the ability of the mechanism successfully to oversee and implement the ODS phaseout process in Article 5 countries.

Collection of Funds   Most of the assessed contributions for the interim phase of the Fund were ultimately received. In 1994, however, the first year of the Fund's second three-year operating phase, only 58 percent of expected contributions had been received by the end of the year (UNEP/OzL.Pro/ExCom/14/4/Rev.1). By the middle of 1995, 17.9 percent of contributions for the previous four years were still outstanding (UNEP/OzL.Pro/ExCom/17/5). Although thus far the delay in contributing has not prevented project approvals, as the IAs streamline their project development and introduce new projects into the pipeline the delay could begin to affect project approval and implementation. Representatives of both developed and developing countries suggest a breach of faith in upholding the spirit of the London amendments: developed countries are faulted for reluctant contributions; developing countries are faulted for slowness of implementation of the treaty and lack of progress in phasing out ODS.

The less than timely collection of assessed funding, particularly in the most recent funding cycle, can be attributed to two main, unrelated factors. First, since the Montreal Protocol was signed and discussions about funding begun, political change in the former Soviet Union and East and Central Europe has thrown the economies of these countries into turmoil and uncertainty. In fact, these "economies in transition" are considered eligible for assistance from the GEF to aid with their ODS phaseout, while they are responsible for contributing to the Multilateral Fund for the ODS phaseout of others. Most of these countries undergoing political and economic transition are behind in their funding, and the ExCom has agreed to work out alternate arrangements for collecting their arrears, such as in-kind contributions, until they can begin to contribute again (UNEP/OzL.Pro/ExCom/10/40).

Of greater consequence are the uncollected funds from economically stable developed countries. At the end of 1994 Austria, Belgium, Canada, Japan, Korea, the Netherlands, and the United Arab Emirates were behind in contributing by at least one-third of their assessed amount (UNEP/OzL.Pro/ExCom/15/4/Rev.1:3). Although these countries agree that they should contribute to the fund and do not dispute the amount assessed, their slowness in contributing sends the message that they do not put a high priority on assisting Article 5 countries with their phaseout. This second factor seems to involve issues of contracting and concern. Given that the initial objective of the developed world was to gain the participation of the developing countries with the ozone depletion objectives, as measured by their adherence to the Montreal Protocol, this initial objective has already been met. It may, therefore, be difficult politically to garner support in national governments for funding when the stated "problem"—participation of Article 5 countries—has been addressed. Slowness in contributing assessed funds may also represent developed countries' concern over how the funding is used, in particular the cost effectiveness of some of the projects undertaken.

Incremental Costs   It is no wonder that the idea of incremental costs, designed to be a straightforward measure of the costs of phasing out ODS compared with doing nothing, has proved to be more complicated to calculate than initially envisioned. Incremental costs are those incurred by firms in shifting to non-ODS technologies with no clear return for the additional expenditure, and constitute the main category of costs to be covered by the Fund. Not all additional costs qualify under the terms of the Multilateral Fund, given that they cannot be used to cover tariffs and taxes a country may impose to promote ODS phaseout. Incremental costs are calculated on a case-by-case basis and typically cover such things as the additional costs of purchasing ozone-safe substitutes for use in manufacturing and the containment and destruction of ozone-depleting substances (UNEP/OzL.Pro/WGIV/6,0, annex 4). The parties developed the concept in a compromise to ensure that the funding would truly be "additional"; it would compensate states for the actual extra costs they incur for joining and implementing the Protocol. But, although incremental costs can be reduced to formula in theory, applying this theory to actual projects exposes both a lack of clarity about some details and a more basic disagreement between developed and developing countries about the fundamental definitions of what costs should qualify.

The first type of disagreement is technical in nature and involves difficulties in applying the incremental costs formula. These types of problems arise simply because of the complexity of applying theory to the real world. For example, with the rising costs of ODS and falling costs of substitutes, the incremental costs of a project may fall between the time it is proposed and the time it is implemented, and the ExCom has yet to figure out a systematic way to address this issue. More dramatic are questions of how to apply the idea of incremental costs to production facilities: it may be relatively simple to calculate the costs of retrofitting to operate without ODS, or to figure out how much more it costs to use substitutes than to use ODS, but it is less clear how to apply the incremental costs idea to the closing of a factory that produces ODS. The costs of technical issues like demolition and dismantling plants can easily be calculated, but issues such as how to compensate or retrain workers and how (or for how long) to figure forgone profits are much more complicated (UNEP/OzL.Pro/ExCom/10/39).

The second type of disagreement is more philosophical in nature and involves questioning the categories of costs that should be considered "incremental." The Fund is only designed to pay for costs associated with switching over to more expensive technology, substitutes, or processes. One difficulty raised involves the question of negative incremental costs: if the completion of a project leads to a net cost savings (or negative incremental cost), the Fund is not supposed to cover it. But such projects may nevertheless require an initial capital investment beyond the resources of the company or country in question in order ultimately to realize the net savings, and may therefore not be undertaken absent funding. The Ex-Com has discussed the possibility of loans, not initially included as a function of the mechanism (in part due to its short time frame and the contracting problems that might arise from having loans due after the Fund has expired) as one way to deal with the idea of negative incremental costs (UNEP/OzL.Pro/ExCom/13/3).

India has often led the call for an expansion of the types of costs to be considered "incremental" and has proposed amendments to the "indicative list of categories of incremental costs" under which projects are supposed to fall. For example, India has argued that incremental costs should be calculated also in terms of future demand. Absent the Protocol, India's demand for controlled substances would increase greatly, so if calculating the amount of funding required to phase out ODS in India does not take into consideration the amount of additional ODS that would have been demanded over time, then it misses some of the actual costs of phaseout (UNEP/OzL.Pro/ExCom/12/36). When first proposed in 1994, the non-Article 5 countries successfully argued that these sorts of philosophical issues should be put off until the Fund is thoroughly evaluated in the context of the Protocol as a whole, and the Secretariat noted that the budget for 1994–1996 relied on continuing the current definition of incremental costs (UNEP/OzL.Pro/ExCom/12/37). These sorts of disagreements over incremental costs can be seen as an attempt by Article 5 countries to receive funding for as much as they can, and by non-Article 5 countries to keep the costs they pay as low as possible. In other words, it is through the category of incremental costs that the more basic unresolved issue of amount of compensation is fought out among the parties.

The concept of funding the incremental costs of action has nevertheless proven more useful for the Montreal Protocol than for the Global Environment Facility, due to the relative clarity of the technological issues and the lists of indicative costs approved by the meeting of the Parties to the

Montreal Protocol (UNEP/OzL.Pro 2/3, annex 4:42). Although there have been disagreements, they have not prevented the general idea of incremental costs to be the guiding force in disciplining funding under the Protocol.

Roles of the Implementing Agencies   The initial delegation of responsibility to the implementing agencies based on their expertise would seem obvious, but the responsibilities of these agencies have since expanded and overlapped. There are both contractual and strategic reasons that the mandates of the IAs have expanded from the original design in which they would specialize. The agencies themselves want as many options as possible to do projects in Article 5 countries, through which they gain funding and work as well as contacts in developing countries for other purposes. The recipient countries want to preserve their options as well, to use whichever agencies are likely to do the sorts of projects they want in a timely manner, and to stay within the good graces of agencies with which they are likely to be working on other issues.

The IAs are given 13 percent of the funding as institutional overhead for each project they undertake; this money essentially pays them for their service, but Fund projects also provide them with an expanding work portfolio. Developing country officials allude to the fact that they are approached by representatives from different IAs, all wanting to do ODS projects in their countries. Agency representatives joke about the competition among them for who can do projects in which countries or which will be asked to assist in preparation of a country program. The World Bank's plan to negotiate "umbrella agreements" with countries, by which a country would negotiate at one time the legal arrangements required for the Bank to do a set of projects, caused concern among the other agencies, which feared that they could be left out of the implementation process in the countries with which the World Bank has such agreements. Given that all of these agencies perform non–Montreal Protocol functions around the world, the links they make or preserve in the course of the Multilateral Fund activity may make other interactions with the recipient countries more productive or more likely.

Developing country representatives, for their part, want to be able to get projects approved and implemented in a timely manner, so as to be able to take advantage of as much funding as possible and gain access to new technology. Given the variation among agencies in ability to implement projects quickly, recipient countries have looked to agencies other than the World Bank with which to do projects. The reason for the varying speed of implementation among agencies is largely contractual: the IAs have different mandated internal processes for project approval and implementation. The World Bank, the agency least accustomed to working with the small grants that constitute most Multilateral Fund activity, has the most complicated approval and implementation process, and as a bank needs to contract with financial intermediaries in the recipient countries in order to implement its projects. UNDP, on the other hand, is a true implementing agency, and its field offices already in place in recipient countries assist in timely implementation of projects once approved. In a number of cases, Article 5 countries have specified their desire to work with either UNDP or UNIDO rather than the World Bank on investment projects. As the agencies and recipient countries have gained experience in working on the ODS phaseout, there has been a gradual spreading of investment project activities. Where the Bank initially accounted for 80 percent of all monies authorized for Fund projects, by the July 1995 meeting of the ExCom the bank's projects accounted for just over 50 percent, and UNDP's share rose to almost 30 percent (UNEP/OzL.Pro/ExCom/17/9).

At the same time, recipient countries acknowledge the need to stay in the good graces of the various agencies that might want to work within their countries—which is one reason they give for using multiple IAs to implement projects. In addition, given the large number of projects that need to be done for complete ODS phaseout, the use of multiple IAs is a reasonable response to a desire on the part of Article 5 countries to get as much funding as possible for phaseout in as short a time as possible, especially if there is uncertainty about whether funding will be forthcoming for all projects. As funding gets tighter it will increase the interest among Article 5 countries in using whatever IAs will do projects in their countries.

Competition among the IAs has both positive and negative effects. On the one hand, activity of multiple agencies in a country may increase awareness of the Protocol across a range of industries and introduce flexibility and creativity in the selection and design of projects. In addition, the IAs have different project cycles and rules governing their contracting abilities, or different agents and experiences, that can influence the pace and timing of project implementation and thus the desire of countries to work with one agency or another. Having a choice among agencies can benefit the recipient countries. On the other hand, however, advantages that might have been gained by specialization are not realized, and duplication of effort is likely. It is the oversight ability of the Fund's bureaucratic structure that determines the extent to which these issues are problematic.

Implementation   The ability of the Multilateral Fund as an institution to phase out ODS use is the real test of its effectiveness. The ability of the organization to ensure that projects of a high quality are approved and implemented is an important first step toward effectiveness. Although there has been a lot of activity at the bureaucratic level for project proposal, development, and approval, implementation of approved projects has remained slow. By the middle of 1995 (halfway through the fund's second phase of operation), the ExCom had committed $292 million and approved more than 830 activities. But few had been executed—two-thirds of the training programs had been carried out, but only thirty-nine investment projects had become operational. Early completed investment projects were in countries that were not the highest producers or consumers of ODS (Egypt, Malaysia, Venezuela, the Philippines, Turkey, and Thailand) (UNEP/OzL.Pro/ExCom/15/7). By mid-1995 only 5,349 tons of controlled substances, adjusted for ozone-depleting potential (ODP), had been phased out (UNEP/OzL.Pro/ExCom/17/9). Implementation of the remaining approved investment projects is expected to lead to the elimination of an additional 44,680 ODP tons, less than one-third of ODS consumption in Article 5 countries (UNEP/OzL.Pro/ExCom/17/9).

The difficulties in implementation have not come because of opposition by the actors involved. There is no incentive for the recipient countries to block implementation of the Fund. The Montreal Protocol gave many of the industries in the recipient countries an incentive to change over eventually to non-ODS technology, given that it requires them to phase out ODS. Also, once the Protocol set in motion the general transformation in developed country industries to non-ODS technology and processes, some developing country industries would necessarily follow. First, any country whose industries buy ODS or manufacture products for export containing or using ODS would need to join and adhere to the Protocol to continue to have access to ODS (unless a set of states operated ODS manufacturing outside the Protocol). Second, corporations with operations in a variety of countries have an incentive to standardize processes and products with non-ODS technology if their operations in one country require them to use such technology or processes. Third, as the newest technology and processes turn away from ODS, it becomes harder and more expensive to obtain ODS and related technology. Fourth, the newest plants and procedures will be non-ODS using, so any company that wants to be the most technologically advanced will turn toward alternative processes and technology. For these reasons there has been no significant industry opposition to the implementation of the Fund. Additionally, industry representatives from recipient countries are involved to some extent in the decision-making process about what gets funded; some serve as members of the country's representation on the ExCom. They are thus involved in decisions to approve projects.

The logjam in implementation happens, however, for several reasons. First is the complexity of the administrative process that must be navigated before ODS is phased out. Implementation entails three phases: project preparation, project approval, and project execution. In general, each country first creates (usually in conjunction with an implementing agency) a country program. The projects identified in the country program must be approved by the country itself; formulated, approved (in whatever multistage internal approval process the agency has), and proposed by the implementing agency that takes on that project; and then reviewed and evaluated by the Fund secretariat before being forwarded to the ExCom itself. Before receiving approval, the agency may be required to revise and/or resubmit the proposal, which in some cases may require returning to every stage in the process. After approval the agency has to make the necessary arrangements to implement the project, including negotiating with various actors in the country in question. Thus, even if most projects receive ready approval, they do not then move immediately to implementation at the firm level. Although some of this complexity may be due to unnecessary or inefficient administrative structures, the fact that the Fund operates through existing organizations means that it has had to contend with existing structures that may not have been optimal for its purposes in the first place.

Second, the lack of capacity in recipient countries also slows implementation, although attention to this issue was not an initial concern of the Fund. The ExCom began to allocate money for "institutional strengthening" because recipient countries showed a lack of awareness of the ozone depletion problem as well as a "weak institutional infrastructure" for addressing phaseout issues (UNEP/OzL.Pro/ExCom/7/20). Funding for institutional strengthening, however, only goes to create and outfit a national ozone office. For large countries especially, the number of industries that need to be reached and transformed is huge, and such an office is unlikely to address much of the problem.

Third, the process of phaseout is complicated. Many of the projects undertaken involve new, rapidly changing technologies, and the IAs and the ExCom must constantly evaluate and attempt to use new methods in new areas. This latter explanation for slowness in implementation is simply the result of addressing a new environmental issue with changing rules and even more rapidly changing technologies. The length of time for implementation that results from this type of complexity is even more likely with other environmental issues that are less well understood or less clearly bounded than the ozone depletion problem is. And this aspect of delay might not be entirely problematic: the lessons learned or the technological advances made in the course of the delay might make subsequent implementation easier or more effective.

Oversight   Although the ExCom and its secretariat were initially designated as oversight mechanisms for the work proposed and carried out by the implementing agencies, these bodies could well have functioned as a rubber stamp to projects proposed by the implementing agencies. This outcome was particularly likely given that the Fund's initial purpose was to persuade developing countries to join the Montreal Protocol; if it functioned to "buy" the participation of the developing countries it could well have ignored the quality of the projects to which the funding would be put. Instead, the ExCom, and to a greater extent the secretariat, have stepped in at various stages in the process of funding projects to ensure that the projects have merit and that the process runs more effectively than it would absent oversight. The high degree of oversight results in part from a common desire of the donor and recipient countries to have funds spent as efficiently as possible, as well as from reactions to the inevitable problems that occur when so many different types of agencies attempt to work together under a common set of guidelines.

The type of oversight exercised by the ExCom and the secretariat can be attributed to the fact that both the Article 5 and non-Article 5 countries want the money that is spent to be spent well. If projects are done poorly or at higher cost than necessary neither donors nor recipients benefit. So the oversight that has emerged in the Multilateral Fund process largely takes the form of the parties disciplining the implementing agencies—a stark contrast to oversight of standard World Bank projects, for instance.

Early in the process the ExCom showed that it was willing to take seriously its oversight responsibilities by rejecting the work programs of the implementing agencies for duplication and confusion (UNEP/OzL.Pro/ExCom/3/18). This initial lack of coordination is not surprising given the complexity and uncertainty associated with instituting a new funding mechanism and given the differences that existed among the implementing agencies. The ExCom's insistence on coordinated work programs and sectoral approaches to organizing projects, however, came as a surprise to the implementing agencies.

By showing early on that it would take its role in oversight seriously, the ExCom set a precedent for challenging the IAs to improve their proposals. The secretariat's role in oversight of proposals has developed at least in part as a way to process the large amount of information that comes from the implementing agencies and needs to be packaged in a way in which the ExCom can address it. Over the course of the first few meetings, the ExCom delegated to the secretariat (whose role is simply to "assist the Executive Committee," according to the terms of reference for the Multilateral Fund [Ozone Secretariat 1993, annex 10]) greater and greater responsibility for assessing and commenting on proposals prior to meetings. The secretariat's role is to increase the quality of the proposals; a high percentage of proposals are accepted at ExCom meetings because, before they are ever proposed at the meetings, the secretariat recommends changes or revisions, or suggests that the projects not be proposed at all. The countries and implementing agencies almost always take these suggestions.

But the heavy involvement of the secretariat is not uniformly appreciated. Tensions between the World Bank and the secretariat over project evaluation have been particularly thorny and have contributed to delays in the approval process. In part, the Bank's structural constraints, internal reporting requirements, and project criteria make its acceptance of secretariat suggestions difficult. For its part, the secretariat has been heavily criticized for its "micromanagement" on a project-by-project basis. The UNEP Economic Options Committee, in its report for the seventh meeting of the Parties to the Montreal Protocol in May 1995, noted the implementation consequences of the design of the Fund. It noted the difficulties inherent in having a "new, small, and relatively inexperienced Secretariat trying to manage large, powerful and established international institutions" (UNEP, Economic Options Committee, 1995).

The weakest link in oversight, however, seems to be project monitoring and evaluation. The implementing agencies insist that they follow up on the projects that they are charged with, and it is unlikely that funding is diverted or used for unintended purposes. But there is no structural process within the procedures of the fund to ensure oversight and evaluation of projects once they are approved. In operating procedures of the IAs, only UNDP shows a formal postimplementation monitoring and follow-up phase. The World Bank relies heavily on the recommendations of a technical advisory committee, the Ozone Operations Resource Group (OORG), that is comprised largely of representatives from Northern industries that stand to benefit from the transition to substitute technologies they may recommend. This reliance has prompted sharp criticisms from the few environmental NGOs (notably Greenpeace and Friends of the Earth) that have continued to attend ExCom meetings to monitor developments in the implementation of the Montreal Protocol. These groups claim too little attention is paid to potential alternatives to traditional chemical solutions to ODS use. Even the outside evaluation of the Fund, required by the protocol, is intended to "assess the feasibility of alternative phase-out schedules" rather than to evaluate the functioning of the Fund as it stands (UNEP/OzL.Pro/ExCom/15/40:ES-1). The Ex-Com has recognized the need for more evaluation, especially in light of increasing financial pressure on the Fund, and has set up a panel to recommend policy guidelines.

Perhaps more important than discerning the reason behind the level of oversight that evolved in the ExCom and the secretariat is noting that neither the type nor amount of oversight was inevitable. It is not clear why the secretariat and the ExCom took the strong oversight role that they did. To the extent that such a role can be seen as valuable, future efforts to create funding mechanisms should seek ways to ensure that such oversight will occur. And although the oversight has helped the IAs to avoid duplication of effort and funding and helped improve the quality and cost effectiveness of the projects, there still is no formal procedure to ensure that all aspects of the process will be scrutinized. The extent to which projects are evaluated against current scientific knowledge is ad hoc, and there is little follow-up on projects that have been completed to ascertain how well, quickly, or cost-effectively they have been done. Because the oversight process was not designed in a systematic way, but has instead responded to a perceived need within the process, it both responds to the immediate concerns of the parties and may not oversee all the parts of the mechanism that could benefit from oversight.

3   Evaluation

The organizational structure of the Fund has evolved to allow it effectively to oversee the collection and disbursement of funds for the phaseout of ozone-depleting substances in the developing world. The ultimate effect of the Fund on the environment, however, remains in doubt. Issues of concern, the contractual environment, and capacity in a variety of actors threaten to undermine the ability of the Fund to accomplish the complete phaseout of ODS by developing countries in a timely fashion. At the same time, it would be premature to suggest that the Fund has been a failure. Although very little ODS has been phased out in Article 5 countries, this lack of progress must be considered in light of the fact that developing countries do have a ten-year grace period in which to implement the Protocol requirements. Developing countries may have an incentive to increase their ODS use or production, because doing so may benefit them industrially and does not currently harm their ability to receive funding to change over some of their industrial practices or to benefit from new technology.

It is also important to remember that the Fund is not the only factor that influences the pace and timing of ODS phaseout, nor was it ever intended to be. Market forces—in particular the increasing cost of ODS and the increasing availability and decreasing cost of substitutes—are certainly a central explanatory variable for the change away from the global use of ozone-depleting substances. Second, the actions of large multinational companies influence policies in developing countries as they implement corporate policies internationally. Third, increasing public awareness and legislative mandates in developing countries also have an effect on phaseout. Many of these factors are difficult to isolate from the influence of the Fund itself, given that the Fund has an impact on them. Nevertheless, there are some aspects of the Fund that can be considered successes or failures in contributing to the global phaseout of ozone-depleting substances.

More important than the small amount of ODS that has been phased out so far in developing countries are the number of phaseout projects under way, and the extent to which these projects actually represent a lessening of developing country reliance on ODS. Has the environment improved because of the Montreal Protocol? Certainly. Has some of this improvement been due to the Fund? Probably, both in terms of gaining earlier accession to the Protocol than might otherwise have happened, and in terms of making possible financing for early shifts to ODS-free manufacturing that would have been financially difficult even for states that wanted to make such a shift.

Successes

Despite the current low return on investment in terms of the amount of ODS that has actually been phased out, the Multilateral Fund has been effective in attracting developing countries to participate in a global phaseout of chemicals that threaten the earth's stratospheric ozone. Two serious impediments to attaining the necessary cooperation—a poor contracting environment in which developing countries viewed with deep suspicion the rationale for global environmental regulations and the low concern on the part of developing countries for saving the ozone layer— were overcome with the assurance that those countries would receive assistance in altering their industrial practices to meet the objectives of the global accord. The creation and structure of the Fund addressed some of the most important concerns about the North-South imbalances that clouded the contracting environment, by providing a new multilateral structure for transferring capital to developing countries that gave joint decision-making control to developed and developing countries. The first success of the Fund could be observed when several major developing countries signed on to the Montreal Protocol with the creation of the Fund under the London amendments, making it clear that their participation in the Montreal Protocol process was a quid pro quo for the creation of the Fund.

The Fund also helped raise concern and the capacity to take action based on this concern. Prior to the existence of the Fund, developing countries had little incentive to acknowledge their role in the problem of ozone depletion or their need to contribute to its correction. Through the Fund negotiations and continuing scientific and socioeconomic assessment panels of the Montreal Protocol itself, the developing countries gained a greater understanding of the problem, the potential negative effects it would have on their own economies, and the need for global action to overcome it. The process helped Article 5 countries discover the extent of their ODS use and what would need to be done to meet their obligations under the Protocol. Thus, even though market forces (as non-Article 5 countries phase out CFCs by 1996) play the most important role in pushing the global phaseout, the Fund helps to ensure that the market transition will be more orderly and equitable than it would be without such assistance.

The use of established agencies to implement the Fund, with oversight from an independent secretariat and the ExCom, provides a compromise between the danger of organizational inertia and the advantages of building on the experience of established organizations. The secretariat does a lot of work behind the scenes, suggesting revisions to projects, work programs, or country programs before these proposals are submitted to the ExCom. Although almost all projects brought before the ExCom are ultimately approved, many are withdrawn, postponed, or modified at the suggestion of the secretariat before they are brought forward. Although cumbersome, the oversight roles of these bodies prevent the Fund from simply being used to buy the participation of developing countries in the Montreal Protocol, and instead allow it to identify worthy projects to which funding can be put. It is this function of the Fund that is the most surprising and worthwhile.

Problems

The Fund, however, has not solved all the potential difficulties that prevent it from having as profound an impact on global transition away from ozone depletion as has been envisioned. In particular, contractual problems with the way the Fund was created, a lack of prioritization of projects, implementation difficulties in balancing the advantages and disadvantages of complex oversight, and both theoretical and practical questions of technology transfer have all plagued its history.

The ten-year lag provides one of the main contractual problems with the funding mechanism. When initially negotiated in the Montreal Protocol, it was seen as a way to encourage developing countries to join the Protocol, by recognizing their needs for further development with ODS before working for phaseout. Because the Montreal Protocol did not initially call for even a full phaseout of ODS by developed countries, the ten-year lag was not seen as a threat to the ozone layer. However, as the scientific process put into play by the ozone negotiations discovered the seriousness of the ozone depletion problem and the amendments to the Protocol called for deeper and deeper cuts (and ultimate phaseout) of ODS, the ten-year lag for developing country compliance came to have more significance. More important, when the Multilateral Fund was created under the London amendments, it created the potential for a contractual dilemma when combined with the ten-year lag. With developing countries currently able to increase their use and even production of ODS, situations can arise in which states receive funding to retrofit plants to phase out ODS while also building identical ODS-producing plants. At its worst, this scenario could lead to a situation in which, at the end of the ten-year lag, the largest developing countries could be in a position to renegotiate financing. With an increased stock of ODS or increased production capabilities, when developed countries have already phased out ODS entirely, developing countries would have the power to demand a much bigger payoff for full phaseout, after the Fund is supposed to have expired. This problem seems to be due to the fact that the whole incentive package for developing countries was not created at the same time; the grace period allowing an increase in ODS consumption and production, which was already in place when the London amendments were negotiated, seems to contradict some of the Fund's intentions.

Also important is the fact that, not only did the negotiating process not consider the possibility of "buying off" the developing country ODS producers, but it also did not consider differentiating among the developing countries in the time line for phaseout. As a result, the funding process is not designed to work for earlier phaseout in the countries whose phaseout would most influence the ozone layer. As the draft review of the Montreal Protocol discussed at the fifteenth ExCom meeting points out, the countries with the highest ODS consumption will probably be among the last to phase out (UNEP/OzL.Pro/ExCom/15/4 Rev 1). The projects that are done are the ones that the IAs put forward, which often means that they are the easiest to carry out rather than the most important. Because there is no structural effort to prioritize phaseout, greater levels of ODS have been released into the atmosphere than would have been if the mechanism concentrated first on transforming industry in the largest Article 5 ODS consumers or producers. The negotiating power of the developing countries as a bloc (and the fact that the largest developing countries pushed solidarity of their cohort on this issue) prevented these distinctions from being made.

Implementation problems remain. The structure of the Fund represents a compromise on the question of whether to create a new mechanism or use existing agencies. In the case of implementation, the advantages of a new structure are clearer than the advantages of the experience of the implementing agencies, especially when combined. Each implementing agency has its own multistage process that its projects must go through before even being proposed to the ExCom; any changes suggested by the secretariat or mandated by the ExCom must go through the process again. Although the experience of the implementing agencies helped get the mechanism implementation off the ground (and provided a certain legitimacy), it is these established processes that also slow it down.

One further problem in implementation is that the Multilateral Fund does not have a regular process for reviewing completed projects, to ensure that they have actually been fully implemented and done in a cost-effective manner, or to learn from the experiences with various processes or technology. Implementing agencies insist that they carry out implementation oversight in a satisfactory manner, and it is unlikely that individual projects that are undertaken will be undermined, given that the firms in question want to acquire new technology. But there are advantages that can be gained from hoarding ODS, and already a black market in illegal ODS is appearing in the developed world (UNEP, Economic Options Committee 1995). It is also possible that agencies will carry over experience gained from one project to the next, but the fact that there is no institutional structure in which to facilitate such learning is disturbing. Cost may also be a problem: once a funding level has been approved, IAs have no incentive to accomplish a project under that cost. The ExCom is aware of many of these problems and will probably begin to address them over time, but there is nothing structural to ensure that these types of oversight will develop.

Finally, the central issue of technology transfer remains problematic. Although viewed as a crucial element in developing countries' advancement to meet the goals of the Montreal Protocol, provisions for it were diluted in the final agreement on the Fund due in part to constraints of domestic legislation in developed countries to protect the proprietary rights of the private sector. Despite much discussion in the ExCom about this issue and significant transfer of technologies to user-industries, the issue of transferring technologies for production of substitutes is a more difficult one. Industries that developed and profit from substitutes are reluctant to relinquish their control over manufacture of these chemicals, and developing countries are concerned about having no control over the chemicals they need for their industrial development. There are still no guidelines for the IAs about how to deal with production processes for substitutes.

There is some disagreement, as well, among various actors involved in the process about what the best technologies are to achieve the overall goals of the Montreal Protocol. Particularly contentious are debates surrounding the approval of projects that introduce the use of intermediate chemicals, such as HCFCs (hydrochlorofluorocarbons), as CFC substitutes. These are chemicals that, because of their ozone-depleting potential, will be phased out in developed countries by 2030 and are likely ultimately to be phased out in developing countries as well. Another area of contention is the use of HFCs, in refrigeration in particular. These substances do not destroy ozone but they do contribute to global warming. Although the participants are concerned about nonozone environmental impacts of substitute technologies, the process is not set up to account for these other environmental impacts, nor are explicit criteria in place with which to evaluate competing claims about technologies.

Conclusion

The Multilateral Fund was used initially to persuade countries to join the Montreal Protocol, and the initial money was committed by non-Article 5 countries before projects were approved by the ExCom that oversees the Fund. This combination of factors could have led to a situation in which the Fund process did not control the content of projects. If the Fund were used as a simple payoff to gain participation, the process might have allowed funding to be given to projects regardless of quality or of how they fit into the overall strategy of ODS phaseout. Instead, the ExCom signaled early in the process that it was not going to automatically approve projects. Implementing agencies work with the secretariat to modify projects to meet the general requirements of the ExCom rules before they are even presented at ExCom meetings; even then the projects are occasionally modified, postponed, or turned down by the executive committee if they do not meet certain criteria. In addition, the ExCom regularly makes new decisions about the requirements that projects or programs must meet in order to be funded. So although the Fund may have been conceived by some as a simple side-payment, it has gone much further than that in its implementation.

The question that remains, given the assumption that much ODS use in developing countries would be phased out without the Fund, is whether the Fund has sped up the process of implementation of the Montreal Protocol. Some states that might have ultimately phased out ODS absent the Fund could not have put up the capital initially without aid. Although the slowness of the disbursement and project completion of the Fund might lead one to believe that it has not helped move ODS phaseout along, the process embodied in the Fund nevertheless has raised awareness and understanding of the ODS-consuming industries and has given information about phaseout and alternatives to the political leaders. This increased capacity, which would not necessarily have required a Fund to achieve, is nevertheless one of the most useful aspects of the Fund. And although it was not the financing that was necessary to accomplish this increased capacity, the funding *is* what brought the relevant actors into the process in the first place.

The successes and failures of the Montreal Protocol Multilateral Fund suggest lessons for designing future multilateral mechanisms to provide funding to address global environmental problems. The main pitfall to avoid is contractual problems from creating ultimately incompatible incentives at different times to encourage developing country participation in the global regulatory regime. Compromises involving symbolic politics, as well as real oversight mechanisms, are worthy objectives.

The main contractual difficulty with the Fund arises because Article 5 countries are allowed to increase their consumption of ODS during the period that the Fund is paying for projects to phase out their ODS use. Given the recent trend toward creating global environmental agreements that are made increasingly stringent over time, this type of contracting problem is likely to arise in other issues. It arose in the Montreal Protocol process because, as scientific evidence increasingly showed the seriousness of the ozone depletion problem, the requirements for phasing out use of ODS were made increasingly severe. When even developed countries were not required fully to phase out ODS, allowing developing countries to increase their consumption did not radically alter the regulatory regime, and was seen as a way to entice these countries into the agreement. But once a regime that involved total phaseout was put into place and a funding mechanism was created, a contradiction emerged: increase in ODS consumption by developing countries could happen at the same time that the Fund was paying for phaseout.

The successes of the Fund also suggest lessons for future agreements. The ability of developed and developing countries to agree on a funding mechanism came in part because of the willingness of both sides to trade off between symbolic and material victories in their negotiating positions. Initially contentious voting rules have not even needed to be used; but, because of productive trade-offs, they allowed the mechanism to begin functioning. Other such productive trade-offs served similar functions. One of these, the use of existing agencies to implement the Fund with oversight from a new organization, proved particularly useful in guaranteeing the quality of the projects funded.

The symbolic message of the existence of the Multilateral Fund as a part of the solution to the ozone depletion problem also provides important lessons. Like it or not, the Fund *has* set a precedent for dealing with global environmental issues with North-South equity problems. It has created expectations that developing countries will be compensated for the forgone development opportunities or the added burdens required by environmental cooperation. And although one can argue that phaseout of ODS in developing countries might have happened absent the Fund, the existence of the Fund facilitated an orderly phaseout, sent a strong message about the willingness of the world community to address equity issues in environmental regulation, and perhaps opens the way for greater participation of developing countries in future environmental agreements.

Note

For support in carrying out the research and writing for this chapter, Elizabeth DeSombre gratefully acknowledges the support of the Center for International Affairs at Harvard University and the Rockefeller Brothers Fund; Joanne Kauffman gratefully acknowledges the Volvo/MIT Award for Environmental Research and the Center for Science and International Affairs, John F. Kennedy School of Government, Harvard University.

5

Nonstate Actors Leading the Way:
Debt-for-Nature Swaps

Cord Jakobeit

Debt-for-nature swaps are a specific financial transfer mechanism for the environment with a relatively long and rich history and a significant track record that has generated both acclaim and controversy.[1] Three mainstream Washington-based international nongovernmental organizations (NGOs) with a focus on the environment and conservation have been the driving forces for the mechanism: Conservation International (CI), The Nature Conservancy (TNC), and the World Wildlife Fund (WWF).[2] Their debt-for-nature initiative attempted to establish an issue linkage between debt and the environment in order to increase funding for environmental activities in developing countries. The swaps became financially attractive due to the reduced prices of commercial debt titles as a consequence of the debt crisis. They were built on the experience gained with earlier debt-for-equity exchanges mainly in Latin America. Debt-for-nature agreements reward developing countries for adopting conservation measures to safeguard their rain forests and biodiversity by writing off a portion of their foreign debt.

This chapter argues that debt-for-nature swaps—although very attractive from a financial point of view—were ultimately limited and short term, and hence had relatively small direct effects on the environmental problems they attempted to address. In fact, their limited role may have contributed to achievement of the agreements. Had their financial scope and potential environmental reach been much larger, the NGO initiatives would have triggered much more interest from the state actors involved, and stronger resistance would probably have formed in the process. The initial issue linkage between debt and the environment made cooperation possible under difficult circumstances. Now, however, the easing of the debt crisis has crowded out many opportunities for the swap mechanism, even though the environmental problems remain as pressing as before. The legacy of the mechanism is less in what it initially set out to achieve and more in what has been learned and demonstrated in the process about more successful ways to build local capacity to achieve lasting biodiversity conservation in the biological "hot spots" (Myers 1988) of the developing world. Even the initial NGO proponents of the nature swap mechanism have by now moved on to other campaigns.

Since the first swap with Bolivia in 1987 and until the end of 1994, thirty-two environmental swaps were completed in fifteen debtor nations, most of them in Latin America. They have reduced the stock of commercial foreign debt by $177 million and generated close to the equivalent of $130 million in domestic currencies for conservation.[3] The conclusion of nature swaps on the basis of commercial debt titles peaked at the beginning of the 1990s. In recent years their number has begun to decline.

After the NGO innovation of the concept (first generation swaps), governments of developed countries have also applied this transfer mechanism in a second phase by writing off public bilateral debt in exchange for environmental commitments from developing country debtors (second generation swaps).[4] This second phase was pioneered by Germany's swap with Kenya in 1989. Since then, second generation swaps have been announced by several other bilateral donors (World Bank 1993c: 115), including Sweden, Canada, Switzerland, and the United States with the Enterprise of the Americas Initiative. The most recent scheme was announced in early 1995 between the Swiss and Bulgarian governments. A major difference between first and second generation swaps is their size. Accumulated from all bilateral sources, second generation debt-for-nature swaps have long passed the $1 billion benchmark[5] and can reduce significantly greater amounts of debt, thus potentially providing greater financial support for the environment.

In addition to bilateral debt, the pressure on multilateral development banks (MDBs)—a third potential source of debt titles so far not used for the swap mechanism—is bound to rise in the future. In the management of the debt crisis over the last decade, commercial banks have successfully sought to reduce their exposure, while bilateral creditors have increasingly opted for partial forgiveness. The most seriously affected and least developed countries were forced to seek financial support from the MDBs, whose share of total outstanding debt has continuously risen over the last few years.

So far, total accumulated financial transfers from the two phases of the swap mechanism pale in relation to total outstanding developing country debt—some $1.8 trillion at the end of 1993 and still growing. The debt-for-nature experience has done very little to alleviate the debt crisis; and its effect on the environment in the developing world, as will be argued in this chapter, is also a cautionary tale in many ways. Nevertheless, the case merits attention for two main reasons: (1) for what it says about the role of NGOs in international environmental politics, and (2) for what can be learned about the usefulness and limitations of linking the environment with another issue area in order to reach agreement and cooperation between states.

Although NGOs do play a role in almost all cases discussed in this volume and are considered an important driving force in international environmental politics, they are rarely the main actors. Their principal role consists of mobilizing public support to put pressure on governments and on MDBs to accept international environmental standards and commitments. They can also be important agents for generating processes of social learning in the field of the environment by creating new linkages between the local and the global arenas and exploring niches that other international actors are ill-equipped to fill.[6] The debt-for-nature case features NGOs as agenda setters, key actors, driving forces for innovative cooperation, and as actively involved proponents for new ways of thinking about venues to stem deforestation and biodiversity loss. The case also provides ample evidence about the limitations and difficulties NGOs encounter when they set out to promote their agenda with more powerful actors.

Issue linkages between two different and seemingly unrelated topics have been suggested as one way to allow for progress in the difficult negotiations over environmental problems.[7] Such linkages can increase the scope of the negotiations and allow for compromise, but they can also increase institutional complexity, bog down the process, and raise the likelihood of failure. The parties may fall victim to the intricacies of an additional major problem and its different dynamics. The debt-for-nature case provides evidence for both the usefulness and the limits of issue linkages for environmental protection.

The analysis below draws heavily on selected country cases in each phase of the debt-for-nature history. Costa Rica was by far the largest recipient in the first generation. The country completed a record six debt-for-nature agreements between 1988 and 1991. A closer look at this exceptional case will permit some preliminary evaluations about the effectiveness of the mechanism and help specify the country-level conditions that must exist for this transfer mechanism to work. Comparing the swaps initiated by Germany and within the EAI will highlight the most significant variations in the development of the transfer mechanism in its second phase and illustrate under what conditions "green" conditionality by governments from the North might work.

In line with the common analytic framework adopted for this research project, section 1 describes the history and development of the debt-for-nature concept. The emphasis is on the first phase from the initial deadlock over the environmental issues at the beginning of the 1980s, continued to the debt-for-nature idea presented by NGOs in 1984 and the peak of the first generation swaps concluded with Costa Rica. The second part of section 1 introduces the two second generation cases discussed, in which bilateral debt was pardoned in exchange for environmental commitments from recipient developing countries. The third part summarizes the defensive position of MDBs reluctant to apply the debt-for-nature mechanism.

Section 2 analyzes the roles of NGOs in their dealing with the other main actors. It analyzes the initial deadlock over the environmental problems, then explores under what conditions interests converge to bring about agreements acceptable to all parties. The section also looks at the solutions devised to overcome contractual problems that could have hindered consummation and implementation of swap agreements. The analysis then returns to the largest recipient in the first phase, Costa Rica, focusing on the role of international and domestic NGOs and the lessons learned from the nature swaps. It also contrasts the German and U.S. experience in the second phase, accounting for the differences by reference to domestic NGO positions and strategies.

Finally, section 3 looks at the effectiveness of the debt-for-nature mechanism. Despite significant successes, major limitations prevented debt-for-nature swaps from having a major impact on problems of forest loss and vanishing biodiversity.

1 Description: The History of the Debt-for-Nature Mechanism

By the beginning of the 1980s, the closely related problems of deforestation and rapid biodiversity loss in many developing countries had attracted some concern and a rising awareness in the scientific and NGO communities of developed countries. A 1981 report by the Food and Agricultural Organization (FAO) of the United Nations (FAO 1981) indicated a rate of deforestation much higher than had previously been supposed. Tropical biologists pointed out that there were more adverse and alarming repercussions when tropical forests were cleared at a seemingly ever increasing rate. Given that tropical forests are estimated to contain at least 70 percent of the world's plant and animal species (Erwin 1988), rapid deforestation is translating into a mass extinction of biodiversity, thereby creating irreversible scientific, aesthetic, ethical, and economic losses (Myers 1992). In addition, tropical forests function as a massive sink for carbon dioxide thus linking deforestation to global warming. When tropical forests are burned, their stored carbon is released into the atmosphere. Estimates put the annual greenhouse effect from the burning of tropical forests to some 30 percent of all carbon dioxide released (Houghton 1990).

Defining Problems and Solutions
Although information and awareness about the implications of these environmental challenges increased during the 1980s, a collective definition of the interrelated problems did not emerge.[8] Conflicting interests and low priorities for environmental protection have slowed down the progress for environmental agreements. With the bulk of tropical forests located in developing countries, the issue has become a major bone of contention between rich and poor nations and a new focal point for North-South polarization.

During the first half of the 1980s, environmental NGOs in developed countries were equally concerned about the ecological implications of a related problem in the world economy. The debt crisis, which had surfaced in 1982 with Mexico's declared debt-servicing inability, put an additional burden on ecological systems in the South that were already very fragile and ill-equipped to stem rapid environmental degradation (Bramble 1986). Structural adjustment programs, the standard remedy prescribed by the International Monetary Fund and the World Bank to guarantee a continued servicing of accumulated debt, tended to exacerbate deforestation and the exploitation of natural resources by trying to increase traditional exports.

After a congressional hearing earlier that year where he had already argued these points, Thomas Lovejoy, then with WWF in Washington, took the initiative and established an issue linkage between debt and the environment in his 1984 op-ed piece in the *New York Times* (Lovejoy 1984). He argued that the financial crisis in developing countries had resulted in catastrophic reductions in their already meager environmental budgets. Because of their economic and financial situation, many developing countries put an additional emphasis on export promotion which led to increased exploitation of natural resources. Given these dismal developments, Dr. Lovejoy proposed that the twin problems of debt and the environment be addressed simultaneously with a bold program of swapping debt for nature.

Although stimulating a debate in the circles interested in environment and development, the NGO initiative was not taken up by the more important actors. Neither private commercial banks nor governments nor multilateral banks—the three different creditors holding on to debt titles from developing country debtors—were willing to subscribe to Lovejoy's suggested problem definition.[9] They were not sufficiently concerned with the problems to engage in negotiations or define new rules to bring about remedies for the environmental implications of the debt problem. The management of the debt crisis continued with an emphasis on case-by-case rescheduling, fresh money, and structural adjustment. The possible negative implications of this chosen abatement strategy, such as environmental or social repercussions, were initially shoved aside or ignored.

The environmental NGOs based in the United States and with a special emphasis on global nature conservation held on to Lovejoy's problem definition and suggested solution. With the broad initiative being blocked by the key and financially potent actors, these NGOs focused on the small subset of creation, maintenance, and conservation of tropical forest reserves, an area where they had been active before and one within their limited capacity and financial means. In this first phase CI, WWF, and TNC became the driving forces behind specifying a solution that was aimed at moving their activities beyond small-scale project work and publicly expressing concern.

Specifying and Implementing the Nature Swap Mechanism   After their initial bet on increased donations and tax breaks for the support of the debt-for-nature concept had failed, the NGOs came up with a different specification of the financial transfer mechanism. As a consequence of the debt crisis, secondary markets had emerged in which commercial bank debts were traded at huge discounts. This gave NGOs an opportunity to piggyback on the concept of debt-equity swaps that had developed earlier. In these commercial transactions, debt obligations by indebted countries are converted into assets or equity within those countries. In contrast, debt-for-nature swaps support conservation projects. In essence, they amount to a reallocation of developing country government spending, whereby hard currency funds, which might have had to go for debt servicing, are allocated instead for environmental projects using local currency.

In a typical first generation debt-for-nature swap (Gibson and Curtis 1990), an international environmental NGO attempts to attract donations to be used for the purchase of the foreign debt title of the developing country in question in the secondary market at a discount on the face value. A commercial bank sells or donates the debt title of the country in question to the NGO. The debt title is then presented to the debtor government and converted into domestic currency, whereby the total amount of outstanding foreign debt is reduced. Finally, the total domestic currency equivalent or an agreed-upon fraction of the debt title is used to finance environmental projects in the debtor country related to conservation activities in or close to national parks, thereby complying with the initial condition that the NGO had attached to the mechanism. Local

NGOs and branches of the national administration from the developing country are the major recipients of the funds. They carry out the environmental activities.

Sorting out the options, convincing the many actors involved to go along, and negotiating the first agreements was no small feat for the three international NGOs. It had taken them more than three years from the launching of the idea to the first swap agreement. Since 1987, however, when the first debt-for-nature swap was concluded in Bolivia, until the end of 1994, thirty-two transactions have been completed in fifteen countries, most of them in Latin America (see table 5.1).

Through such financial mechanisms, about $128 million have been raised for environmental projects at an initial cost of $46 million (or an average discount of 63 percent), enabling NGOs to leverage their funds by a factor of 2.3, while developing countries reduced their stock of foreign debt by $177 million.

Nevertheless, the mechanism had a bumpy start. The first swap with Bolivia even threatened to discredit the entire concept (Patterson 1990; Page 1990; Dogsé and Droste 1990; Sarkar 1994). The agreement was aimed at changing government policy with respect to buffer zone management for an existing biological reserve as a quid pro quo for the cancellation of a commercial debt title. The Bolivian government delayed the provision of funds for the implementation of the agreement because it was concentrating on economic reforms instead. It took about two years after the agreement was signed to get actual implementation started. In the process, it had become clear that NGO-negotiated nature swaps are an inappropriate vehicle for policy reform, however well intended (Rubin, Shatz, and Deegan 1994: 25).

After the near failure with Bolivia, the mechanism was successfully redesigned following the example set by the WWF's first swap with Ecuador. Most subsequent swaps were bond based and relied on strengthening local NGOs that share the same environmental concerns as their international counterparts. In this redesigned approach, donated debt is converted to bonds issued in the name of a local NGO that receives the interest paid by the central bank over the life of the bond, usually only a few years (Patterson 1990: 8). This concept was later applied in other Latin American countries, in Asia (the Philippines), in Africa (Madagascar, Zambia, Ghana, Nigeria), and even in Eastern Europe with a small nature swap in Poland.

Since all first generation swaps were carried out subject to independent audits, one can conclude that the funds did flow into conservation projects and were largely used as specified in the detailed agreements of the mechanism.[10] Contrary to what many lawyers had predicted, compliance or enforceability problems did not arise to a significant degree. The reasons for this are discussed in more detail in section 2.

Implementation: Evidence from Costa Rica   In many ways, Costa Rica was an ideal candidate for the nature swap initiative.[11] From the international NGO perspective, the country was a safe bet for a concept still struggling to prove its worth, especially after the less than encouraging first experience with Bolivia. At the same time, Costa Rica was an exception when compared to most other potential recipients. It had established national parks by the end of the 1960s and managed to safeguard close to one-third of its territory to some degree by the end of the 1980s. Many observers ranked the country's conservation commitment as "one of the most effective in the world" (Wallace 1992). Costa Rica entered the process with a "worldwide reputation as a leader in conservation" (Umaña and Brandon 1992: 85) and with a nondogmatic orientation toward the outside world.

Despite its commitment to conservation, however, Costa Rica faced both serious environmental problems and a debt crisis; thus it was a natural country for the debt-for-nature concept. It had one of the highest rates of deforestation on earth for much of the 1980s, and also suffered from the highest per capita foreign debt in the developing world. The commercial banks had lost confidence in its ability to pay, and as a consequence, prices of commercial debt titles from Costa Rica had plummeted on the secondary market, to below 15 percent of face value. High inflation rates reduced the value of the small amounts budgeted for conservation efforts, while the country's fiscal and economic strains exacerbated the threats to nature parks and protected areas from cattle ranching, mining, migration, and logging pressures. The standard structural adjustment program with the IMF and the World Bank resulted in disproportionate budget cuts for government agencies responsible for managing protected areas, leading

Table 5.1
First generation debt-for-nature swaps, 1987–1994
(U.S. millions)

Country	Year	Cost	Face value	Conservation funds	Purchase price (percent)	Redemption price (percent)	Leverage	Organization/country
Bolivia	1987	0.10	0.65	0.25	15.4	38.5	2.5	Conservation International
Ecuador	1987	0.35	1.00	1.00	35.4	100.0	2.8	World Wildlife Fund
Costa Rica[a]	1988	5.00	33.00	9.90	15.2	30.0	2.0	The Netherlands
Costa Rica	1988	0.92	5.40	4.05	17.0	75.0	4.4	National Park Foundation of Costa Rica
Costa Rica	1989	3.50	24.50	17.10	14.3	69.8	4.9	Sweden
Costa Rica	1989	0.78	5.60	1.68	14.0	30.0	2.1	The Nature Conservancy
Ecuador	1989	1.07	9.00	9.00	11.9	100.0	8.4	World Wildlife Fund/The Nature Conservancy/Missouri Botanical Gardens
Madagascar	1989	0.95	2.11	2.11	45.0	100.0	2.2	World Wildlife Fund/U.S. Agency for International Development
Philippines[b]	1989	0.20	0.39	0.39	51.3	100.0	2.0	World Wildlife Fund
Zambia	1989	0.45	2.27	2.27	20.0	100.0	5.0	World Wildlife Fund
Costa Rica	1990	1.95	10.75	9.60	18.2	89.3	4.9	Sweden/World Wildlife Fund/The Nature Conservancy
Dominican Rep.	1990	0.12	0.58	0.58	19.9	100.0	5.0	Conservation Trust of Puerto Rico/The Nature Conservancy
Madagascar	1990	0.45	0.92	0.92	48.5	100.0	2.1	World Wildlife Fund
Poland	1990	0.01	0.05	0.05	24.0	100.0	4.2	World Wildlife Fund
Philippines	1990	0.44	0.90	0.90	48.8	100.0	2.1	World Wildlife Fund

Country	Year							Organization
Costa Rica[c,d]	1991	0.36	0.60	0.54	60.0	90.0	1.5	Rainforest Alliance/Monteverde Conservation League/The Nature Conservancy
Madagascar[e]	1991	0.06	0.12	0.12	49.6	100.0	2.0	Conservation International
Mexico[f]	1991	0.18	0.25	0.25	72.0	100.0	1.4	Conservation International
Mexico[f,g]	1991	0.00	0.25	0.25	0.0	100.0	n.a.	Conservation International
Ghana[h]	1991	0.25	1.00	1.00	25.0	100.0	4.0	Debt for Development Coalition/Conservation International/Smithsonian Institution
Nigeria	1991	0.07	0.15	0.09	43.3	62.0	1.4	Nigerian Conservation Foundation
Jamaica	1991	0.30	0.44	0.44	68.6	100.0	1.5	The Nature Conservancy
Guatemala[d]	1991	0.08	0.10	0.09	75.0	90.0	1.2	The Nature Conservancy
Mexico	1992	0.36	0.44	0.44	80.5	100.0	1.2	Conservation International/U.S. Agency for International Development
Philippines[i]	1992	5.00	9.85	8.82	50.8	89.5	1.8	World Wildlife Fund
Guatemala	1992	1.20	1.33	1.33	90.0	100.0	1.1	Conservation International/U.S. Agency for International Development
Guatemala	1992	1.20	1.33	1.33	90.0	100.0	1.1	Conservation International/U.S. Agency for International Development
Brazil	1992	0.75	2.20	2.20	34.0	100.0	2.9	Conservation International
Panama	1992	7.50	30.00	30.00	25.0	100.0	4.0	Conservation International
Bolivia[i]	1992	0.00	11.50	2.76	0.0	24.0	n.a.	World Wildlife Fund/Conservation International
Philippines	1993	13.00	19.00	17.70	68.4	93.2	1.4	World Wildlife Fund
Madagascar	1993	0.91	1.87	1.87	48.7	99.9	2.1	World Wildlife Fund
Madagascar[i]	1994	0.00	1.34	1.07	0.0	79.9	n.a.	World Wildlife Fund
Total average		46.30	177.56	128.77	37.2	86.3	2.3	

n.a. Not applicable.

Table 5.1
(continued)

Note: *Cost* = Expenditures by environmental agency to acquire the sovereign debt. *Face value* = Face value of the sovereign debt acquired by the environmental agency. *Conservation funds* = Value in dollars equivalent to the local currency part of the swap (either face value of the environmental bond or local currency equivalent). For bonds, the figure does not include interest earned over the life of the bonds. Overhead fees charged by government are not deducted. *Purchase price* = Price at which the debt was acquired (cost or face value). *Redemption price* = Conversion rate from foreign debt to local debt (conservation funds or face value). *Leverage* = Redemption price or purchase price.

a. Includes $250,000 donation from Fleet National Bank of Rhode Island.

b. Total amount of agreement is $3 million.

c. World Wildlife Fund contributed $1.5 million to this deal on top of the swap.

d. Purchase of CABEI debt.

e. Total amount of program is $5 million.

f. Total amount of program is $4 million.

g. Debt donated by Bank of America.

h. Involves buying blocked local currency funds from multinational. Also includes Midwest Universities Consortium for International Activities and U.S. Committee of the International Council on Monuments and Sites.

i. Face value of debt includes $200,000 debt donation by Bank of Tokyo.

j. Debt donated by JP Morgan.

Source: World Wildlife Fund, The Nature Conservancy, and World Bank, 1994g: 165.

to fewer funds and lower staff morale. Prior to the infusion of funds from the debt-for-nature mechanism, the country's ability to manage the existing system of parks and protected areas was in serious jeopardy.[12]

A change in government in 1986 paved the way for the country's leading role in the debt-for-nature concept. With the endorsement from the top political level, the newly created Ministry of Natural Resources reached agreement with the country's central bank about the general provisions of the swap mechanism, paving the way for a relatively smooth implementation of the concept. Over the next five years, Costa Rica was able to attract substantial resources and thus generate close to the equivalent of $45 million in local funds for natural resource management from a wide variety of donors. As the star performer in the first phase of the transfer mechanism, the country received one-third of all conservation funds mobilized for all fifteen recipient countries. The debt-for-nature swaps effectively reopened the dialogue with the commercial banks and helped set the country's debt-servicing reputation back on track.

With a new change in Costa Rica's government in 1990, the debt-for-nature mechanism was discontinued. Fears of inflation voiced mainly by the central bank and a focus on debt rescheduling along the lines of the Brady Plan are to be held responsible. With this change in the recipient's interest, new funding for the national park system shifted largely back to other bi- and multilateral funds and domestic sources. Meanwhile, as a consequence of the country's participation in Brady Plan rescheduling at the beginning of the 1990s and due to better economic prospects, the price of Costa Rican debt titles has risen sharply in the secondary market, making a reemergence of the debt-for-nature mechanism unlikely. The international NGOs started to concentrate on other potential recipients where the leverage of the swap mechanism could still be brought to bear.

What is the legacy of the nature swaps for Costa Rica? As the analysis in section 2 will attempt to demonstrate, the country has been ultimately unable to finance more than a fraction of its ambitious conservation and national park activities from this source. The funds mobilized were not sufficient to stem the many root causes of deforestation and rapid loss of biodiversity on the national scale. High-resolution satellite photographs from a research project at the University of New Hampshire estimated the forest cover to have declined to 14 percent of national territory by

1991, down from 17 percent in 1983. More alarming, although park administration and park rangers usually get good marks for their effective protection of the park's core areas, deforestation is also slowly progressing in the buffer zones of at least one national park.[13] Nevertheless, the country case has some valuable lessons to teach about the interplay of international and local NGOs in mobilizing financial transfers, about the merits and limitations for local NGOs as agents for environmental activities, and about the impact of new thinking about how to achieve effective conservation.

Second Phase: Swaps with Public Bilateral Debt

The innovative leadership of NGOs helped pave the way for the involvement of official creditors in what I call the second phase of debt-for-nature swaps. The mainstream Washington-based NGOs were especially successful in influencing and shaping the environmental component of the Enterprise of the Americas Initiative.[14] Their own action and lead had provided the showcases to demonstrate that the swap mechanism worked. The earlier commercial swaps initiated by the NGOs had also helped to sharpen their own understanding about what was needed to let the instrument succeed.

The Dutch and Swedish governments were early leaders in the field. They made public funds available in 1988 to purchase discounted private commercial debt from Costa Rica to be used in debt-for-nature swaps. They acted just as NGOs had before. Public debt titles, however, were initially considered sacrosanct. After debt forgiveness had been considered as an option at the 1988 G-7 Toronto Summit and after the dangers of tropical deforestation had reached G-7 communiqués in 1989 (Myers 1992: 440), Germany started a debt-for-nature program with Kenya later that year, forgiving about $500 million (817 million deutsche marks) in exchange for additional environmental project commitments from the Kenyan government (BMZ 1990). This single transaction more than quadrupled the face value of all prior debt-for-nature transfers, thereby demonstrating the much larger financial potential of second generation swaps. Other bilateral donors followed suit after an agreement had been reached in the framework of the Paris Club in 1990 allowing for different types of swaps, including debt-for-nature swaps with bilateral debt (Kloss

1994). The mechanism was extended to Eastern Europe, where Poland proposed the world's largest debt-for-nature swap program in 1991, which would potentially generate up to $3 billion over eighteen years (Sarkar 1994: 128).

The United States joined the second generation swap program which the EAI announced in 1990.[15] Up to $12 billion in government loans would potentially be available for reduction or cancellation, whereby enhanced environmental cooperation figured as one of the stated goals of the initiative for Latin America and the Caribbean. Due to legal and congressional budget constraints, however, the EAI had a slow start, despite favorable views from the Bush administration and from large parts of (the Democratic) Congress. In view of the slow and relatively small disbursement applied by the U.S. Treasury as a consequence of strict financial criteria, it appears to be too early to evaluate the implementation of funds from the EAI in the recipient countries. By the middle of 1992, the EAI had reduced the official debt of Bolivia, Chile, and Jamaica to the United States by $263.3 million. In 1993, Chile, Colombia, El Salvador, Uruguay, Argentina, and Jamaica benefited from an additional $90 million debt reduction. However, since the arrival of the Clinton administration and with the stubborn resistance of a single subcommittee in Congress,[16] the EAI has effectively been put on hold with no additional funding made available.

Potential Third Phase: No Swaps with Multilateral Debt

Although private commercial and public bilateral debt titles have been used in debt-for-nature transfers, the use of multilateral debt has been rejected for the time being.[17] MDBs, such as the World Bank, fear for the contamination of their portfolios and for their triple A rating critical for the refinancing of the organization's future lending (Cody 1988: 16). Debt forgiveness is perceived as incompatible with the statutory agreement making it mandatory for debtors to repay in full and on time. To the present day, the NGO initiative has faced insurmountable obstacles to include debt titles controlled by MDBs, whereby the latter are at least indirectly backed by their major state shareholders.

This has not excluded, however, indirect forms of MDB support for the debt-for-nature concept, such as World Bank advice to developing countries on how to integrate swaps with the country's environmental action plan. After all, swaps—not just nature swaps but the many off-springs from equity swaps, such as debt-for-health, debt-for-education, and debt-for-development swaps—are a market-based instrument to deal with the debt problem in combination with other pressing issues and therefore, at least theoretically, in line with the major development philosophy advocated by MDBs.

Partial involvement went even beyond technical support. The World Bank played with the idea to use funds flowing back from older IDA loans for debt buybacks but dropped the suggestion when no agreement over the selection of potential recipient countries could be reached. The International Finance Corporation (IFC), another affiliate of the World Bank group, sold a parcel of subtropical forest in Paraguay to an international environmental NGO below market value in 1990 (Brand 1990). The property was acquired by the IFC in a foreclosure in 1979. The Inter-American Development Bank (IDB) initiated a $400 million pilot program in 1990 that would enable the IDB to provide loans to its member countries to purchase nonofficial debt in the secondary market. The central bank of the participating country would then convert the purchased debt into local currency at an agreed conversion rate between the country and the IDB to be deposited in an inflation-indexed environmental fund (World Bank 1993c: 115). The last two examples, however, do not include multilateral debt titles. MDBs remain the laggards in the swap mechanism.

2   Analysis: NGO Initiatives and Limitations

When the closely related problems of deforestation and biodiversity loss emerged on the international agenda at the beginning of the 1980s, the degree of consensus among the key actors was extremely low, repeating the confrontational stance of North and South over global versus national priorities. Because a common definition of the problem did not exist—some actors even refusing to see a problem at all—the likelihood of quick, successful, and coordinated action appeared to be very small. While the North wanted the South to stem the alarming trends, hardline developing country governments, such as Malaysia, India, and Brazil (Hurrell 1992), insisted on their free-riding incentive and on their sovereign rights to repeat the environmentally destructive action of today's developed countries. They were eager to use the more pronounced concern expressed by the North as a most welcome bargaining chip and as a new opportunity for blackmail to attract additional financial and technological transfers. The environment was perceived as a new chance to alter the perceived imbalance between North and South after similar attempts had failed in the debate over a new international economic order a decade ago. Sovereignty over their environmental resources became a fiercely defended principle for developing countries.

Genuine concern for the environmental problems at hand was notoriously underdeveloped, with many developing country governments often directly connected to logging, cattle ranching, or other potentially destructive activities. The repercussions of the environmental problems were potentially endangering the livelihood of an increasing number of local and indigenous people who lived in the forests or were forced to move there for survival. They lacked the political clout to make their concern felt. The major domestic and global repercussions of the environmental problems would only be felt more dramatically by future generations. Developing country–based domestic interest groups for the environment were nonexistent or in their infancy at best.

Developed countries, on the other hand, disposed of both money and skills. They controlled the technology necessary to transform the long-term economic and financial potential of a rich pool of biodiversity into medium-term gains via data collecting, pharmaceutical prospecting, and genetic and biological engineering. Developing countries insisted on their fair share from the benefits of such endeavors based on the biodiversity of the tropical world (WRI 1994: 158). However, whereas the price of timber could be calculated more or less exactly, the benefits from conservation and/or sustainable use of biological diversity were and still are notoriously hard to measure and allocate. As the position of the United States on the biodiversity issue at UNCED shows, concern in the North is not strong enough to accept international agreements at any price.

The NGO Initiative: Innovation and Negotiation

Given the low or missing congruence of actors' interests to the present day and the absence of preexisting institutional networks/regimes to address the specific problem, a solution did not present itself after the problems had surfaced. Even worse at the time, the existing aid network/regime of the North contributed more to the problem rather than doing something helpful about it. World Bank loans for forestry projects and infrastructure in tropical forest areas, for instance, were notorious for their often devastating impact on the environment (Rich 1994). Until awareness rose by the end of the 1980s, aid to developing countries often did more to destroy species and habitats than to save them. The combined "efforts" of aid and organized (logging companies, cattle ranchers) and unorganized (miners, landless farmers, poachers) local actors put the remaining tropical forests under increasing pressure.

Thomas Lovejoy's 1984 initiative, described above, attempted to overcome the apparent deadlock over the issue. By linking the debt problem and the environment in developing countries and by suggesting a specific solution (the debt-for-nature concept), the initiative proposed a new broad problem definition. Developing countries would much rather talk about debt relief than about their environmental problems. The initiative tied the problem to a specific financial dimension, thereby opening the way for negotiations. The initiative, however, lacked specificity as to detailed mechanisms, actors involved, and particular objectives. It also left ample room for the involvement of other actors and for a wide variety of responses. Lovejoy's initiative was deliberately intended to stimulate a broad debate that would move, or so it was hoped, beyond the more narrow confines of the NGO community (Lovejoy 1994).

Due to the predominant concern with the debt crisis on the part of the major actors in the North, the broad NGO initiative was initially not taken up. The debt issue was equally polarized. Governments in the developed countries and the preexisting global financial institutions (IMF and World Bank) were primarily concerned with the prevention of a possible global financial crisis. Structural adjustment promised the best way for the restoration of financial credibility and for a continued debt-servicing ability. Improving macroeconomic conditions and getting the price signals right would also be the best medicine to help the environment, or so it was argued. Many developing countries, on the contrary, questioned the legitimacy of the debt, thereby also rejecting the suggested linkage in the NGO problem definition. They played for massive debt forgiveness instead.

Commercial banks—the third possible source for the suggested response to the problem—insisted on repayment in full for fear of contamination. Nevertheless, right after the beginning of the debt crisis, a secondary market for the interbank trade of commercial debt titles emerged, where discounts on the face value reflected the fact that the heralded principle of full repayment would probably never be met. This gave banks an opportunity to rebalance their portfolios.

Given the conflicting interests of the major actors over the two issues, it is not surprising that the broad NGO initiative appeared quickly to be equally stalled. The major sources for debt titles were rejecting the causal belief and the linkage expressed in Lovejoy's initiative.[18] Left with this reluctance of the major actors to go along, the three Washington-based environmental NGOs settled for a small subset of the initial problem definition. Instead of broadly combining debt relief with environmental projects, they concentrated on the subset of creation, maintenance, and conservation of tropical forest reserves that were in dire need of financial support. This strategy was based on the premise that successful biodiversity conservation requires a comprehensive program to establish a network of protected areas that are representative of the earth's biological regions (Klinger 1994: 236). The NGOs had firsthand experience from years of small-scale project activities in developing countries. They knew how quickly the pressure on these areas was growing from a variety of intruders. The administrative capacity in developing countries to maintain and guarantee park protection, if present at all, was suffering in many cases from financial cutbacks and political neglect in the wake of structural adjustment policies.

The focus on conservation was still very much congruent with the initial problem definition but ultimately only a small subset thereof. As conceptualized by the NGOs, and given their limited means unless bigger players could be convinced to join the effort, debt-for-nature swaps were neither going to solve the debt crisis nor significantly affect the root economic causes that lead to deforestation outside of small restricted areas.

Pushing the debt-for-nature concept further, however, the NGOs perceived of a number of additional advantages. Showing that something could be done would bring increased recognition and media attention.

The negotiations would bring the NGOs in closer contact with the "big" financial players and the other key actors, while potentially raising their organizational scope and reach and allowing them to tap into new sources of funding (Patterson 1990: 31; Dogsé and Droste 1990: 26). Talking to developing country governments would bring opportunities to raise concern for the environment and to support domestic interests willing to promote the issue. Debt-for-nature swaps could be perceived as a policy-leveraging tool. They would help to shift resources to meet legal obligations that were—at least in cases like Costa Rica—already there. In addition, the mechanism was ideal for the public in the developed world. It helped to raise donations by pointing at a specific number of hectares being protected in the South by the incremental dollar from the pockets of concerned citizens in the North.

Perhaps a bit too naively, the international NGOs had initially hoped for a much larger willingness from the commercial banks to come forward with donations for their new idea and "good cause." The NGOs also intensified their lobby activities to improve the tax treatment of donations. But despite some notable exceptions,[19] private banks have by and large preferred to either hold on to their debt titles or sell them rather than to give them away (Chamberlin 1992: 19). From their point of view, the public relations factor of donations for debt-for-nature swaps did not outweigh other factors, like portfolio considerations or future expectations for the recovery of certain country risks. The banks ended up treating the environmental NGOs like any other potential buyer in the secondary market. In this newly emerged market, there was no preexisting central authority or international organization to mount any resistance against the use of newly purchased debt titles for whatever purpose the buyer had in mind. Commercial banks had no reason not to sell to an additional source of demand. To the contrary, this would potentially drive up the price of an otherwise dubious asset. Increased demand and a higher volume of transactions would also bring additional fees and commissions.

The NGOs realized that they had to raise donations from sources other than commercial banks for their initiative. The first swap was ultimately facilitated by the announcement of the Connecticut-based Frank Weeden Foundation to make $300,000 available for the first purchase of debt titles to be used for environmental activities. Raising funds to purchase the debt has remained a major bottleneck and challenge for the NGOs.

Once the funding was secured, NGOs could provide a financial incentive to developing countries while also being able to advance their environmental objectives. A relatively small amount of hard currency would buy a significantly larger amount of local currency to be used for conservation and the improvement of environmental quality. The debt-for-nature mechanism comes with a tempting financial multiplier or leverage for the donor, making the swap more attractive than a straight transfer of hard currency. However, being able to purchase debt still left the NGOs with the task of convincing potential developing country recipient governments to enter into agreements with specific design features, appropriate for solving contracting problems.

To get the negotiations started, a recipient developing country government had to see its financial benefit in the endeavor, while also accepting the issue linkage. In its financial dimension, every swap is ultimately the result of striking a deal between the interests of the international NGO and the developing country government. Whereas the international NGO would like to purchase the debt at the lowest price and to convert it to as high a price as possible into local funds for the environment, the developing country recipient would like to see a lower obligation in local funds in order to maximize its savings when compared to the initial debt-servicing obligation in hard currency. Within this wide range for the specific negotiations, flexibility is made possible via the daily price fluctuations on the secondary market, the conversion rate of the funds, and the choice of exchange rates, that is, the decision whether the local currency equivalent is calculated at the daily exchange rate or at a different rate agreed on by the parties.

From the point of view of the environmental NGOs, concluding a swap agreement is still not necessarily an easy matter. An observer correctly called a specific swap "a result of connections, hard work and blind luck" (Chamberlin 1992: 19). Although there are common features to all debt-for-nature swaps, every individual swap is to a large extent an ad hoc activity depending on current conditions in the secondary market, on whether there is already a debt swap program in place in the developing country, on the outcome of negotiations over technical and financial details and environmental conditions, and on existing capabilities and capacity of the implementing NGOs and/or government branches—all differing widely from country to country. On average, it took the NGOs up to half a year to negotiate an agreement (Rubin, Shatz, and Deegan 1994: 26). Transaction costs in terms of administrative and organizational efforts for the international NGOs are significant in the debt-for-nature transfer. The international NGOs have therefore worked under the premise that there must be at least a discount of 50 percent on the face value of the debt title to make their effort worthwhile (Moltke 1988).

Another potential problem of the nature swap mechanism that could have posed difficulties to reaching any agreement has been reduced by the redesign of the mechanism after the first experience with Bolivia. The Bolivian swap did not involve a domestic constituency and it raised real questions concerning fairness: either Conservation International got more than it had paid for or Bolivia had no intention of doing what it promised. These two issues bedeviled Conservation International in the implementation phase and helped pave the way for the bond-based approach. Even this improved approach, however, is not without contractual problems. After the transfer of the debt title to the debtor, the latter has—at least in theory—every opportunity to ignore his initial commitment. Other than moral pressure, there is little the international NGO can do to enforce the environmental part of the agreement in court. In practice, however, enforcement has not been an issue. The gains of agreement for all sides involved in the potential win-win situation outweighed the potential contracting problems.

The negotiations and the close contact to the international NGOs and their agenda might have altered recipient government concern at the margin. In addition, the contractual innovation of the bond approach facilitated compliance as well as improved critical local capacity. The contract spells out the roles, responsibilities, and obligations of the implementing agents in great detail, thereby limiting the likelihood of disagreement later. The main implementing agents, local NGOs, share the environmental concern of the international NGOs. Given that they are directly affected in their operations by the detailed provisions of the agreement (recruiting and paying of NGO park rangers, purchase of land and equipment, setting up of facilities, etc.), they are in an ideal position to monitor disbursements. The swaps were intended to upgrade the role and mission of local NGOs.

Lessons from the Swaps in Costa Rica

Contrary to a number of other swap recipients, where local environmental NGOs did not exist prior to the mechanism, Costa Rica already featured a vibrant environmental NGO community. Much of the swap funding went to the National Parks Foundation (NPF), a private nonprofit foundation that had already started to operate in 1978 with a mandate to plan, manage, protect, and develop national parks and reserves. As an established institution, the NPF could play out its good connections to Washington's international NGO community. Being able to present a well-established and connected local NGO for receiving contributions and assuring responsible financial accounting and management has definitely helped foster donor generosity.

Strengthening local capacity within the environmental branches of government has been advocated as an alternative option by some observers (Patterson 1990). However, the NGO option presented a number of advantages. All recipient countries, not just Costa Rica, had to operate under constraints imposed by the MDBs. Strengthening the environment via financial transfers to local NGOs was consistent with the standard structural adjustment condition to reduce government bureaucracies. Moreover, a change in government would not directly affect the operations of the transfer mechanism. In addition, the long-term interests of the environment might ultimately be better served by helping to form a civil society. Enhancing independent institutional capacity can empower local self-interest, thereby gradually reducing the dependence on continued external support.

Did the local NGO option work? Optimizing the government-NGO interplay is no easy task, as the case of Costa Rica illustrates. Decentralized decision making on regional and local levels created new bureaucratic hurdles, and in some of the swaps, substantial delays in the disbursement of funds occurred due to bureaucratic infighting.[20] In an evaluation of their swap at the end of 1993, the Swedish government criticized the National Parks Foundation for not being able to account for all the funds received from debt-for-nature swaps, thereby hinting at some diversion and/or waste and at serious managerial problems (Lewin et al. 1994).

Evidence from other recipient countries highlights the fact that local NGOs need not necessarily be portrayed as the ideal agents for implementation. With a vibrant local environmental NGO community, as in some countries in Latin America or as in the Philippines, the selection of a particular local NGO counterpart for a swap can easily lead to new tensions and accusations in the NGO community. NGOs can also be overwhelmed by their new role. They often receive relatively huge inflows of funds compared to the status quo ante, which can pose serious problems for weak institutions. Without a careful look at the specific conditions in a recipient country, local NGOs should not be seen as the new panacea for channeling funds with environmental or larger developmental objectives into developing countries (Wegner 1993).

Did the swap transfers to Costa Rica meet their financial and environmental objectives? The transfers managed to more than triple the external aid for nature conservation by the end of the 1980s when compared to the situation prior to the swaps (Kloss 1994: 146). They provided urgently needed funding for an ambitious park system that was threatened to exist only on paper at a specific phase of the country's development. Only a few years later, however, their financial impact on the national system of conservation, as a proportion of all expenditures for the country's nature parks, had become negligible.

The bond-based approach has spread the actual disbursement over many years—as long as two decades in the case of the 1989 swap using Swedish funds. As a result of the initial bargaining with the donors over the detailed terms of the agreement, most bonds came with relatively low nominal interest rates. The net present value of such a long-term bond is driven down significantly, however, in the wake of double-digit inflation rates in later years (Kloss 1994: 132–135). As a result, the contribution from debt-for-nature funds for the total expenditures in the national system of conservation will fall to only 2 percent for the years 1992 to 1996 (MIRENEM 1993: 2). Public funds and bilateral assistance are expected to cover 27 percent each. The principal from the debt-for-nature bonds is or will be paid into the newly established national trust fund for the environment (*fondo patrimonial*) once the bonds have matured. And yet, the contributions from the trust fund will only finance 17 percent of the total expenditures for the period 1992 to 1996.

Such government estimates for the present and the near future must be interpreted with a grain of salt. They often serve as "shopping lists" for potential donors. Being able to point to the financial gaps and shortcomings of a particular national strategy or system is a good way to solicit funding from abroad. As with its active role in welcoming and promoting the debt-for-nature concept, the Costa Rican government has undoubtedly developed a fascinating creativity in this respect.

When the debt-for-nature concept was first discussed, the Costa Rican government made a deliberate decision to let potential donors and international NGOs identify the specific park or activity they wished to fund. Rather than channeling all the receipts into a general, country-wide conservation fund, it was believed at the time that contributors would be more likely to make long-term commitments to specific lands, "where they could see both the challenges and the yearly progress" (Umaña and Brandon 1992: 97). Naturally, this favored areas where Western biologists, scientists, and NGOs had already worked before, such as Guanacaste in the north of the country. A few conservation areas attracted the lion's share of funds and are today almost financially self-sustained. Other areas and parks, equally rich in biodiversity and not less threatened, have been more or less left aside by the debt-for-nature transfers. The funding of the national conservation effort is characterized by serious imbalances. But this was one of the implicitly intended results of the mechanism. Debt-for-nature swaps were also meant to work as "the bait to multiply the fish caught" (Ugalde 1994), as one key actor in the country's environmental policies put it. The country was outstandingly successful at that.

The remaining gaps in the ambitious national park system, the experience gained from the transfers, and the demonstrated absorptive capacity—all this made for a convincing case to attract additional bilateral and multilateral funding. Not surprisingly, Costa Rica was among the first recipients of grants from the Global Environmental Facility for biodiversity protection. These funds went to a nature park in the southern part of the country that had received very little from the nature swap initiative.

The limited reach of many swaps and the unsolved problem of meeting the recurrent costs of conservation projects once the short-term funds from the debt-for-nature transfer had been exhausted also helped to alert NGOs and policymakers to the need for environmental trust funds as more effective means to provide the necessary long-term, reliable, and steady support for projects promoting conservation and biodiversity protection (Wright 1992: 2; Resor and Spergel 1992; Rubin, Shatz, and Deegan 1994). With the creation of the national trust fund, this insight was included in the later swaps and subsequent strategies of international environmental NGOs.

The Costa Rican case has also contributed to alter the state of knowledge about viable conservation strategies. The initial focus of some NGO-initiated swaps was on purchasing remaining private lands within the legal confines of protected areas and on paying rangers and other staff to manage and protect the parks. The experiences of Costa Rica and other recipients of swap funds, such as Madagascar, for instance, has shown the limitations of this "fences-and-fines approach" (*Economist*, 2 April 1994: 44). Legally securing and policing the parks will only work if local people surrounding them and depending on these areas for survival are provided with meaningful alternatives. A viable conservation strategy must include broader issues of sustainable development and poverty alleviation. As a consequence, education and the provision of job opportunities for local people in the buffer zones, beyond the recruiting of park rangers, as well as the creation of infrastructure for nature tourism and scientific research have figured as core activities funded by nature swaps and other conservation projects. The Dutch swap in Costa Rica, for instance, sought to provide jobs and rural income by financing reforestation outside of park areas.

Costa Rica has confirmed its role as a magnet for innovative environmental activities and creative financing to the present day, whereby the debt-for-nature concept has played its part. Among many other sources, funds from the second swap with Sweden from 1990 helped to establish INBio, the country's national institute for biodiversity (WRI 1993). In its subsequent agreement with Merck and Company, the U.S.-based pharmaceutical firm, INBio went on to demonstrate the opportunities for the creative and much-needed inclusion of private actors to protect the environment.

In its quest for international financial transfers to protect the nation's environment, Costa Rica has skillfully managed to follow every trend in the international debate. Earlier campaigns, such as the debt-for-nature concept, highlighted deforestation as the main threat. In later years, the country's role as the biodiversity "hot spot" at the isthmus between the two parts of the American continent figured more prominently. As the most recent trump card, the country has actively advocated "carbon bonds" for joint implementation as an abatement strategy against climate change (Orlebar 1994). With these proposed financial mechanisms, companies in developed countries can offset greenhouse gas emissions by financing forest preservation.

The question, however, is whether Costa Rica has done enough to set its domestic record straight and to mobilize domestic sources of funding. If some of the obvious creativity in attracting funding from abroad had gone into such efforts, the country would look less like an example of environmental aid addiction. For instance, the close to 300,000 foreign ecotourists who visited Costa Rica's national parks in 1991 were only charged entrance fees of less than $1 each.[21] A still modest $5 entrance fee for foreigners would have been enough to pay for the entire current staff and operational budget of the national park service. It took another change in government in 1994 before steps in this direction were taken. In the meantime, the number of foreign visitors to Costa Rica's parks has continued to rise sharply, warranting even higher fees.

In conclusion, the Costa Rican case—although an exceptional and "best case scenario" in many ways—taught the three international NGOs a number of lessons: the importance of trust funds and local counterparts, the necessity to move beyond parks, and the need to get the private sector and the more important actors involved. In the second phase of the debt-for-nature mechanism, the final form of the Enterprise of the Americas Initiative reflected many of these lessons.

Contrasting the German and the U.S. Experiences in the Second Phase
Looking back at their innovative lead in the first phase, the international NGOs had to live with a financial scope and reach of their initiative that was certainly below some of their initially high hopes and expectations. It became clear in the process that the ultimate test for the concept's success lies in the NGOs' ability to reach out to the overarching institutions and more important actors—the world of trade and finance, the much larger aid budgets and strategies of governments and MDBs—to

respond more adequately to debt-aggravated deforestation and loss of biodiversity.

Undoubtedly, the NGO initiative had paved the way and demonstrated that the mechanism can be successfully implemented. This opened the door for potentially much larger flows of financial transfers for the environment, in which taxpayers in developed countries are shouldering part of the financial burden. Falling subject to various domestic and foreign policy interests, however, the largely excellent record of the first generation swaps on implementation has not continued in all cases. Some governments have been significantly less successful in achieving compliance with the mechanism than NGOs. Compared to the first generation, second generation debt-for-nature swaps differ more widely in the way they are designed and implemented.

The German swap with Kenya, merely dressed in environmental garb, comes close to outright debt forgiveness by inviting reneging. Early in 1988, the German chancellery had commissioned a study to suggest ways that would combine debt relief with environmental protection for tropical forests (Oberndörfer 1989). After a brief evaluation of the first generation nature swaps, the study argued in favor of applying the initiative to bilateral debt titles. The study also suggested creating a new international institution to supervise and coordinate debt relief and environmental activities (Oberndörfer 1989: 39). The German government played an active role during the G-7 summits in 1988 and 1989 to bring these issues to the international agenda. However, apart from vaguely supportive sentences in the official G-7 communiqués, nothing much followed. The existing international institutions gave the environment a higher organizational profile after the World Bank/IMF annual meeting in Berlin in 1988, but a new oversight institution did not develop. Management of the debt crisis with the existing multilateral institutions—eager to defend their new-found importance and wary of new issues—remained the priority over environmental considerations. Early bilateral activities applying the swap mechanism lacked international coordination and reflected domestic interests.

Contrary to the domestic situation in the United States, German environmental NGOs and observers close to the German Green Party were highly critical of the debt-for-nature concept and advocated outright debt forgiveness instead (Radke and Unmüssig 1989). The generous treatment West Germany had received from its creditors in 1953 was seen as the model for the way out of the debt crisis. A debtor should only shoulder that part of the obligations which the country is able to carry.[22] This would necessarily imply more generous forms of debt forgiveness. Germany has been the leading Western country in applying this concept primarily to least developed countries. Germany's total debt forgiveness from official development loans has amounted to more than $6.4 billion[23] since 1978 (Taake 1994: 228). This represents more than a third of total Western debt forgiveness during that period.

Germany's comparatively greater propensity for debt forgiveness has tainted its pioneering role in the second generation of debt-for-nature swaps. The swap agreement with Kenya from 1989 had obliged the Kenyan government to make additional funds available for environment and resources protection "to the extent possible" (BMZ 1990: 53, 119) as a consequence of the bilateral debt forgiveness. Not surprisingly, the Kenyan authorities took this rather vague formula as an invitation to renege on its commitment. They broke the letter but apparently not the intent of the agreement. Instead of additional projects, the Kenyans used the newly available funds to continue existing projects threatened by cuts from structural adjustment measures. This did not cause a stir—either in the German public or in the German Ministry of Cooperation charged with supervising the implementation of the agreement. Applying sanctions would only stop projects that were considered to be in the interest of the Kenyan environment (Böhmer 1994b). There was no serious effort to enforce the swap agreement.[24] More recent German swaps planned with Honduras, Jordan, and Vietnam will include funding for environmental projects from multilateral donors and NGOs in addition to bilateral German projects (Böhmer 1994b). It remains to be seen whether this modification of the German approach to include local constituencies, as well as the recipient country's government, can alter the existing rather dismal record on compliance and implementation.

In contrast to Germany's early attempt of a second generation swap, the EAI is a mechanism in which the donor country keeps the debt title allowing for drastic reactions to cases of noncompliance by reactivating the old obligation to service the debt in hard currency. Apart from legal differences and the strong role played by the U.S. Treasury in the EAI, this is to a large extent also a reflection of the strong focus on compliance in the discussion of the nature swap concept in the United States. In addition, the EAI as it emerged from Congress is deliberately built around the detailed experience of the international NGOs with the first generation swaps.

The three Washington-based environmental NGOs played an active role in lobbying for the debt-for-nature initiative to be expanded to bilateral credits. They would describe the final shape of the EAI as a major success story of productive NGO influence on government policies. Within the Bush administration, William Rielly, head of the Environmental Protection Agency (EPA) and previously with the World Wildlife Fund, was arguably the most active proponent of the mechanism. After the changes to the executive draft legislation, the EAI reflected a number of lessons learned by the three environmental NGOs: local NGOs—not just one but as many as possible—can actively participate in the process of determining what environmental benefits should flow from debt reduction (Weiss 1992: 923); a portion of the payments in local currency are deposited in an environmental fund to address the long-term recurrent cost problems of conservation projects; the United States keeps the title to the debt and is involved in institutionalized oversight mechanisms for the environmental projects (Sarkar 1994: 131). The international NGOs could make a powerful case on the basis of their own project experience with the mechanism. They also benefited from the fact that there was not enough money involved to make it worthwhile for special interests to figure out how to control the allocations.

It is too early to evaluate the implementation of these regulations in specific projects on the ground. The ultimate financial dimension of the EAI for the environment, however, and the seemingly endless battles over congressional authorization for additional funds, make it difficult to escape the conclusion that the EAI has its fair share of symbolic politics as well. Compared to the financial contribution other bilateral creditors were willing to make to the swap mechanism, and in contrast to the more than $120 million mobilized by private environmental NGOs that were ultimately limited in their capacity to raise donations, the $154 million in local currency that the EAI made available for environmental projects over a number of years in the participating countries (U.S. Department of the Treasury 1993: 1) seems to be a relatively meager financial result of intensive NGO lobbying.

For different reasons—the limited financial reach of the EAI and the greater propensity for debt forgiveness in the case of Germany's early activities—the record of the initiative's second phase left much to be desired from the perspective of the international NGOs.

3 Evaluation: A Window of Opportunity Is Closing

Debt-for-nature swaps were an attempt to bring together the broader problems of deforestation, biodiversity loss, and the environmental implications of the debt crisis in developing countries, while simultaneously outlining a solution. However, the main actors at the beginning of the debt crisis were following different priorities and agendas. They rejected the debt-environment linkage expressed in the 1984 initiative. This left the international environmental NGOs as the only actors willing to follow up on their idea.

In addition to a number of other advantages, piggybacking the debt-for-equity concept had an appeal that NGOs found difficult to resist given one of their main deficiencies: a notorious shortage of funding. This transfer mechanism came with a high financial efficiency that potentially could reduce by half the initial amount of hard currency needed to purchase a debt title. This made for a powerful incentive when presenting the case to potential donors. Private commercial banks, however, were reluctant to donate debt, preferring instead to hold on to an asset that was hoped to gain in value over the years. The initially higher expectations of NGOs for massive resource mobilization failed to materialize. It proved to be a lot more difficult to mobilize funds from foundations or individual donors. Once funding for a specific swap was secured, however, the transfer mechanism continued to demonstrate its appeal to the parties involved. The swap reduced the problem to a specific financial negotiation, whereby the outcome could be presented as a win-win situation to all parties. The remaining differences between a recipient developing country government and an international environmental NGO could be settled in the bargaining over the specific financial terms of the agreement. The debt crisis opened up a new way for the NGOs to fund their on-the-ground priority for developing countries, that is, conservation projects.

Did the issue linkage and the funds that were transferred keep up with the scope of the environmental problems? The initial issue linkage between debt and the environment was a precondition to get the flow of funding started under very difficult circumstances. Its main drawback has only developed over time. As countries outgrew their debt problems, the secondary market price of their debt titles has also gone up sharply, thereby reducing the attractiveness of the debt-for-nature linkage. Given these dynamics in the debt crisis, it is no coincidence that the number of first generation agreements has declined sharply since 1992. The April 1994 refinancing agreement between Brazil, the largest debtor in the developing world, and its commercial foreign creditors led many observers to conclude that "the international debt crisis is over" (Gilpin 1994). This might be true from the bankers' point of view. With the Brazilian deal, roughly 80 percent of the money lent on commercial terms has been restructured. The supposed end of the debt crisis, however, has in many cases not really affected the persisting pressures on tropical forests and biodiversity reserves. The initial argument the NGO initiative had started out with—debt-servicing obligations as the major cause for environmental destruction—did not fully stand the test of time. In fact, three case studies commissioned by the NGOs themselves argued that even though the debt crisis and subsequent structural adjustment policies had often exacerbated the environmental problems, more deeply rooted domestic and international causes had preceded this phase and were also likely to outlast it (Reed 1992).

There were other limitations built into the issue linkage and the design of the transfer mechanism from the very beginning. The first generation's reliance on available commercial debt titles from the secondary market reduced the major recipients to those in Latin America. Other parts of the tropics with substantial rates of deforestation but with a different or much less of a debt problem, such as large parts of Southeast Asia, fell almost completely out of the picture (Kraemer and Hartmann 1993). The secondary market consisted mainly of debt titles for the four big debtors in Latin America, that is, Brazil, Mexico, Argentina, and Venezuela (Tammes 1990), whereas other country titles had a much smaller market share.

Heavy demand would drive up the price, thereby reducing the possible leverage of the instrument. Depending on the specific country conditions, all debt swap instruments—not just debt-for-nature swaps—were ultimately unable to surpass this limitation of relatively narrow markets.

Apart from the exceptional case of Costa Rica, larger funds could only be generated for Panama and the Philippines. The conservation funds mobilized by the first generation debt-for-nature mechanism for the remaining twelve recipient countries have remained relatively small. How do the recipients compare to a list of countries most in need of environmental transfers to stem the loss of biodiversity and deforestation? In terms of ten "hot spots" that (1) are characterized by exceptional concentrations of species with high levels of endemism and (2) are experiencing unusually rapid rates of depletion (Myers 1988), the mechanism's reach has been partial at best. Out of this list, only four areas received funding from the first phase of the nature swap mechanism: Madagascar, Western Ecuador, the Brazilian Atlantic coast, and the Philippines. The record becomes even worse when one looks at those areas where about two-thirds of the recent destruction of tropical rain forests occurred: Brazil, India, Indonesia, Burma, Thailand, and Colombia (MacNeil, Winsemius, and Yakushiji 1991: 82). Only Brazil received nature swap transfers, albeit only with a relatively tiny swap in 1992. The swap that could have given the concept a whole new dimension, a major initiative to protect the Brazilian Amazon, never did really get off the ground (Bramble 1989; Hecht and Cockburn 1990; Hurrell 1992). The strength of the sovereignty principle, persistent questions as to the legitimacy of the debt, and fights among the different interest groups within the Brazilian administration provide some clues why this critical campaign ended in a deadlock.

Given the unabated North-South polarization over the environmental issues and the notoriously challenging and difficult business of raising donations for their initiative, the limited financial reach and scope of the nature swap transfer mechanism should not be held against the international environmental NGOs. Moreover, even if the big Brazilian swap had worked, the NGOs had already begun to reconsider their role in the process. The ultimate success of their initiative would depend on their ability to influence the more important actors—not to replace them. The NGOs realized that they would be stretching their role by multiplying biodiversity and conservation projects worldwide. In addition to the increased prices on the secondary market, this might help explain their limited attention span. From the end of the 1980s onward, the GEF promised to be the new multibillion dollar game in town. Trying to shape its structure and design held a number of advantages over the nature swap campaign that merited moving on to this new issue. The GEF would be of global reach and not limited to countries with serious debt problems. The funding would hopefully be much larger. Moreover, the GEF opened up an opportunity to get a foot into the door of the World Bank, something the nature swap concept had failed to deliver. The upcoming UNCED was yet another major event worthy of increased NGO attention and activities.

The swap mechanism was ultimately only of limited reach even in the case of its major recipient. With recipient concern exceptionally high in Costa Rica, the transfers helped to address domestic priorities and policies that had already existed prior to the first agreement. Costa Rica received a much-needed infusion of funds which have contributed to strengthening the ambitious national park system during a critical phase of its development. The direct financial impact of the transfers, however, will be insignificant in a few years. In comparative terms, other recipients with less of a track record on environmental concern and with more limited or nonexistent administrative, legal, political, and technological capacity for the implementation of an effective conservation strategy have probably benefited to a larger extent from the creation of nature parks and the strengthening of local environmental NGOs through the mobilization of funding from abroad. It became clear in the process that projects that depend on continued external support are in constant jeopardy. The factors leading to deforestation and rapid loss of biodiversity outside the confines of restricted areas must be addressed as well.

In sum, the greater value of the debt-for-nature concept lies in the insights gained in the process of implementing conservation and biodiversity projects in developing countries. These include: the need for long-term financial support, whereby environmental trust funds can play a significant role; a broader strategy of sustainable development work in areas surrounding the parks; a strengthening of both local environmental

NGOs and government capacity to create the necessary domestic conditions for viable, long-term solutions. There is some clear evidence in the process that the three mainstream international environmental NGOs have shifted the focus of their work due to organizational learning.

Evidence from the debt-for-nature mechanism also confirms that neither local nor international NGOs should be glorified. Just like states or any other political actor, they are subject to competition and other pressures that can easily get in the way of effective transfer design and policy implementation. For instance, it must be pointed out that the relations among the three mainstream Washington-based international environmental NGOs in the history of the swap mechanism were not always easy. Jockeying for position is characteristic of the behavior of not just states when new issues reach the international agenda. In the 1987–88 race to conclude the first swap, coordination among the three NGOs left much to be desired. As a new organization, Conservation International— founded in 1987 mainly by former members of The Nature Conservancy's international department—was most eager to benefit from the $300,000 offer of the Frank Weeden Foundation. This competition provoked a great deal of implementation problems. It was only after the successful redesign of the mechanism to the bond-based approach pioneered by the World Wildlife Fund that the three NGOs were brought back in line.

The NGOs eventually realized that their attempts to influence the actions of the more important actors had failed to turn into a real success story. Commercial banks made little more than symbolic gestures. MDBs have strictly refused to let the NGO initiative affect their newly upgraded role in the management of the debt crisis, although the World Bank has been made increasingly responsive to environmental concerns in the process. Governments have tried on their own, but the German swap with Kenya, for example, was little more than debt forgiveness dressed up in environmental garb. Again, the attention spans of donors have been quickly dominated by other issues. The EAI could be interpreted as a major exception. Nevertheless, three NGOs were even unhappy with the EAI, when the funding for its trade and investment component reduced the debt reduction and environmental component to little more than a symbolic gesture.

4 Conclusion

A decade since the debt-for-nature idea was launched from within the Washington-based international environmental NGOs, it is difficult to escape the conclusion that the concept is now beyond its prime. In a few years, the concept may be seen as a limited NGO initiative during a certain phase of the debt crisis. As of 1996, the combined effects of lowered pressure—and hence leverage—from the debt crisis and of short attention spans of funders have already reduced interest in the mechanism. The window of opportunity from the debt crisis is about to be closed. Even the initial NGO proponents of the mechanism have reduced their staff working on these issues and moved their emphasis on to other, more promising campaigns and activities.

If the debt crisis worsens again or MDBs reconsider the mechanism, commercial debt swap conversions could be quickly reactivated.[25] Yet the conditions in those parts of the world where some limited potential for commercial swaps might still be exist—sub-Saharan Africa and some parts of the former Soviet sphere of influence, in particular—do not lend themselves easily to the concept. Environmental concern tends to be low or overridden by much more pressing economic issues. Local NGOs are either nonexistent or have to operate under much more difficult circumstances. In addition, the quest for debt forgiveness has begun to overwrite the leverage that conditionality, green or otherwise, might have held in the regions unable to overcome their debt problems. The nature swap experience has confirmed that conditionality as a means to achieve lasting government policy reform is a risky business. Strengthening local NGOs might be the better long-term strategy, albeit depending on local conditions.

As with a number of issues discussed in this volume, the debt-for-nature transfer mechanism can hardly be called a success story. It has helped to generate some valuable insights about meaningful ways to finance nature conservation and biodiversity preservation, while also teaching both international and local NGOs a lesson about their capacities and limitations when dealing with the more important actors and institutions. NGOs, with limited funds, can only become more significant actors under very specific contextual conditions, and there is certainly no guarantee that governments and MDBs will follow suit. Even situations that seem to be win-win are often only temporarily available. In addition, such situations are vulnerable to opposition by actors able to veto national policies, as illustrated by action taken by the Foreign Operations Subcommittee of the U.S. House Appropriations Committee in defining the financial scope of the Enterprise of the Americas Initiative, or by national players in the collapse of the initial Brazilian swap. There can also be little doubt that the broader agenda at the heart of the initial NGO concern is still as valid as when it was first brought to the world's attention. Deforestation and the rapid loss of biodiversity will be on the international environmental policy agenda for some time to come.

Notes

I want to thank all members and invited guests of the Harvard University Financial Transfers Project for their many comments and continuous support during the research and writing of this chapter. As ever, Hildegard Bedarff provided the most valuable encouragement and comments. I am also particularly indebted to the Thyssen Foundation, the Center for International Affairs and the Center for European Studies, both at Harvard University, for allowing me to stay as a Thyssen Fellow and as a Kennedy Fellow at Harvard during the academic year 1993–1994. I would like to gratefully acknowledge the very helpful written comments on earlier drafts I received from Barbara Connolly, Peter Haas, Ronald Mitchell, Konrad von Moltke, Miranda Schreurs, and from two anonymous referees. In addition to the sources cited in this chapter, important background information was provided from many members of the international environmental NGO community, from people working for commercial and multilateral banks, and from people in or close to the government in the United States, Germany, and Costa Rica. Of course, all remaining errors are entirely my own and should not be attributed to any of the aforementioned or cited individuals or institutions.

1. The legal, financial, environmental, political, and economic merits and pitfalls of this transfer mechanism have been debated at length. Lawyers have mainly complained about the limitations in the contractual design of the mechanism, which put limits on enforceability and compliance (Hrynik 1990; Greener 1991). Economists have taken issue with the inflationary potential inherent in the swap mechanism (Panayotou 1992), with the high transaction costs involved, and with the suboptimal allocation of resources in this case (Kraemer and Hartmann 1993). Third World activists have brushed the swap mechanism aside because it diverts attention from a supposedly more important objective, that is, large-scale debt forgiveness (Mahony 1992; George 1992). For an initial overview and a more balanced and mostly positive assessment of the concept, see among others,

Hawkins 1990; Patterson 1990; Gibson and Schrenk 1991; Sher 1993; and Sarkar 1994.

2. Other, smaller international NGOs have also been involved with the mechanism, albeit on a more selective and partial basis. Some local NGOs from developing countries provided smaller amounts as well. See table 5.1 for details about the different institutions involved. No doubt, however, that the three United States-based NGOs spearheaded the concept and must be seen as the driving forces.

3. For an overview of all first generation swaps, see table 5.1.

4. The term "second generation" swaps was introduced by Conservation International, the pioneer with the first debt-for-nature swap in Bolivia in 1987 (Sher 1993: 151).

5. Although complete records on first generation swaps are kept by environmental NGOs and the World Bank, detailed figures for all bilateral swaps in the second phase are much harder to come by. There is no central institution that would distinguish between bilateral debt forgiveness in general, debt forgiveness for reasons other than the environment, and second generation swaps for the environment. However, the combined two bilateral nature swaps from the second phase discussed in more detail in this chapter, Germany and the United States, are almost reaching the aforementioned benchmark on their own.

6. On the role of environmental NGOs, see Princen and Finger 1994; and Haas 1992. For an analysis of the debt-for-nature mechanism that highlights the role of NGOs, albeit only in the first phase of the concept's history, see Klinger 1994.

7. For an early discussion of the debt-for-nature concept from the perspective of negotiation analysis, see Dawkins 1990. Susskind (1994: 84–86) is also following an analysis along these lines.

8. The initiative from 1985 that led to the Tropical Forestry Action Plan (TFAP) might be seen as a partial exception with regard to deforestation.

9. As formulated by Thomas Lovejoy in an interview with the author: "People laughed at me and thought I was crazy."

10. Evidence to the contrary could neither be found in the author's interviews with the environmental NGOs nor via the study of several of the aforementioned audits, delays in the spending of some funds due to technical, political, or unforeseen events notwithstanding, as notably in the first swap with Bolivia (Sarkar 1994: 126; Rubin, Shatz, and Deegan 1994: 25). It would have been surprising, however, to have the international NGOs admit that fraud or waste occurred on a massive scale given that they are the main proponents of the swap transfer. To be honest, one must admit that it would almost be impossible to closely check thirty-two transactions in fifteen countries. I am indebted to Konrad von Moltke for pointing out to me that the later NGO emphasis on trust funds rather than on the bond-based financial transfer from the debt-for-nature mechanism can also be interpreted as an admission of problems. Relying on weak local organizations

for the distribution of relatively large sums has created a number of problems, as will be pointed out in later sections of this chapter.

11. For an introduction to Costa Rica's environmental efforts and conservation policies, see Wallace 1992; and Umaña and Brandon 1992. For a more critical assessment, see Carrière 1991.

12. One of the most active proponents for the debt-for-nature concept in the country summarized Costa Rica's situation by the middle of the 1980s: "Overlapping and unclear jurisdictions, serious funding and personnel shortages, an inability of environmental agencies to coordinate with each other or with other agencies, and environmental agencies with insufficient power (compared to that of other government ministries) were major factors inhibiting the government's ability to manage protected areas" (Umaña and Brandon 1992: 99–100).

13. E-mail communication from Arturo Sanchez, Complex Systems Research Center, University of New Hampshire, 23 September 1994. The annual rate of deforestation for Costa Rica between 1981 and 1990 is estimated at 2.6 percent, the third highest in the hemisphere after Jamaica and Haiti (WRI 1994: 307).

14. Interviews with representatives from the Washington-based international environmental NGOs, see Rubin 1993; Curtis 1993; and Resor 1993.

15. For the environmental aspects of the EAI, see Gibson and Schrenk 1991; Weiss 1992; U.S. Department of the Treasury 1993; Sarkar 1994; and Rubin, Shatz, and Deegan 1994.

16. The Foreign Operations Subcommittee of the House Appropriations Committee.

17. As the World Bank put it: "While debt-for-nature swaps are not part of the World Bank's lending operations, the objectives of those swaps fit with Bank support for conservation and debt reduction." *World Bank News*, special issue, May 1992: 24.

18. Given this predominant resistance to the causal belief in the debt and environment linkage, it is somewhat surprising that the NGO community waited until the end of the 1980s to prove their case with a detailed comparative study (Reed 1992). This study also made it clear that debt servicing is by no means the only responsible factor for environmental degradation in the developing world.

19. See table 5.1 for names of private banks that did donate some of their debt titles in particular cases. In addition, some banks waved fees and commissions and donated their time and expertise to assist NGOs in the financial engineering required to close the transactions (Rubin, Shatz, and Deegan 1994: 26).

20. Umaña and Brandon (1992: 98). For a stronger critique of the environmental bureaucracy in Costa Rica, see Carrière (1991: 194).

21. Personal communication from staff of the Servicio de Parques Nacionales (SPN), San José, Costa Rica, 17 February 1994.

22. Kampffmeyer (1987) and Böhmer (1994a) discuss the concept in detail. On NGO advocacy for debt relief, see World Bank (1994g: 34).

23. DM 9 billion calculated at the DM/$ exchange rate from the middle of 1995: DM 1.4 for $1 U.S.

24. Similar symbolic politics have been pursued by France in its dealings with the former French colonies in West and Central Africa. Germany continued to use the environmental and resources protection option to the tune of $35.7 million (DM 50 million) in 1993, and of $57.1 million (DM 80 million) in 1994 (Böhmer 1994a: 38). Again, the DM/$ exchange rate applied is at DM 1.4 for $1 U.S.

25. Mexico's recent financial troubles after the massive devaluation of the peso in December 1994 is a case that comes to mind. However, financial experts point to the difference between the debt crisis at the beginning of the 1980s and the situation by the middle of the 1990s: "Unlike the debt crisis of 1982, when a liquidity squeeze in several big Latin American countries rapidly turned into a genuine solvency problem, most emerging markets are now in good shape. They are growing rapidly; their exports are booming; and their budget deficits are under control" (*Economist,* 18 January 1995: 90 ["Going with the flows"]).

6

Conditionality and Logging Reform in the Tropics

Michael Ross

No global environmental issue has raised more concern, to less avail, than tropical deforestation. The tropical forests of Asia, Africa, and Latin America are receding at a rate of 0.8 percent, or 15.4 million hectares, per year (FAO 1993); their loss contributes to soil degradation, global climate change, and diminished biodiversity. Efforts to conserve these forests through international agreements have largely failed. As a result, a handful of bilateral and multilateral funders—principally the World Bank—has taken up the job of promoting rain forest protection.

The World Bank and other funders have long used their forestry aid to promote commercial logging in the tropics. In the 1980s, under pressure from the environmental movement, funders began shifting their attention to the problems of forest loss and forest degradation. By the late 1980s funders were using their most coercive tool—conditionality, the practice of making aid conditional on policy reform—in an attempt to force developing states to better protect their forests.

At first glance, conditionality looks like an attractive tool for donors. It seems like an easy way to persuade aid recipients to rewrite their most noxious policies; it is typically used with loans, not grants, making it cheap to use; it can be "piggybacked" on to existing aid programs, giving donors the promise of more clout for the same amount of money; it circumvents the complicated shortcomings of project-oriented environmental assistance; and it offers aid agencies a more defensible rationale for their programs, a clear quid pro quo for the money they send overseas.

Yet conditionality's impact in hard cases, when powerful domestic interests oppose reforms, is unclear. In the late 1980s and early 1990s funders tried to use conditionality to stop unsustainable logging practices in Indonesia and the Philippines. Both countries have been major exporters of tropical hardwoods. Both have lost much of their forests to environmentally destructive logging, and both have had powerful logging industries that are closely tied to, and subsidized by, their governments. In Indonesia, the World Bank's attempt to impose conditionality failed when the government heeded the interests of the timber industry and rejected the proposed loan. In the Philippines, however, funders used conditionality to help raise logging taxes, protect virgin forests, enforce long-dormant forestry regulations, and break the logging industry's grip on government policy.

This chapter uses these examples, and the wider literature on conditionality and contract theory, to make two arguments. First, when a recipient government relies on its logging industry for political support—as was the case in Indonesia—the type of conditionality used by the World Bank and other major funders is likely to fail. Second, when a government is committed to reform but lacks the political capacity to overcome the influence of pro-logging interests—as was the case in the Philippines—conditionality can play a critical role in helping the reformers prevail.

The first part of this chapter reviews both the problems of commercial logging in the tropics and international efforts to address them. It explains the role of logging in the degradation of tropical rain forests, the common political obstacles to logging reform, the failure of earlier multilateral initiatives to protect tropical rain forests, and the evolution of World Bank forestry lending. It then describes the World Bank's failed attempt to promote logging reform in Indonesia and the successful campaign of the World Bank and other funders for logging reforms in the Philippines. The next section draws on the broader literature on conditionality and contract theory to assemble a simple model of conditionality, highlighting the variables that influence its effectiveness. It then uses this model to explain conditionality's failure in Indonesia and its success in the Philippines. The final section revisits the central question of this chapter: can conditionality promote sustainable logging in tropical forests? It also speculates about the ability of conditionality to foster other types of environmental reforms.

1   Description

The gradual devastation of the tropical rain forests of Latin America, Africa, and Asia has caused a profusion of environmental problems: the eradication of a large fraction of the world's biodiversity; increased soil degradation; siltation in waterways and coastal areas; fluctuating bouts of flooding and drought; and the atmospheric release of stored carbon dioxide that may contribute to global climate change (WRI 1994; Houghton 1993). Deforestation has also led to new hardships for hundreds of millions of people who dwell in or near tropical rain forests and depend on them for their livelihood (Ascher 1994).

In those tropical rain forests with dense stands of commercially valuable trees (which are found principally in Central and West Africa and Southeast Asia), logging can be an important source of deforestation and forest degradation. In ideal settings, logging can create jobs, profits, and government revenue while causing few environmental or social problems. But more commonly, logging contributes to deforestation or forest degradation in two ways: first, unsound logging practices often damage forests so severely that their ecological functions are impaired for decades or centuries; and second, logging enables settlers to enter and cultivate logged-over areas, which are then unable to regrow (Bruijnzeel and Critchley 1994). Environmentally sound logging practices are rare in the tropics. A 1987 study of forest management in seventeen tropical states (covering over 70 percent of all tropical forests) found fewer than 1 million hectares being "sustainably" managed—less than 0.2 percent of the forest area surveyed (Poore et al. 1989).[1]

Policies, Politics, and Unsustainable Logging

Government policies often help establish or perpetuate unsound logging practices in three ways. First, many governments sanction logging at unsustainable rates, even in forest areas set aside for permanent timber production. Second, governments throughout the tropics regularly fail to enforce their own silvicultural and environmental regulations, even those that require little administrative capacity. Finally, governments tend to collect only a small fraction of the rents (or windfall profits) accrued by loggers, which both fosters corruption and deprives the government of a large pool of revenue (Repetto and Gillis 1988).

These perverse government policies often have a political basis. Almost all of the forests in tropical areas are government-owned, enabling public officials to preside over the distribution of logging permits. The logging permits are commonly distributed as a form of political patronage by public officials, who thereafter undertax and underregulate the clients who retain these licenses. Changes of government often lead to the redistribution of timber licenses, rather than systemic reform. Hence the timber industries of developing states are almost always in the hands of government allies; the government, in turn, is loath to act against their interests.

The domestic pressures for logging reform in tropical states may be widespread, but their impact is usually small. Even when environmental groups face no political restrictions, their efforts to mobilize support for forestry reform must overcome a daunting collective action problem. Those who benefit from unsustainable logging, underregulation, and undertaxation are a concentrated and politically influential constituency. Those most harmed by it—typically forest dwellers who lose their resource base, the public at large who lose tax revenues, and future generations who lose their natural resource endowment—are dispersed, underenfranchised, and often impoverished.

Hence government leaders rarely have much concern about unsustainable tropical logging. Even when they do, their ability to regulate the timber industry may be weak. Most forestry institutions in tropical states are chronically short on human resources, training, information, funding, and often administrative authority (Vollmer 1993; Miranda et al. 1992; World Bank 1991b). Because tropical forests have a distinctive ecology, the experience and techniques of foresters trained in temperate zones is often of little use. Tropical rain forests are also geographically extensive and usually far from the infrastructure of government authority, which makes logging sites difficult to monitor. Finally, the logging industry's profits tower over the tiny salaries of forestry officials in developing states, rendering the officials vulnerable to corruption. Even when confronted with a reform-minded government, the harmful practices that favor the industry—unsustainable logging, underregulation, and undertaxation—

may themselves generate enough rents to allow loggers to purchase further regulatory immunity.

International Initiatives on Tropical Forestry

Since the mid-1980s Western funders and environmentalists have tried to promote forestry reform in the tropics by pressuring recalcitrant governments. They have launched three initiatives; none of them has had a significant impact.

The first initiative was the Tropical Forestry Action Plan, a program begun by the FAO in 1985, and later co-sponsored by the World Bank, the UNDP, and the World Resources Institute.[2] The TFAP pursued a twofold strategy, seeking both to boost international funding for forestry projects and to use this new money to prod recipients toward reform. Although international funding has risen sharply, most observers agree that it has produced little, if any, reform (Ullsten, Nor, and Yudelman 1990; Winterbottom 1990; Colchester and Lohman 1990). Moreover, TFAP offers no real funding of its own and only seeks to coordinate the aid flows of others—a strategy that has left it with little leverage. Since 1990, support for TFAP has collapsed among major donors and NGOs, including two of TFAP's original cosponsors, the World Bank and the World Resources Institute.

The second initiative was the negotiation of the 1985 International Tropical Timber Agreement, which led to the formation of the International Tropical Timber Organization (ITTO) in 1987. The ITTA was originally designed as a commodity agreement under the UN Conference on Trade and Development; its goal was to stabilize international markets for tropical timber products. Yet the ITTA also contained a brief environmental clause that unexpectedly made ITTO the only intergovernmental body with the authority to address the problems of sustainable logging in the tropics.

The ITTO has nonetheless lacked both the authority and the resources to have a discernible influence on logging practices in the tropics. The ITTO's aid budget is small and relies on voluntary contributions, which for the years 1989, 1990, and 1991 together totaled $46 million (International Tropical Timber Council 1990, 1992). Moreover, ITTO's temperate and tropical members disagree over the organization's mission.

Temperate members favor a more aggressive approach to rain forest protection, while tropical members regularly veto restrictions on the use of their forests, arguing that any regulations should apply equally to temperate forests. The ITTO's single attempt to influence the logging policies of a specific member—the Malaysian state of Sarawak—has had little impact (Pearce 1994). When the ITTA was renegotiated in early 1994, temperate states were forced to dilute an already tepid provision, initially agreed to in 1990, that called on tropical states to sustainably manage those forests that produce wood for export by the year 2000.

The third initiative was an attempt to negotiate a forest convention during the preparations for the UN Conference on Environment and Development in 1992. As at ITTO, developed and developing states were bitterly divided over whether a treaty should cover only tropical forests or temperate forests as well. For developing states, this is not only a matter of principle, but a matter of self-interest: environmental restrictions could raise the cost of logging in the tropics, making tropical timber less competitive in world markets. In the end, the developing states decided no treaty would be better than one that seemed to discriminate against them. In place of a treaty, a statement of seventeen nonbinding principles on world forests was released. Delegates could not agree whether a forest convention should be created in the future (Parson, Haas, and Levy 1992).

Conditionality and Forestry Lending in the World Bank

This string of diplomatic failures left the task of promoting tropical forest protection to a handful of bilateral and multilateral funding agencies, principally the World Bank. World Bank funding for forestry projects has risen steadily since the late 1970s. Between 1982 and 1984, the World Bank funded nineteen new forestry projects worth $380 million; from 1992 to 1994, the Bank made twenty-three new forestry loans worth $1.6 billion (World Bank 1991b; World Bank 1994d). As the Bank has assumed the central role in international forestry funding, it has gradually adopted a more pro-environmental stance. In the 1970s, the Bank viewed virgin tropical rain forests as an "inefficient use of tree-growing space" and often urged states to replace them with tree plantations. By 1991, the Bank had reversed its position and would no longer finance logging in old-growth forests (World Bank 1991b: 7–8).

Since the late 1980s, the World Bank has become increasingly prone to attach policy conditions to its forestry loans (World Bank 1994f). Although the use of conditionality to promote environmental reforms is part of a larger trend in World Bank lending (World Bank 1994c), it also reflects the Bank's frustration with the politics of forest protection. A 1991 internal evaluation of the Bank's forestry loans found that successful projects at the local level were often undermined by perverse national policies. A project in Côte d'Ivoire, for example, boosted the country's tree planting capacity from 1,000 to 20,000 hectares a year and was considered a success. Yet the project's replanting efforts were dwarfed by the national rate of forest depletion—which was ten to twenty times higher than the planting rate—and which the Bank project failed to address (World Bank 1991b: 32). This and other cases persuaded the Bank to pay more attention to reforming national policies, if for no other reason than to prevent its local projects from being undermined. To do this, the Bank often relied on conditionality.

The World Bank recently reviewed thirty-three completed forestry projects that had carried a total of eighty-seven conditions. It found that 56 percent of the conditions had been fully complied with, 25 percent had been partially complied with, and 18 percent had not been complied with at all (World Bank 1994f: 5). These figures almost certainly overstate the Bank's influence. Some of the activities mandated in the conditions would have been carried out by the governments anyway. The sample is also marked by a selection bias. The figures do not measure compliance with all conditions demanded by the Bank, but rather, only with those that recipients have agreed to fulfill. Recipients often object to some conditions initially requested by the Bank; the more important their status as a borrower, the more likely they are to prevail and still obtain the loan. In some cases—such as Indonesia—the Bank's call for reforms have led to the collapse of loan negotiations. The figures exaggerate the Bank's leverage by excluding data on proposed loan conditions that recipients rejected and including data on conditions they would have enacted anyway.

Still, the review contains insights into the effectiveness of World Bank conditionality. The Bank's figures suggest that compliance was most common when the conditions were transparently in the government's political interest. Mandates to strengthen the government's Forest Service, for example, were always adhered to. On the other hand, when the Bank called for reforms that could weaken the government's political authority—by increasing public participation in forest management, privatizing the forest industry, or altering land tenure—compliance was much lower (World Bank 1994f: 6, 19, 20).

The study also implied that the effectiveness of mandated logging reforms varies according to the political stature of the recipient's timber industry. While noting that the Bank has pushed for logging reforms in six tropical countries (Gabon, Lao PDR, Zimbabwe, Kenya, Tanzania, and Haiti), the review finds, "Despite this effort, it must be acknowledged that Bank attempts to support policy reform in this area have been more successful in small countries than in large countries with significant forest resources and export-oriented forest product industries (such as Malaysia and Indonesia)" (World Bank 1994f: 28). Indeed, even in small countries a well-placed logging industry can foil World Bank efforts. In Haiti, for example, politically powerful timber interests prevented the government from complying with a World Bank mandate to raise logging fees (World Bank 1994f: 15).

The review also suggests, however, that in a different set of states conditionality has been effective despite the presence of influential timber industries. In Kenya, Cameroon, Liberia, and Nigeria, the review found that the Bank's mandates to raise logging fees were successful, even though they were opposed by politically influential logging interests (World Bank 1994f: 18).

If the logging interests were influential in each set of states, why did the World Bank's efforts fail in the first set and succeed in the second? The following section explores this problem by examining a single, representative case of conditionality failure and a single case of success. In Indonesia, the World Bank's efforts to promote logging reform were blocked by the government, which regards the timber industry as part of its governing coalition. In the Philippines, however, the government welcomed the help of funders in shackling and dismantling the powerful logging industry, which was largely excluded from the governing coalition.

Logging Policies and Politics in Indonesia

Indonesia has the second largest array of closed-canopy tropical rain forests in the world. Since 1973 it has used these forests to dominate international timber markets, becoming the world's leading exporter of hardwood logs from 1973 to 1980 and the leading exporter of wood-based panels (including plywood, veneer, and particle board) since 1983. Wood products have been Indonesia's most important export, after oil, since 1972 (FAO Yearbook, various; Central Bureau of Statistics 1992).

Commercial timber production, combined with agricultural encroachment, has taken a heavy toll on the Indonesian forests. The FAO estimates that deforestation in Indonesia rose from 300,000 hectares a year in the early 1970s to 1.3 million hectares a year by the late 1980s. Part of this jump is due to logging that is both unsustainably extensive and harmfully intensive (Revilla 1994; Sutter 1989: 12–17; Fakultas Kehutanan 1989).

Many government policies have contributed to the rapid loss of the forests. Some flow from planning decisions to convert forest land to agricultural land. Others reflect the government's use of timber concessions for political patronage and the perverse forest policies that result.

The Indonesian government's logging policies have three particularly harmful facets. First, the government has sanctioned logging at unsustainable rates. The most comprehensive survey of Indonesia's forests, conducted by the FAO, estimates that the "sustainable" level of timber production in Indonesia is roughly 25 million cubic meters a year; the same estimate was offered by an earlier UN study in the 1970s (Revilla 1994: 13; U.S. Department of Commerce 1977). Yet current timber production is well over 40 million cubic meters a year, and may be as high as 55–60 million cubic meters a year.

The government has also failed to take even simple measures to enforce basic logging regulations. Between 1980 and 1988 just over 10 percent of all timber concessions complied with government rules; in 1989–1990 a different government survey found 4.2 percent in compliance (Djamaludin 1991).[3]

Finally, the government has taxed only a fraction of the timber industry's rents (windfall profits). Since logging firms are cutting trees from state-owned forests, the government is entitled to collect 100 percent of the resource rents.[4] Yet in the late 1980s, the government captured no more than a quarter, and perhaps as little as one-tenth, of the potential rents generated by logging (Haughton, Teter, and Stern 1992; WALHI 1991; Ingram 1989).

The Indonesian government's support for large-scale commercial log-ging began in the late 1960s, when the Suharto government used the distribution of timber concessions to consolidate its own political foun-dation (largely among military officers) and to boost hard-currency ex-ports (see Manning 1971). Since then, control of the Indonesian timber industry has gradually shifted away from the military and toward a small group of ethnic-Chinese entrepreneurs. But the government still finds a large, lightly regulated, and undertaxed timber industry to be in its po-litical and economic interests (Schwarz 1994; Robison 1986). Today pro-environmental reformers are scattered throughout the central gov-ernment, notably in the National Development Planning Agency (BAP-PENAS), the environment ministry, and the forestry ministry. Yet their influence within the government is consistently outweighed by the logging industry, particularly with President Suharto. Moreover, the incipient components of a popular constituency for reform—small farmers, indige-nous peoples, taxpayers at large, and future generations—are dispersed and disenfranchised and have little influence.

International Funds and Conditionality in Indonesia

In the late 1980s, the importance of Indonesia's forests as a biodiversity site, combined with the government's reputation as a reliable aid partner, began to attract a flood of international funds slated for rain forest pro-tection. By September 1992, Indonesia had received funding for sixty-three forestry projects—forty supported by grants, twenty-three by loans—worth a total of $300.4 million (TFAP 1993: 9).

Of all the funders, the World Bank made the most concerted effort to influence Indonesia's logging policies. Beginning in 1988, the Bank made two forestry loans to Indonesia, worth a combined $53 million, largely to fund conservation projects and forestry research and planning. By 1993, the Bank was prepared to offer Indonesia a third loan, worth $120 million; attached to the offer was a series of conditions that could trans-form the logging industry.

The terms for this third, proposed World Bank loan reflected the Bank's frustration with past attempts to encourage reforms by funding forestry research and planning projects. Bank officials argued that even though the government learned from these earlier projects, they had not yet applied this knowledge to their policies.

Among the conditions in the proposed loan were a series of hikes in timber royalties; a heightened role for local communities in forest management; reform of the way concessions for timber plantations were allocated; and the establishment of an inspection service—independent from the forestry ministry and potentially independent from the entire government—to monitor the logging industry.

Unwilling to accept such sharp restraints on the logging industry, in 1994 the Indonesian government rejected the loan. Although some of the reforms called for by the Bank were supported by officials high in the government, their influence was limited; even with the Bank at their side, they could not prevail. The government had no need for new forestry loans. They were already deluged with aid projects for "rain forest protection" that were often funded with grants, not loans. Nor was the government running low on its own forestry funds. The forestry ministry's reforestation fund, which was collected from a tax on timber, contained some $1.6 billion at the end of 1993. The fund was also growing more quickly than it could be spent: from 1989 to 1993, only 16 percent of the annual receipts were disbursed (Ministry of Forestry 1994). For all of these reasons, the Indonesian government suffered little by rejecting the World Bank's forestry loan and the conditions affixed to it.

Timber Politics and Deforestation in the Philippines

From 1951 to 1965, the Philippines was the world's leading exporter of tropical timber; from 1966 to 1973, tropical timber was the leading commodity export of the Philippines (FAO, various; National Statistical Coordinating Board, various). Since the late 1970s, however, Philippine timber production has steadily declined, as the nation's valuable hardwood forests have receded. In 1948, 59 percent of the country was covered by forests; by 1987, forest cover had dropped to 22 percent, one of the quickest rates of deforestation in the world. The twin causes of forest loss have been unsustainable logging and the encroachment of small farmers on logged-over land (Kummer 1992: 46, 56).

The end of the timber boom and the onset of a painful "timber famine" had been widely anticipated in the Philippines since the 1950s. But the gains reaped by firms that acquired logging rights, and the gains reaped by politicians who dispensed these rights, created a political juggernaut that neither popular disapproval nor foreign pressure could slow before the late 1980s.

Like the Indonesian government, the Philippine government abetted the harmful practices of the logging industry in three major ways. First, it licensed logging at grossly unsustainable rates in spite of repeated warnings from experts. In 1955, Luis Reyes, the "father of Philippine forestry," issued one of the first warnings that the forests were being cut too quickly (Reyes 1955). In 1958, Tom Gill, a prominent American forester, reported to the Philippine government that "at the present rate of timber removal, disastrous inroads are being made on the forest capital" (Gill 1958). When Reyes issued his warning at the beginning of 1955, 3.6 million cubic meters of wood had been cut the previous year; in 1958, the year of Gill's warning, 5.4 million cubic meters of timber were cut. By 1968 timber production had risen to 13.5 million cubic meters. It remained near this level until the late 1970s, when it began a slow descent due to forest loss (FAO Yearbook, various).

The government was also arbitrary and ineffectual in enforcing forestry regulations. Penalties were more often levied on concessionaires for political reasons than legal ones. This contributed to forest degradation in two ways: it enabled well-connected loggers to violate cutting regulations with impunity; and it made the tenure rights of all concessionaires insecure, which encouraged them to log as quickly as possible, before they fell out of favor with the faction in power. Shortly after the Aquino government took office in 1986, it found that 90 percent of all loggers were in violation of forestry regulations.

Finally, the government taxed only a small fraction of the rents accumulated by logging firms, which led to widespread corruption and the loss of vast government revenues. One study found the government collected an average of 11.4 percent of the timber rents between 1979 and 1982 (Boado 1988: 184). Another estimates rent collection ranged from 4 to 30 percent for most timber products before 1981, with a slight increase thereafter. In some years, one tax (the silvicultural fee) went entirely uncollected (Bautista 1992: 43–49). According to a 1989 World Bank report, "Failure of Government [sic] to collect a significant rent from licensees largely explains the rapid depletion of timber resources the last two decades" (World Bank 1989: 18).

When the government of Corazon Aquino took office in February 1986, any concern it had for the nation's forests paled beside more urgent tasks, which included reviving an economy ransacked by former President Ferdinand Marcos and fending off a series of armed rebellions. Aquino was immensely popular at first. Yet she followed a cautious political strategy, working hard to retain the support of major business firms and members of the traditional aristocracy. This strategy was reflected both in the appointments she made and in the halting pattern of her economic reforms, which removed subsidies from sectors of the economy dominated by Marcos supporters while protecting sectors populated by her own backers (Hutchison 1993; Mosley 1992; Haggard 1990).

Aquino's cautious tactics were evident in her February 1986 appointment of Ernesto Maceda to head the Department of Environment and Natural Resources (DENR), which held authority over the Forest Management Bureau (FMB).[5] Maceda was a longtime opposition politician who had formerly worked for Marcos; allegations of corruption had trailed him since the 1950s. Maceda's forestry policies were contradictory. He banned the export of unprocessed logs, which pleased conservationists. But he also seemed to approve and cancel forest concessions arbitrarily. Maceda distributed scores of logging permits that lacked expiration dates, volume limits, and specific boundaries of operation, giving these license holders extraordinary lenience. He also issued timber licenses in provinces where logging had been banned. Accused of corruption, he was fired from his post in November 1986, when Aquino began a drive to clean up her cabinet (Vitug 1993: 40–42).

After a four-month delay, human rights lawyer Fulgencio Factoran was appointed to fill the DENR post. Factoran soon embarked on an ambitious plan to reform both his department and the logging industry: he restructured DENR to downgrade the status of the corrupt Forest Management Bureau, installed pro-environment allies in key positions, and began to cancel timber licenses held by close associates of Marcos[6] (Remigio 1992).

Many loggers had been closely linked to the Marcos government, and measures to curtail the timber industry were consistent with Aquino's strategy of reforming sectors of the economy controlled by her opponents. Yet Factoran faced heavy opposition from members of Congress who backed the timber industry and from bureaucrats within his own department; together they delayed his confirmation in Congress for one year. Although he appeared to have the backing of President Aquino herself, some of Factoran's reforms found little support within her administration. Several of the measures he advocated—such as suspending the licenses of firms that were violating the logging rules—hurt members of the Aquino administration who were tied to the logging industry, including Fidel Ramos, the armed forces chief of staff and later Aquino's successor as President (Vitug 1993).

One of the early priorities of the Aquino government was meeting its international debt repayment schedule, which by 1988 consumed 32 percent of the nation's foreign exchange earnings (Korten 1994: 973). The timber industry had once been the nation's leading source of foreign exchange. In 1973, the forest sector produced 18.4 percent of the Philippines' exports. By 1987, it produced 5.3 percent of all exports (National Statistical Coordinating Board, various). The demand for foreign exchange was so great, however, that economic planners might have opposed Factoran's reform agenda, which was certain to reduce timber exports further still. Instead they supported Factoran, because the forests had become a magnet for funders who hoped to help the government through its balance-of-payments crisis while promoting environmental conservation.

Conditionality and Reform in the Philippines

Partly due to its reputation for corruption, the Forest Management Bureau had received little foreign assistance in the past. From 1979 to 1987, FMB had secured just $60 million in foreign loans. From 1988 to 1992, however, the government received forest sector loans from the Asian Development Bank, the World Bank, and the Japanese government totaling $726 million, plus an additional $130 million in grants from ADB and the U.S. Agency for International Development (Korten 1994: 973).

Four of the ADB loans and one Japanese loan (totaling $397 million) went to fund reforestation efforts. The World Bank loan was divided in two. Sixty-three million dollars went toward establishing protected forest areas, enforcing logging regulations, and developing community-based projects in watersheds on forest land. The remaining $166 million (joined by an additional $100 million loan from the Japanese government) was not to be spent on forest projects at all. Instead, it was sent directly to the Philippine Central Bank to ease the balance-of-payments crisis.[7] These funds were nonetheless treated as part of the "natural resources sectoral adjustment loan" because their disbursement was conditioned on a series of forest policy reforms, whose frankly stated goal was to "preserve what remains of the biological diversity of the Philippines" (World Bank 1991b: i). The USAID grant followed a similar strategy, sending $75 million in conditional aid to the central bank (Hermann et al. 1992).[8]

Some of the conditions attached to the World Bank and USAID loans were routine. But others were politically stringent, requiring the Aquino government to boost logging fees, ban logging in virgin forests, and use the Department of Justice and the Department of Defense to crack down on logging violations. The logging industry and its allies in Congress fiercely opposed these measures.

The reformers at DENR and the funders collaborated in drafting these conditions. Factoran and his allies realized that to change Philippine forest policies they would have to fight the logging industry and its proponents, particularly in Congress; and to win in Congress they would have to round up public support and gain more leverage within the Aquino administration. Many members of Congress supported forest reforms, but others were closely tied to the logging industry. Factoran's nemesis in Congress was Representative Jerome Paras, chairman of the House Committee on Natural Resources, who oversaw the work of DENR. To win a major fight in Congress, Factoran needed strong backing from the rest of the Aquino administration. But DENR had little sway within the cabinet and received only passive support from President Aquino. Moreover, outside the network of environmental NGOs (which at the time was tiny), popular support for forestry reforms was at best diffuse and unmobilized. Commercial logging caused environmental problems that led to terrible hardships for many rural dwellers. But for others it provided jobs, which in rural areas could be scarce (see Broad and Cavanagh 1993).

The proreform faction at DENR turned to international funders for help in Congress, within the administration, and among the public, partly through the use of conditionality.[9] By claiming that the measures they favored—such as raising the logging fees and banning logging in old-growth forests—were essential to getting international funding, DENR officials deflected the wrath of the timber industry and its allies toward the funders, who were impervious to their pressure. The DENR reformers also gained from issue linkage, tying the prospects for economic assistance, which had wide backing, to logging reforms, which were more controversial.

The use of cross-conditionality—making funds needed by one sector or institution (in this case, the central bank) conditional on reforms in another (DENR and the timber industry)—helped Factoran win the support of economic planners and the central bank, giving forestry reforms much wider backing within the administration.[10] DENR also gained leverage in Congress from the conditions, because the conditions mandated congressional approval of measures, such as higher logging fees, that DENR alone had been unable to get passed.

The ADB loans had a different and equally important political function. Politicians had long used the allocation of timber licenses to build a network of political supporters in rural areas. Factoran and his allies now decided to use the ADB reforestation funds in the same manner, allocating them selectively to bring members of Congress into the proreform camp. Reforestation funds not only eased the pain of rural communities that lost jobs when logging concessions and processing plants closed, it also eased the pain of members of Congress who lost their ability to use timber licenses as a form of patronage.

The conditions attached to the foreign funds not only helped DENR outmaneuver the logging industry, it helped them attack the foundations of the industry's economic and political power. Some of the conditions had been crafted to make logging much less profitable and force many firms out of business. The World Bank, for example, mandated a large hike in logging fees and competitive bidding for new logging licenses, while both the Bank and the ADB required loggers to post a reforestation bond before they began cutting. These measures were meant to remove most of the government subsidies from the logging industry and to force from business any firms uncommitted to long-term sustainable management of their concessions. The ban on logging in old-growth forests, which was tied to the USAID grant, was not simply designed to protect the forests, but to undercut the industry's economic strength by denying loggers access to the country's most valuable timber. Finally, the World Bank loan required DENR to enforce the logging regulations more vigorously, giving the administration a stronger legal basis for canceling the timber licenses of errant loggers. With the timber industry's income drastically trimmed, their political influence—and ability to block further reform—began to subside.

Conditionality had one final benefit for the reformers: it helped them "lock in" the reforms so that future governments could not rescind them. This had special importance in the Philippines. President Aquino decided not to run for reelection, giving reformers little time to implement their new policies. Even if the reformers kept control of DENR in future administrations, their influence could evaporate: circumventing the bureaucracy was a time-honored tradition among Philippine heads of state. Most ominously, a series of coup attempts made it clear to the reformers that not only the Aquino government but the Congress and the new constitution could disappear at any time. The conditions attached to the international funds might carry little weight with a new government. But they also might endure long after the Aquino government had been dislodged by a new leader.

The victory of the reformers over the logging industry slowed the destruction of the Philippine forests. When Factoran was appointed to the DENR in 1987, 154 licensees were permitted a combined annual allowable cut of 8,118,000 cubic meters of wood. By mid-1994, there were just thirty-one licensees remaining, whose combined annual cut was 878,583 cubic meters—barely one-tenth of the previous total (Forest Management Bureau 1993, 1994). Moreover, the activities of these remaining firms were better monitored, and their profits were taxed at a higher rate, than before the loans were arranged. Supported by anticipation of the forthcoming loan, in 1989 DENR raised the implicit forest charge on the most common species of timber from 35 pesos per cubic meter to 535 pesos per cubic meter; it also required loggers to post a bond of 12,500 pesos per hectare before logging, which they would forfeit if they failed to reforest the area (Bautista 1992: 31).[11]

Ironically, the substance of the aid programs—apart from the conditions attached to them—has had a more ambiguous impact on the Philippine forests. The funders boosted the government's capacity to overcome political obstacles and institutionalize reforms. But the scale of the reforestation programs badly outstripped DENR's administrative capacity, and a USAID review found that the implementation of many programs in rural areas was poor (Korten 1994; Hermann et al. 1992). Illegal logging may have risen as legal logging has slowed. One observer argues that the eventual repayment of the loans may put new pressure on Philippine natural resources (Korten 1994).

Nonetheless, due in part to the help of international funders, the rudiments of sustainable forest management have been put in place in the Philippines. Funders did little to boost concern for forest protection within the Aquino administration. But they supplemented the political capacity of a preexisting reform faction in four critical ways. First, they helped reformers overcome inertia within the administration, through both issue linkage (tying the desired balance-of-payments assistance to the more controversial reform package) and cross-conditionality. As a result, the reformers were able to get support across the administration for measures to shrink the timber industry, an important source of foreign exchange, during a severe balance-of-payments crisis. Second, funders helped the administration overcome opposition in Congress by absorbing part of the responsibility for unpopular measures, and again, through issue linkage that mandated congressional action. Third, funders helped the administration both in Congress and with the public by providing reforestation funds that could be used to buy off opponents to reform and to finance a new, pro-reforestation patronage network. Finally, by attaching the conditions to long-term loans and grants, the funders helped protect the reform program in case an unsympathetic government later came to power. Although the projects actually financed by the loans have had a mixed record, the conditions attached to them helped the government "get the policies right"—a critical step toward sustainable forest management that few timber-exporting states ever take.

2   Analysis

What can these cases tell us about the variables that determine conditionality's effectiveness in promoting sustainable logging? To understand the factors that distinguish cases of failed conditionality (like Indonesia) from cases of successful conditionality (like the Philippines), this section constructs an analytical framework that describes conditionality's broader problems and limitations.

The section begins by using a simple theory of contracts to explain how conditionality, in an ideal setting, should work. It then draws on the experience of funders with structural adjustment conditionality to explain why conditionality usually fails to lead to reform in recalcitrant states, but how in a limited set of circumstances it may be effective. Finally, drawing on both the model of "conditionality as contract" and the analysis of its real-world limitations, it attempts to explain why green conditionality failed in Indonesia and similar cases, while succeeding in states more like the Philippines.

Conditionality as a Contract

Conditionality—the exchange of financial aid for policy reforms—has been a part of international aid programs since the Marshall Plan.[12] Until the late 1980s, it was used almost exclusively by the International Monetary Fund and the multilateral development banks to impose fiscal discipline on developing states and promote market-oriented economic reforms.[13] Since the end of the Cold War, however, many funders have tried to use conditionality to promote a broader set of goals, including democratization, reduced military spending, and environmental reform (Nelson and Eglinton 1993).

For conditionality to work, a funder and a recipient must accomplish two things. First, they must reach an agreement on the terms of the bargain; second, they must fulfill them. The failure of the two parties to reach an agreement is usually easy to explain. The funder and the recipient each go to the bargaining table with a set of acceptable outcomes (or a "win-set") in mind. A wide range of factors determines the contents of each side's win-set, including the recipient's need for funding and the funder's determination to impose conditionality. If the two win-sets fail to overlap—as was the case in Indonesia—then no agreement can be reached. The win-sets of the two parties will overlap only if the funder's package can make the recipient better off than they would otherwise be: the benefits of the loan must outweigh the costs of the conditions. A loan with no policy conditions is attractive because it offers a government immediate economic benefits while delaying the costs of repayment (or better still, placing these costs on the shoulders of a future government). Policy conditions are distasteful because they impose immediate political costs while, at best, promising only long-term benefits. In a successful loan package, the allure of the loan will compensate for the sting of the conditions. If it does not, the package will be rejected.

If the funder and the recipient reach an agreement but fail to keep it, the explanation grows more complex. Given that conditional aid is a transaction between two parties and is governed by a set of implicit and explicit rules, it can be seen as a kind of contract—a contract in which funders trade concessional loans or grants for policy reforms in a recipient (Milgrom and Roberts 1992). Contracts normally must be enforced by a third party, who can penalize either party if they fail to meet their obligations. Because the conditionality contract crosses international boundaries, however, there is no third-party enforcement. The contracts are hence designed to be unilaterally enforceable by the party most at risk, the funder.

To make this possible, conditional loans and grants are usually disbursed in a series of sequential blocks or "tranches"; the release of each tranche is conditioned on the recipient's progress toward reform. To keep them from abandoning their reforms after the final tranche has been given out, funders may insist that the recipient install policies that are hard to reverse. Funders also rely implicitly on the recipient's desire to obtain future funding to keep them from reneging on their obligations. When funders bargain with a state that has failed to adhere to earlier contracts or whose commitment to reform is in doubt, they may even demand costly policy changes before the first tranche is disbursed. Recipient states may nonetheless find it in their interest to ignore the conditions on their loans if they expect to need no funding beyond the point at which their cheating becomes known to the donor, and if they discount the long-term consequences of damaging their reputation (see Mosley 1987).

Conditionality for Economic Reform

For conditionality to work, three prerequisites must be met. First, the recipient must have the political and administrative capacity to fulfill the conditions; second, the funder must know if the recipient is violating the conditions; and third, the recipient must expect to be penalized if they are caught cheating. When using conditionality to promote structural adjustment reforms, funders have had problems with all three prerequisites.

The first prerequisite—that a recipient can fulfill the conditions—is self-evident, but shapes the conditionality bargain in important ways. A government that is administratively weak, politically divided, or faces strong opposition to reform may have trouble implementing complex or unpopular measures. Researchers have found that recipients of structural or sectoral adjustment loans will have difficulty meeting conditions that have the following attributes:

1. they are not politically sensitive (Kahler 1992: 97);
2. they can be put in place by a small number of officials, and do not require the cooperation of large numbers of bureaucrats or agencies;
3. they do not require extensive institutional change to implement;
4. they "break bottlenecks," opening the way for further actions by the government or private actors;
5. there is a strong technical consensus on their utility (Nelson 1989; Nelson and Eglinton 1993: 69).

Funders often insist on conditions that fail to meet these criteria, despite their lower success rate. The most important reforms of any type, for example, may be politically controversial or require extensive institutional change. Still, placing difficult-to-implement reforms in the conditionality contract erodes its effectiveness, because it opens the door to cheating by recipients who falsely claim that they are unable to implement cumbersome reforms, when they are actually unwilling to do so.

The second prerequisite—the ability of funders to monitor compliance—depends on both the administrative capacity of the funder and the conditions in the contract. Funders find it easier to monitor conditions that are unambiguous, easily observable, can be quickly fulfilled, and cannot be circumvented. When monitoring requires extensive interpretation or investigation, it can tax the funder's administrative capacities; it may also invite cheating by the recipient. When a condition requires many months or years to fulfill, it can lead to three problems: it opens the door to foot-dragging and deception; it gives opponents of reform time to mobilize; and it makes it easier for the recipient to evade punishment for cheating, if they can hide their noncompliance until the loan is fully disbursed (Nelson 1989; Mosley 1987: 9). Finally, when a recipient is able to circumvent a condition—for example, by meeting its provisions but nullifying its impact with countervailing measures—there may be little a funder can do to salvage the reform, at least until the recipient requests a new loan.

The efforts of the MDBs to promote economic restructuring have also been hampered by their reluctance to penalize cheating. The MDBs' aversion to punishing their own borrowers is largely the result of their own internal institutional conflicts (Brechin 1996). As advocates of policy change, the MDBs have an interest in promoting reforms in recipient states. But as banks, they must continue to make loans—regardless of whether their borrowers have enacted reforms—in order to remain solvent. They also have institutional incentives to leave no part of their budgets unspent. The larger the recipient and the more important its status as a borrower, the greater the incentive MDBs have to offer them loans.[14] Hence some of the most important borrowers are given the "softest" conditions. Although funders may penalize recipients by delaying the allocation of a tranche, they virtually never cancel a loan once disbursement has begun, no matter how poorly the recipient is complying with the conditions (Mosley, Harrigan, and Toye 1991: 48, 71–72, 165).

Moreover, if a borrower has fulfilled at least half of the conditions in their previous loan (and if they are an important borrower or have a close relationship with the bank, fewer than half), the MDB is almost certain to offer it a new loan once the previous one has been dispensed. There is now so much evidence of the World Bank's laxity that recipients feel no obligation to adhere to more than half the conditions they initially agree to (Mosley, Harrigan, and Toye 1991: 170–171).

The contrast between the successful use of conditionality by the IMF and the less-successful record of the MDBs can largely be explained by variations in these three prerequisites—the ability of the recipients to implement the conditions, the monitoring problems faced by funders, and the funder's proclivity to penalize cheating. When the IMF sets conditions for its short-term loans, it creates a tight contractual environment. The conditions can generally be fulfilled by a handful of officials without delay. They are also relatively precise and easy to monitor, because they are expressed in quantitative, not qualitative terms. This setting makes cheating by the recipient difficult and unambiguous, which in turn makes punishment by the funder swifter and more certain. But the structural and sectoral adjustment loans arranged by MDBs are made in a much more porous contractual environment: the conditions are typically complex, irreducibly ambiguous, take years to fully implement, and require the cooperation of many public officials. Those recipients with a strong incentive to cheat find it easier to do so while escaping any penalties.

Because of this loose contractual setting, MDBs have generally been unable to use conditionality to persuade recipients to take economic measures they would otherwise oppose. Scholars have found that MDB conditionality has had little or no influence on the decisions of borrowers to support or oppose reform. Even when there are internal battles between proreform and antireform factions, the outcome is usually clear and has little to do with any enticements offered by funders (Mosley, Harrigan, and Toye 1991: 141). Governments that wish to obtain loans and cheat on their conditions have been able to do so, partly by taking advantage of the limited enforcement capacity of MDBs, and partly by fulfilling their nominal obligations while nullifying their impact with countervailing measures. Because the MDBs have been eager to disburse their loans and reluctant to punish their customers, those governments lacking a strong commitment to reform have few incentives to take painful measures.

But the experience of MDBs with structural adjustment lending also shows that conditionality can help proreform governments in at least five ways, even when the contractual environment is porous:

1. it can give public officials a more defensible rationale for taking politically unpopular measures;
2. it can raise the political authority of proreform technocrats who work closely with the funder, or are responsible for carrying out the policy changes;
3. it can provide the recipient with the resources to buy out opponents to reform or build a new, proreform constituency;
4. it enables the recipient to use the technical skills of the funder to implement complex reform programs;

5. it can build reformist alliances within the government and strengthen the hand of key players by using cross-conditionality—making funds for one sector dependent on reforms in another (Mosley, Harrigan, and Toye 1991: 119–120, 160).

Conditionality in Indonesia and the Philippines

This model of conditionality-as-contract suggests that "straight conditionality"—the use of funds to "buy" policy reforms in a recalcitrant recipient—is unlikely to succeed in promoting logging reforms. The conditions in the loans offered to Indonesia and the Philippines were complex, difficult to fulfill, time-consuming to implement, and probably easy to circumvent. Moreover, the most stringent conditions were attached to funds from the World Bank, which often fails to penalize borrowers who renege on the conditions in their loans. There was little chance that straight conditionality could work in this or any similar setting.

On the other hand, the financial transfers to the Philippines might be seen as an example of "reverse conditionality," when a proreform government covertly invites a funder to "impose" conditions on its aid. The conditions affixed to the World Bank and USAID funds—conditions that were largely welcomed by the Philippine government—helped proreform technocrats in DENR gain leverage both within the Aquino government and with Congress; gave the government a more politically defensible rationale for dismantling the timber industry's privileges; and helped bind future governments to a course of reform. In addition, the ADB loan helped the government assemble a proreform constituency by establishing a rural patronage network built on the dissemination of reforestation contracts. The new patronage system also supplanted an earlier patronage network based on the distribution of timber concessions, which had anchored the opposition to reform.

There were two major differences between the contractual environment of the proposed Indonesian loan and the contractual environment of the Philippine loans. First, the funders had more leverage over the Philippines than over Indonesia: the loan offered to the Philippines was larger; the Aquino government had a greater need for external funding than the Suharto government; and the Indonesian government had alternative sources of nonconditional funding, whereas the Philippine government did not. Nonetheless, the experience of the MDBs with structural adjust-ment conditionality suggests that the amount of leverage held by the funder matters little. Even under desperate circumstances, governments unwilling to go through with funder-sponsored reform programs have re-fused conditional loans or accepted the funds while evading the condi-tions. If the World Bank had the same leverage with Indonesia that it had with the Philippines, straight conditionality was still unlikely to work. The contractual environment was too porous to force the government to make politically difficult reforms.

The second difference between the Philippine and Indonesian cases is both more obvious and more critical. The level of concern about unsus-tainable logging was greater in the Philippines, where the timber industry was not part of the governing coalition, than in Indonesia, where it was. In the Aquino administration, support for forestry reform was strong among some actors and weak among others. Nonetheless, a proreform sentiment prevailed within the cabinet by mid-1987, apart from the in-ducements offered by international funders. The strength of the reformers reflected the position of the loggers, who—tainted by their affiliation with the Marcos government—had largely been excluded from the gov-erning coalition.

In Indonesia's Suharto government, support for forestry reform was scattered and weaker. While President Suharto has made tentative over-tures to environmental interests, the logging industry exists wholly under his patronage and constitutes part of his political coalition. When envi-ronmental concerns clash with the interests of the timber industry, Su-harto's decisions consistently favor the industry. In this setting, the loose contractual environment of the World Bank's proposed forestry loan could do little to raise the Indonesian government's level of concern. Yet without a higher level of concern, a loan to help implement reform pro-grams would be of little use. Moreover, a Bank loan that hurt the interests of the Indonesian timber industry, a valued political constituency, would not make the Suharto government better off.

The programs of outside funders may be able to improve Indonesian forest management in ways that do not conflict with the interests of the logging industry. But the most important source of forest degradation in Indonesia today, unsustainable commercial logging, cannot be curtailed without a greater level of concern in the Indonesian government, a concern that conditionality cannot induce.

The conditional funding offered to the Philippines and Indonesia to promote sustainable logging occurred in a porous contractual environment: the conditions were difficult and time-consuming to fulfill; they were possible to circumvent; and their principal sponsor, the World Bank, often failed to punish states that reneged on their commitments. Under these circumstances, recipients are seldom persuaded to embark on reforms they would otherwise oppose. Yet if the recipient has the concern for reform, but lacks the political or administrative capacity to act on that concern—which was true in the Philippines—conditionality can play a key role.

3   Evaluation

This section draws on the case studies of Indonesia and the Philippines, and the preceding analysis of conditionality, to revisit this chapter's central question: can conditionality help promote sustainable logging in the tropics? It also speculates on the ability of conditionality to foster other types of environmental reform.

Viewed in isolation, the success of conditionality in the Philippines might raise the hopes of environmentalists who seek to protect the rain forests of other countries. It should not. The Philippine political setting in the late 1980s was unusual in two ways. First, after thirty years of unsustainable logging, the industry's political influence had waned along with the forests themselves. Second, many longtime concession holders had been closely tied to President Marcos and were effectively disenfranchised by his fall from power. In 1986, for perhaps the first time since Philippine independence, the timber industry was left out of the governing coalition. For both of these reasons, the Aquino government had few incentives to protect the timber industry from reform.

But most states with large timber industries look less like the Philippines and more like Indonesia. In Indonesia, the logging industry is controlled by members of the governing coalition, which protects their ability to amass rents from unsustainable logging. The analysis of conditionality suggests that funders cannot force governments to enact forestry reforms that they would otherwise oppose. For most major timber producers in the tropics, conditionality can do little to advance reform, if the reforms hurt the interests of the logging industry.

In some settings conditionality may still be an important tool. When a timber industry is large and flourishing, concern for sustainable forest management may be overwhelmed by the allure of logging rents. But there are two settings in which the industry should be smaller, less profitable, and less influential politically. The first is at the end of a timber boom, when (as in the Philippines) the industry has shrunk along with the volume of standing timber in the forests. Some might think it pointless to promote sustainable forest management in countries whose virgin forests have disappeared. In fact, protecting a nation's forests after a timber boom has crested has special ecological significance. Once all or most of a country's virgin timber stands have been cut, loggers are more prone to reenter logged-over stands before the forests have fully recovered. Early reentry can produce the most enduring type of logging damage to tropical rain forests, and is critical to stop.

Conditionality may also work before a timber boom has begun. Over the last forty years, the tropical timber industry has migrated through Central America, West Africa, the Pacific island states, and most of Southeast Asia, inducing unsustainable boom-and-bust cycles in a long string of developing states (Ooi 1990). Other states whose tropical forests still remain commercially unexploited—largely in Central Africa and tropical South America—will feel the lure of world timber markets within the next few decades. International funders would be wise to use conditionality (and other tools) to help these states prepare the laws, institutions, and personnel they will need to sustainably manage their forests in the face of world market pressures.

What does conditionality's record of promoting forestry reforms, and structural adjustment reforms, suggest about its ability to promote other types of environmental reform?

Conditionality can certainly promote environmental reforms in "soft" cases, where reform is impeded only by lack of concern or information, not by a politically influential constituency. In these cases, conditionality can raise the salience of environmental issues that would otherwise be neglected, through issue linkage and cross-conditionality.

In "harder" cases, when reform is opposed by powerful domestic interest groups, conditionality may be useful if the government is willing but not politically able to carry out environmental reforms. In these settings, green conditionality can strengthen the hands of pro-environment reformers—buffering them against political discontent, buying out their opponents, and providing technical assistance.

But if a government is strongly opposed to environmental reform, green conditionality is likely to suffer from the same three contractual problems that plague structural adjustment conditionality. Environmental reforms may be hard for recipients to implement if they require extensive institutional change, intersectoral cooperation, or harm the interests of members of the governing coalition (as they often do). Progress in meeting environmental conditions will often be difficult for funders to monitor, particularly when the conditions are imprecise and easy to circumvent, and when implementation is time-consuming and difficult to observe (as they frequently are). And those funders most likely to use green conditionality—the MDBs and the major bilateral donors—are loath to punish cheaters, which makes compliance even less likely. Some environmental reforms may have few or none of these characteristics and work relatively well in conditionality bargains. The reduction of energy subsidies, for example, may encounter political opposition, but it is technically simple to implement and easy to monitor. But in general, a leaky contractual environment—one that enables recipients to cheat with impunity—places sharp limits on the effectiveness of green conditionality.

Conclusion

Conditionality is the sharpest utensil in the toolbox of major funders. To environmental advocates, it looks simple to use and well suited to a world where badly indebted states have few incentives to reform their environmental policies. This chapter takes a more skeptical view.

A review of contract theory, structural adjustment conditionality, and green conditionality in the Philippines and Indonesia suggests that the type of conditionality used by multilateral development banks usually fails to compel states to take measures they would otherwise oppose. Conditionality's ineffectiveness in these hard cases, where governments oppose reform, can be explained in part by the quality of the contractual environment. When the funder uses conditionality to promote reforms that are difficult for the recipient to carry out, hard for the funder to monitor, and when the funder is loath to penalize cheaters, it creates a porous contractual environment. In this type of setting, recipients can cheat with impunity and thus have few incentives to implement measures they dislike.

Conditionality can also fail at a more fundamental level, if the funder and the recipient cannot agree on a "price"—how much the former should pay for reforms in the latter. To strike a bargain, the funder must offer the recipient a package of conditional aid in which the benefits of the loans outweigh the costs of the conditions. Unfortunately, governments commonly find unsustainable logging booms intensely beneficial, because they produce an abundance of rents, foreign exchange, and political leverage. States that exploit their forests rapidly are therefore likely to resist reforms, even when they are sweetened by large concessional loans. This was the crux of the problem in Indonesia, where the government found unsustainable logging too useful to curtail.

Even when the contractual environment is poor, however, conditionality can still help governments that favor reforms but lack the capacity to implement them. This is illustrated by the case of the Philippines, where a coalition of funders used conditionality to help a reform faction overcome a series of political obstacles—including inertia within the Aquino administration, opposition in the Philippine Congress, and mixed public support—and enact a major set of forestry reforms.

In developing states, few types of environmental reform are as difficult to bring about as the reform of a major logging industry. Governments in tropical states almost always own the forests. With few exceptions, they distribute the right to harvest timber to their allies, and thus oppose reforms that might hurt them. Conditionality will seldom influence logging policies in these states. But in cases where the logging industry is politically weak—because it is at the beginning or end of a timber boom, or has been excluded from the governing coalition—funders may be able to forge alliances with reformers, and use conditionality to push forward politically onerous measures.

Conditionality is at best a complex, highly specialized tool for promoting environmental reforms. But in the field of environmental assistance, even limited tools can be essential.

Notes

Special thanks are due to the Center for Energy and Environmental Studies at Princeton University, the Center for International Forestry Research in Indonesia, the Institute for the Study of World Politics, the Harvard University Financial Transfers Project, and the Social Science Research Council for their magnanimous support during the research and writing of this chapter. For their contributions, the author is indebted to Hildegard Bedarff, Jay Blakeney, Steven Brechin, Jim Douglas, David Fairman, Delfin Ganapin Jr., Cord Jakobeit, Frances Korten, Molly Kux, Thomas Schelling, and Marites Vitug. The comments of Lisa Curran, Tamar Gutner, Robert Keohane, and Thomas Wiens on an earlier draft were especially valuable.

1. There are broad disagreements over what constitutes "sustainable" logging in the moist tropics, and whether it is even possible (Johnson and Cabarle 1993). This study was based on a modest definition of sustainability as "Forest use in which nothing is done to irreversibly reduce the potential of the forest to produce marketable timber" (Poore et al. 1989:5). Although this chapter loosely uses the term to refer to logging that results in a sustained level of timber production and causes no long-term damage to the forest's ability to provide environmental services, the precise definition of sustainability is unimportant for the arguments here.

2. In 1993, TFAP was renamed the Tropical Forests Action Program.

3. Since the appointment of Forestry Minister Djamaloedin Soerjohadikoesoemo in early 1993, enforcement has grown stricter. For 1993, the ministry reported that compliance with logging regulations had risen to 15 percent (*Jakarta Post* 1994).

4. This would still leave loggers with a "normal" rate of profit.

5. At the time of Maceda's appointment, DENR was still named the Department of Natural Resources, and FMB was the Bureau of Forest Development. The names were changed when the Department was reorganized under Secretary Factoran.

6. Factoran assembled what John Waterbury has called a "change team"—a group of technocrats, insulated from political pressures, who are assigned to carry out a politically hazardous reform plan (Waterbury 1993: 27).

7. The Global Environment Facility contributed $9.5 million to the World Bank package. Although strictly speaking the GEF thereby contributed to the bank's use of conditionality in this instance, in general the GEF's projects have thus far been too small to exercise the leverage needed for conditionality.

8. The rest of the USAID grant was divided between technical assistance ($25 million) and an endowment for an independent environmental organization ($25 million).

9. This was not the first time that a Philippine government invited foreign funders to "impose" conditionality to help it implement politically difficult reforms. In

the late 1940s, a U.S. aid package negotiated by the Bell Mission served the same purpose (Schelling 1957: 136).

10. Even the funds sent to DENR, not the central bank, were meant to help the Philippines through its balance-of-payments crisis. The DENR projects could largely be paid for with local currency, allowing the foreign currency sent by funders to be used to repay foreign debts and purchase imports.

11. Factoran and his undersecretary, Lito Monico Lorenzana, were sued by seventy-four logging firms for raising the logging fees without a congressional mandate. Eventually Congress—under pressure from the conditions in the World Bank loan—passed legislation to support a sharp hike in logging fees.

12. For a review of the problems of conditionality and coercion under the Marshall Plan, see Schelling (1955; 1957).

13. The MDBs include the World Bank Group, the Asian Development Bank, the Inter-American Development Bank, and the African Development Bank.

14. In addition, the World Bank has sometimes been coerced to make loans to states that are poor credit risks but strategic allies of the bank's major benefactor, the United States—states such as the Philippines under Marcos (Mosley 1992).

III

Financial Transfers in Europe

7

Protecting the Rhine River against Chloride Pollution

Thomas Bernauer

On its journey from the Swiss Alps to the North Sea the Rhine River drains some of the most populated and industrialized areas of four European countries: France, Germany, the Netherlands, and Switzerland. The Rhine is the most important inland waterway in Western Europe. It produces drinking water for around 20 million people; and it receives the waste water of around 50 million people, large chemical industries, potash and coal mines, and other sources. By the mid-1970s, pollution of the river had developed to the point where the once romantically described home of the Lorelei received the nickname "sewer of Europe" (Le Marquand 1977).

In 1963, in response to the rapid degradation of the Rhine's water quality, the riparian countries established the International Commission for the Protection of the Rhine Against Pollution (ICPR). One of the principal issues on the ICPR's agenda has been pollution by chlorides. Collaborative solutions to this environmental problem have involved international financial transfers in the order of $100 million.

Germany, France, the Netherlands, and Switzerland have shared the costs of reducing chloride emissions of a state-owned potash mine near Mulhouse, France: the "Mines de Potasse d'Alsace" (MdPA). This mine, as well as coal mines in the German Ruhr area, are the largest sources of chloride pollution. The principal victims of chloride pollution have been waterworks and farmers in the Netherlands. The four riparian countries have also agreed to jointly finance a project that will reduce chloride pollution of the IJsselmeer. The latter is a large artificial lake in the Netherlands that is fed by the Rhine and serves as an important freshwater source for the Dutch province of North-Holland.

Given the narrow framing of the issue (one substance, few polluters, and few victims), a favorable contractual environment, and no significant capacity problems—all of which are usually thought to make financial transfers easier to design and implement—the outcome is surprising: international financing of environmental protection in the chloride case involved protracted bargaining. Moreover, the outcome is characterized by "too little, too late." This chapter examines the circumstances that have led to this outcome. It analyzes the possible sources of ineffectiveness of financial transfers as a facilitator of international environmental protection. Being the only West-West financial transfers case examined in this study, it also highlights that, contrary to conventional wisdom, environmental cooperation among Western countries, based on financial transfers, is not necessarily easier to achieve than North-South or East-West cooperation based on financial transfers. Western countries seem to have learned this lesson at least implicitly: they tend to use strategies other than financial transfers, such as issue linkage, to overcome environmental cooperation problems. Where asymmetries of capacity are involved, however, financial transfers are usually indispensable. To understand the financial transfers discussed below, it is worthwhile to indicate the central collective action problem posed by chloride emissions into the Rhine, and to outline possible solutions to it.

When the Rhine reaches the German-Dutch border, it carries an average annual load of up to 400 kilograms of chloride ions ($Cl-$) per second (kg/s), with peaks of up to 700 kg/s. The average chloride concentration at the German-Dutch border since the early 1970s has ranged from 150 to 200 milligrams per liter (mg/l), with peaks of up to 400 mg/l (IAWR 1988: 10).[1]

Chloride concentrations in this range do not pose significant health problems for most humans and animals. They also do not have a major impact on the ecosystem of the river. Since the 1930s, Dutch farmers and waterworks have persistently claimed, however, that chloride pollution of the Rhine causes crop damage and damage to water supply systems. Consequently, they have demanded that upstream polluters reduce their emissions. Unlike the Netherlands, the riparian countries upstream, Germany, France, and Switzerland, do not suffer from their own or from other countries' chloride emissions along the Rhine. Not surprisingly, therefore, these countries have been reluctant to engage in costly emission reductions from which they do not benefit directly.

The principal collective action problem emanates from the strong asymmetry of the demand by the riparian countries for pollution reductions. Contrary to many North-South or East-West environmental issues, however, this asymmetry of concern does not stem from differential income or capacity, but rather from differences in vulnerability in regard to the pollution problem and from different marginal costs of pollution reduction. International financing of chloride reductions has served principally as a strategy to increase the demand for pollution reductions on the part of upstream polluters. Other obstacles to effective international environmental cooperation, which can often be mitigated by financial transfers, such as a low technological or administrative capacity of polluters, or an unfavorable contractual environment at the international level, have been of minor relevance in the Rhine case.[2] In other words, financial transfers in the chloride case, to the extent that they have been effective, have influenced environmentally relevant behavior by increasing concern.

Given the collective-action problem identified above, several types of solutions would seem possible: financial transfers, liability rules, issue linkages, supranational solutions, and private good substitutes. This section describes each of these solutions and outlines those that have actually been adopted.

Financial transfers. To homogenize the otherwise asymmetric demand for pollution reduction, the victims (the national, regional, or local governments in the Netherlands, or Dutch waterworks or farmers) could compensate the polluters (MdPA, German coal mines, or their governments at various levels) up to the point where the marginal benefit from pollution reduction is equal to the marginal cost of paying the polluter. This is the Coasian solution.[3] It requires the definition of property rights and the establishment of a compensation mechanism. In the chloride case, property rights were assigned to MdPA in terms of specific emission reduction targets. Property rights were also defined for Germany, France, and Switzerland as part of stand-still agreements. The property rights of the Netherlands, on the other hand, were only vaguely specified. The financial transfers by the four riparian countries involved buying some of MdPA's (France's) property rights.

The reductions at MdPA and the associated international financing have been organized into two projects. These projects have been implemented since 1987 and 1992 respectively. In the first project, MdPA is curbing its chloride emissions by 15 kg/s. The cost of 132 million French francs (FF) is shared according to a formula that was agreed to in 1972; Germany and France pay 30 percent, the Netherlands 34 percent, and Switzerland 6 percent. Germany, the Netherlands, and Switzerland transferred their shares as lump-sum payments to the French government. The latter added its own share and has financed the reductions at MdPA.

The second project consists of modulated reductions; MdPA reduces its emissions by up to 56 kg/s (in addition to the 15 kg/s constantly retained since 1987) whenever the chloride concentration of the Rhine exceeds 200 mg/l at the German-Dutch border. In both reduction projects, the retained salt waste is stockpiled in land-based dumps. The costs of the second project are limited to FF 400 million and are shared according to the 1972 cost-sharing formula. The money is transferred through the same channels as in the first project. In contrast to the first project, however, payments in the second are made on a yearly basis, and the exchange of money for pollution reduction is marked by stricter mechanisms of reciprocity. Both types of reductions are accompanied by a stand-still agreement, under which all four riparians are commited to stabilizing their chloride emissions according to specific quotas.

The internationally financed project in the Netherlands, agreed to in 1991, will reduce chloride pollution of the IJsselmeer. The cost of this project is limited to FF 100 million and is to be shared according to the same cost-sharing formula as the other two projects. The contributions are transferred as lump-sum payments to the Dutch government. The latter adds its share and issues payments to the Dutch waterworks that will carry out the project.

The assignment of property rights and the nature of financial transfers in the Rhine case deviate from the Coasian solution that one may have expected: a Dutch-French or Dutch-German exchange. The solutions to the chloride problem involve financial transfers from all four riparian governments to the polluter as well as the pollutee, not just transfers from the Netherlands to France or Germany. The principal alternative in terms of assigning property rights, which was not adopted, would have been to determine the homemade share of the Netherland's chloride pollution in those Dutch river and lake systems that are fed by the Rhine, and to define the pollution shares of the three upstream countries. On this basis, a stand-still agreement for all four countries and financial transfer arrangement could have been reached.

The financial transfers are bilateral and based on individual country targets, but they are internationally coordinated. They do not involve an international fund, unlike most other transfers examined in this book. The most important alternative to the decentralized approach to financial transfers would have been to establish a joint trust fund; let upstream polluters bid for project funding in exchange for chloride reductions; and allow suppliers of environmental protection know-how and technology to compete for contracts to supply conceptual and technical solutions for these projects.

Liability rules. To solve the chloride problem, the countries involved could have established liability rules. In contrast to the Coasian solution, which is based on the purchase of entitlements through voluntary transactions, liability rules specify how much an actor must pay for the destruction of an entitlement. No concerted international effort has been made to establish liability rules for chloride pollution along the Rhine. However, Dutch farmers and waterworks initiated lawsuits against MdPA and the French government at a time when international negotiations on a Coasian solution were already under way. These lawsuits have resulted in one instance of small-scale compensation payments by MdPA to Dutch farmers. But they did not have any wider impact in the sense of creating a strong legal precedent or leading to generally applicable liability rules for Rhine pollution (Kiss 1985; Romy 1990). International financing of the project in the Netherlands does not involve any formal recognition of liability on the part of upstream polluters.

Issue linkage. The actors involved in the chloride case could have linked the chloride issue to other issues. Instead of exchanging money for chloride reductions, they could have exchanged chloride reductions for some other type of action. Chloride pollution is only a small part of the overall pollution problem the Rhine riparians have faced. The other major item on the ICPR's agenda has been chemical pollution (particularly pollution by heavy metals). In principle therefore chloride pollution could be dealt with in the context of a larger package deal for the Rhine pollution problem. No issue linkage has occurred. Despite its relatively small significance in environmental terms, chloride pollution has persistently been framed as an individual issue, not as an integral part of Rhine pollution as a whole.

Supranational solutions. A fourth solution would be to resort to existing supranational institutions (for example the European Union) or establish new institutions that could overrule upstream polluters—for example through majority voting in the ICPR. Indeed the Netherlands repeatedly sought to replace the ICPR's consensus rule by a majority-vote rule (Dieperink 1992). But the upstream countries persistently blocked all such proposals. The European Union (formerly the European Community) has been only marginally involved in the chloride issue, even though it has been a member of the ICPR since 1976. Moreover, the Netherlands has not made any visible attempts to initiate EC/EU directives or regulations that would impose chloride reductions on France and Germany (Switzerland is not a member of the EC/EU).

Private good substitutes. Finally, the victims of upstream emissions can unilaterally adapt to increased pollution; that is, instead of a collective solution to the problem, the victims employ private good substitutes at their own cost. While international negotiations on financial transfers and transnational lawsuits were under way, Dutch farmers and waterworks introduced technical measures to mitigate the effects of chloride pollution. Farmers improved their irrigation techniques and waterworks diversified their water intake and invested in technologies to protect water supply systems against corrosion. The extent of these investments, as motivated by upstream pollution, is largely unknown.[4]

Section 1 examines the process that has led to the framing of the chloride problem, the specification of solutions, and financial transfer mechanisms, and the implementation of mechanisms and solutions. Section 2 explains four key features or outcomes of these five phases of the problem-solving process: the persistently narrow framing of the issue; the selection of financial transfers as the principal solution; the decentralized form of financial transfers; and the protracted bargaining and its implications. Section 3 evaluates the effectiveness and efficiency of solutions to the chloride problem and summarizes the principal findings of this chapter.

1   Description

Since the 1930s, MdPA has been the largest single source of chloride pollution of the Rhine, contributing about 35 to 40 percent to the total chloride load (MdPA 1988: 13).[5] From 1910 to 1931, MdPA dumped its salt waste directly on the ground at several sites around the potassium mines. Much of the salt has since leached into the ground and has caused serious groundwater contamination that affects the water supply of the Alsace region to the present day. In 1931, in response to increasing groundwater contamination, the French government authorized MdPA to discharge its salt waste into the Rhine. This exportation of pollution costs to downstream actors did not remove the already existing salt waste dumps, which continued to contaminate the Alsatian groundwater. But it prevented the problem from becoming worse. Despite protests by Dutch waterworks and the Dutch government, MdPA constructed an open canal by which the dissolved waste was transported from the mines to the Rhine (IAWR 1988: 21). Well into the 1970s, the French government permitted MdPA to gradually increase its emissions along with its production (IAWR 1988: 21–22). Since the mid-1970s, however, MdPA has implemented an expensive clean-up program. This program is partly financed by the French government and will continue well into the next century (Schreiber and Dufond 1992).

Preferences and Institutional Context

As the above account suggests, France has been quite willing to impose the costs of chloride pollution on Holland, the major downstream country; and France has had the capacity to do so. Not surprisingly, the Dutch government, and Dutch farmers and waterworks in particular, have had opposite preferences, although their bargaining power has been relatively low. The Dutch have persistently tried to persuade upstream polluters, especially MdPA, to reduce their chloride emissions. The Netherlands draws around 65 percent of its freshwater supply from the Rhine. Because its groundwater has become more and more polluted, it is increasingly dependent on Rhine water. The IJsselmeer, a large artificial freshwater lake fed by the northern arm of the Rhine, the IJssel, provides a large share of the water that is consumed by the northern part of the province of North-Holland, which includes Amsterdam. In addition, around 50 percent of the water for the province of North-Holland is drawn from surface water of the southern arm of the Rhine, the Lek (CCE 1973).

The chloride problem in the Netherlands is partly homemade, partly imported.[7] Fifty percent or more of the chloride load in the Netherlands' groundwater and canals allegedly stems from saltwater intrusion from the North Sea. In their constant struggle against salination, Dutch farmers and waterworks have relied on flushing the polders with less salty Rhine water. When chloride concentrations in the Rhine increased, this beneficial effect decreased (MdPA 1988: 6–7). In other words, the conflict between France and the Netherlands has resulted from attempts by both countries to appropriate the benefits of a scarce natural resource in a situation where property rights are only vaguely specified.

Dutch farmers have claimed that they suffer from chloride pollution because chloride concentrations greater than 100 mg/l cause crop damage in certain forms of horticulture and greenhouse vegetable production. These two forms of agriculture, which are widespread in the Netherlands, depend on irrigation and sprinkling with Rhine water. Dutch waterworks have argued that they suffer from chloride pollution in two respects. First, they find it difficult to comply with water quality guidelines and regulations set by the World Health Organization (WHO), the EU, and national authorities. These guidelines and regulations stipulate that the chloride concentration of drinking water should not exceed 200 mg/l. The actual chloride concentration often exceeds 200 mg/l and sometimes even 400 mg/l. Second, the waterworks have claimed that a high chloride concentration causes corrosion in water supply systems. Corrosion leads to increased iron content in drinking water and results in higher maintenance costs.

Germany has been the second largest chloride polluter of the Rhine. Coal mines in the Ruhr area contribute around 15 percent to the total chloride load of the river. German industry other than the coal mines adds around 25 percent (ICPR 1974; ICPR 1991). Germany has been unwilling to reduce its own emissions for two reasons. First, it has remained largely unaffected by chloride pollution of the Rhine because it lies upstream and draws its water supply from river bank filtered water, smaller wells that are hydrologically connected to the Rhine, and other sources (ICPR 1991; Goppel 1991). Second, curbing German chloride emissions would be very expensive and difficult to implement because the salt waste occurs in dissolved form. Germany has, however, supported the Netherlands' demand that MdPA reduce its emissions. It has also been willing to finance chloride reductions in France to avert potential demands by Dutch or other actors (for example the German Green Party) for reductions in Germany, and to mitigate the conflict between the Netherlands and France.

Switzerland's contribution to chloride pollution has been relatively small (around 4 percent of the total chloride load) (ICPR 1991). Its interest in chloride reductions is similar to that of Germany. It does not suffer from chloride pollution and reducing its own diffused emissions would be very costly. It has thus been unwilling to reduce its emissions. But it has supported reductions at MdPA and has been willing to contribute financially to this end. Until the mid-1980s, its support for chloride reductions in France also enabled Switzerland to somewhat improve its poor environmental track record. Until 1984, for example, the city of Basel discharged its (and its large chemical industry's) largely untreated waste water into the Rhine.

The preferences just described, as well as the envisaged and adopted solutions to the chloride problem, have been influenced by the institutional context. Since the 1930s the Netherlands had tried to reach a bilateral solution with France. But the Dutch demands were ignored by France. The issue was also raised in the Central Commission for the Navigation of the Rhine and international Rhine conferences on salmon fishing—to no avail. As the water quality of the Rhine continued to deteriorate rapidly throughout the 1950s, the riparian states started discussions at the expert level. Out of this process emerged the ICPR in 1963. At that point, the option of financial transfers was not on the negotiating table and no lawsuits over liability for pollution damage had been initiated. The ICPR focused merely on fact finding. A Dutch proposal to freeze the Rhine's chloride load at the 1954 level of 225 kg/s, or an average concentration of 250 mg/l, was rejected by France and Germany. They argued that the Dutch problem was not caused by upstream pollution but by seepage of seawater in the Rhine delta (Dieperink 1992; Lammers 1989).

During the 1960s and 1970s, the Netherlands became more and more interested in a multilateral approach for several reasons. A collective solution to the chloride issue could serve as a catalyst for negotiations on other Rhine pollution issues. The problem of mustering enough domestic political support in the Netherlands for financial transfers to an upstream polluter would be smaller if other countries paid the polluter as well. By playing the upstream countries against each other, the Netherlands would not have to pay the entire cost of pollution reductions. It seems to have succeeded in doing so to the degree that one of the German and Swiss motivations for paying for reductions in France was to avoid potential demands for reductions on their territory. The upstream countries reluctantly agreed to address the chloride issue multilaterally. But they made every effort to prevent majority voting in the ICPR and to avoid equipping the secretariat of the ICPR with authority to conduct research into the issue.

First Phase of Reductions

In 1972 the four riparian states agreed that a jointly financed solution to the chloride problem should be found. The Netherlands would pay 34 percent, Germany and France 30 percent each, and Switzerland 6 percent of the total cost of reducing the chloride emissions of MdPA (Communiqué 1972). This cost-sharing formula, which has persisted until the present day, reflects the relative contributions of the individual countries to the pollution problem and the intensity of their demand for chloride reductions. The starting point for negotiations on this formula was the assessment scale for administrative costs of the ICPR.

In exchange for international financing, MdPA was to reduce its emissions by 60 kg/s (annual average) and stockpile the salt waste in land-based dumps. Reductions were to begin by 1 January 1975. MdPA was selected because it was the largest point source of chloride pollution of the Rhine. Moreover, reductions at MdPA could be achieved at lowest marginal cost because MdPA's salt waste occurs in solid form. No specific reduction project was adopted, however, because of disagreement in France over the location of the salt-waste dumps. Proposals to clean the salt waste and sell the salt were rejected due to the high costs of cleaning the material and the difficulties of selling the product on the already saturated European market for salt.

The parties also agreed to stabilize their chloride emissions. The ICPR would monitor compliance with this commitment. The stand-still agreement was a precondition for the French consent to reductions on its territory. The goal of reduction and stabilization measures was to achieve a chloride concentration of 200 mg/l or lower at the German-Dutch border. This goal was influenced by national water quality regulations and recommendations by the WHO and other international bodies.

In 1974 France submitted a detailed plan for a 60 kg/s reduction at MdPA. It proposed that the project be financed on an annual percentage basis, including contingency and inflation costs. This proposal was rejected by the other governments because there was no explicit cost limit and the total cost was estimated at around five times the previously envisaged FF 100 million. In addition, the proposed stockpiling of retained salt waste in land-based dumps evoked local opposition in the Alsace region. This opposition feared that the proposed reduction method would worsen local groundwater contamination. After a series of tests, MdPA and France proposed to inject the dissolved salt waste into the subsoil at around 1,800 meters depth. Twenty kg/s Cl− could be disposed of in a first phase with this method and another 40 kg/s in two subsequent phases if no technical problems arose and if international financing was assured (Hartkopf 1976; Dupont 1993).

An international treaty reflecting the French proposal was signed on 3 December 1976.[8] The parties maintained their goal of reducing MdPA's emissions by 60 kg/s. But the reduction was to be implemented in two phases. In the first phase, MdPA would inject 20 kg/s Cl− into the subsoil for ten years, which was equivalent to a 7 percent reduction of the total chloride load of the Rhine. The ten-year time frame was chosen because MdPA's production was expected to decrease afterward, thus reducing its emissions irrespective of the international effort. Negotiations on the second phase were postponed.

In exchange for its consent to the treaty, Germany required the closure of a buffer basin that was used by MdPA to modulate its discharges. MdPA used this basin to increase its discharges at high flow of the Rhine and reduce them at low flow. Lacking proper insulation, the use of the buffer basin resulted in groundwater contamination beyond the French border into neighboring Germany. Closure of the basin undid much of the expected effect of the planned 20 kg/s reduction by MdPA, especially at times of low water flow, thus increasing the benefits of Germany at the expense of the Netherlands.

The four parties agreed to share the projected cost of FF 132 million according to the 1972 formula. Germany, the Netherlands, and Switzerland would pay their entire contribution at once to the French government after the agreement had taken effect. France would then pay MdPA. It also undertook to report regularly on the implementation of the project.[9]

In December 1979, after a long period of hesitation, the French government broke off the ratification debate in its parliament, arguing that it was unable to muster enough support for the agreement. Again, it pointed to local opposition in the Alsace region. It remains unclear whether local opposition was the genuine cause of the French failure to ratify the agreement; whether this argument served as a disguise for a general lack of interest on the part of the French government in chloride reductions; or whether it was part of an attempt to wrest more concessions from the other parties. The Dutch government responded by taking the highly unusual step of ordering its ambassador to France back to The Hague for consultations. But both Germany and the Netherlands left their lump-sum payments, which they had made even before the treaty was enacted, with the French government to underline that they were not willing to give up the matter. Switzerland, on the other hand, retrieved its payment in 1981 (*Neue Zürcher Zeitung*, 26 March 1987).

In 1983 France resubmitted the chloride treaty to its parliament. Ratification was completed by the end of 1983. On 5 July 1985, after more than twenty years of negotiation, the chloride agreement went into effect. This change in the French position can be attributed to the following circumstances. The three other countries had maintained their political pressure on France to live up to its commitment. Dutch farmers and waterworks had initiated lawsuits against MdPA and the French government, and had reached favorable court decisions in principle (Romy 1990; Kiss 1985). Finally, after elections in June 1981 a new socialist government came to power that put environmental issues higher on its agenda than previous French governments.

The 20 kg/s reduction at MdPA was set to begin on 5 January 1987. A few weeks before this date, the French government suddenly decided to stock the salt residue on land instead of injecting it into the subsoil (MdPA 1988: 15; *Neue Zürcher Zeitung*, 4 June 1986). The other governments were surprised because land-based dumping had met with local opposition before. This change in the French position made earlier arguments for nonratification somewhat suspect. But none of the other riparian countries objected. They were interested in chloride reductions at MdPA, no matter how they were achieved, as long as the costs did not exceed the agreed limits, or as long as France would pay for additional expenses.

On 5 January 1987, implementation of the 1976 agreement began, albeit at a reduced rate. France had argued in the meantime that a soda factory along the French part of the Moselle, a tributary of the Rhine, had closed a few years earlier. This development had reduced France's total chloride emissions. The French government requested that this reduction be taken into account by lowering MdPA's reduction from 20 kg/s to 15 kg/s. The extent of funding, on the other hand, would remain the same. The other governments gave their consent to this increase in the marginal cost of the project because they feared that France would otherwise further delay implementation of the agreement.

Second Phase of Reductions

In 1988 France proposed a project that would reduce MdPA's emissions by the 60 kg/s total agreed to in 1972. The estimated cost of FF 816.3 million was to be shared according to the 1972 formula. Until 1998, MdPA would stock 60 kg/s and, after that date, discharge the stockpiled salt into the Rhine at an agreed upon rate. France insisted on the latter provision because it feared that after 1998 it might find itself with the stockpiled salt and no one to pay for its disposal (into the Rhine or elsewhere).

The Dutch government rejected the French proposal (*Neue Zürcher Zeitung*, 12 October 1988; *Financieele Dagblat*, 13 October 1988). It argued that chemical pollution had in the meantime replaced chloride pollution as the top priority on the environmental agenda for the Rhine. This perception was enhanced by a major spill in November 1986 at the chemical firm Sandoz near Basel, Switzerland. This spill had disastrous environmental consequences for the river and disrupted the drinking-water supply downstream (Schwabach 1989). Hence the Dutch government declared itself unwilling to finance reductions at MdPA that were, in its view, excessively costly and of small benefit. It proposed to spend the money for the Rhine Action Program. The program was initiated by the ICPR shortly after the Sandoz accident in 1986 and serves to coordinate the cleanup of Rhine pollution by substances other than chlorides.

The Dutch government also expressed dissatisfaction with the proposed discharging of the stockpiled salt after 1998; it did not wish to pay MdPA for the retention of pollutants that the Netherlands would receive in smaller doses later on. The Dutch incentive to pay for chloride reductions upstream had been further eroded because in the meantime several Dutch victims of chloride pollution had initiated lawsuits against MdPA and had obtained favorable court decisions in the Netherlands and France. Finally, Dutch horticulturists had improved their irrigation techniques, thus making them less vulnerable to chloride pollution. Germany and Switzerland were reportedly willing to subscribe to the French plan; but deprived of support by the principal victim of chloride pollution, the negotiations collapsed.

The Dutch government's withdrawal from the second phase of chloride reductions was only temporary. It had invested a lot of time and effort in the matter and was not willing to abandon the issue for domestic political reasons. The chloride agreement contained a stand-still provision that was regarded as beneficial. Most important, however, lobbying by Dutch waterworks, which had privileged access to decision makers in the ministry of transportation and public works, intensified.[10] In addition, the drinking-water company of North-Holland had recently started construction of a water-softening installation. This installation would use technology compatible with the anticipated lower chloride concentration in the IJsselmeer that would come as a result of reductions upstream. Without further chloride reductions, this investment would be in jeopardy.[11] In coordination with France, the Dutch government elaborated a new project in 1989.

The treaty that was signed as a result on 25 September 1991 departs significantly from the earlier approach. It provides for modulated instead of constant reductions at MdPA, a project in the Netherlands that is to reduce the chloride concentration in the IJsselmeer, and a stand-still agreement. It downgrades the chloride issue to a problem of drinking-water quality and does not make reference to agriculture.[12] Finally, it is designed to settle the chloride issue once and for all: it states that reductions beyond those provided for in the accord would be impossible from an economic and technical standpoint, and unnecessary from an ecological perspective.

Modulated reductions at MdPA. In addition to the constant reduction by 15 kg/s, MdPA reduces its emissions whenever the chloride concentration of the Rhine exceeds 200 mg/l Cl− at the German-Dutch border.[13] International financing for the project is limited to FF 400 million, with FF 40 million allocated for investment costs and 360 million for operational expenses. This amount covers the preparation of land-based storage sites, the preparation, loading and transportation of the salt waste to these sites, and the costs of discharging the stockpiled salt into the Rhine at a later stage. The project and the cost estimates were largely developed by the French government and MdPA (ICPR 1991, ICPR 1992). Switzerland received a 50 percent discount on its financial contribution. It had argued that a soda factory on its territory had closed in the meantime, eliminating its only major source of chloride emissions. To avoid upsetting the 1972 cost-sharing formula, this discount was factored into the accounts as a FF 12 million payment.

The financing of reductions at MdPA proceeds in phases throughout 1998 according to a strict schedule.[14] The reasons for the step-by-step cooperation and strict reciprocity are as follows. Whereas the FF 40 million in investment costs could be estimated with reasonable accuracy, the amount of salt that will have to be stockpiled (operational expenses) could not. The extent of stockpiling depends on the chloride concentration of the Rhine. The latter is a function of water flow, which depends on climate, and the behavior of other sources of chloride pollution. Moreover, the second phase of reductions involves larger-scale financing than the first phase. In particular the problems and delays experienced in the first phase had made the parties aware that the incentives to renege on commitments remained significant and that the agreement could break down at some point along the way. The relatively low asset-specific investments in the project and the step-by-step payment process are common mechanisms for protecting cooperating actors against opportunism (Williamson 1985).

Diversion of brackish water in the Netherlands. One of the major concerns of Dutch waterworks has been the chloride concentration of the IJsselmeer. Most, but not all, chloride pollution of the IJsselmeer stems from the Rhine. Another source is brackish water that is pumped from the Wieringermeer polder into the IJsselmeer. To reclaim land from the sea, the Wieringermeer polder, originally a part of the North Sea, was sealed off by a dam and is being drained by pumping the brackish water into the IJsselmeer. To reduce chloride pollution of the IJsselmeer, the internationally financed project will divert the brackish water input from the Wieringermeer to the Wattenmeer (the North Sea). This project was elaborated by the Dutch government in cooperation with Dutch waterworks. It requires the construction of a pipeline, the installation of new pumps, and the enlargement of some canals. When fully implemented, it will reduce the chloride concentration of the IJsselmeer by around 30 mg/l (ICPR 1991: 2–3).

Financing of the project is limited to FF 100 million. Additional costs, including operational expenses, will be paid by the Dutch government or the waterworks. International payments are due to the Dutch ministry of transportation and public works within three months after the treaty is signed, which occurred in October 1994. The Netherlands has to report annually on the implementation of the project as well as on the chloride concentration of the IJsselmeer.

This project was adopted only after considerable dispute. The opponents contended that it would curb a pollution problem that the Dutch themselves had created. Ironically, the brackish water input into the IJsselmeer lies very close to the major point of freshwater intake. The opponents also argued that the project would only benefit a few Dutch waterworks and would have no impact on the water quality of the Rhine. In addition, they pointed out that the modulated reductions at MdPA and the measures in the Netherlands, taken together, would result in a higher chloride load of the Rhine than the previously envisaged fixed reduction of 60 kg/s.

The proponents argued that the plan would reduce the chloride concentration of the IJsselmeer beyond 1998. Reductions at MdPA, by contrast, imply that the stockpiled waste will probably be discharged into the

Rhine at some point in the future, albeit in smaller doses. The Dutch government opined that the Dutch chloride input into the IJsselmeer would have no significant impact on water quality if the Rhine did not carry so much salt into the IJsselmeer. It also argued that Dutch waterworks were the principal victims of upstream chloride emissions. Consequently, measures that improved their water quality were justified, even if they did not have a direct impact on Rhine water quality. The combination of modulated and fixed reductions at MdPA and the diversion of brackish water in the Netherlands, so the Dutch argument went, would produce a similar effect on the quality of Dutch drinking water than the more expensive French proposal of 1988. Hence this combination would be more cost efficient. There were also two additional (unstated) reasons. First, a large part of Dutch payments under the 1991 treaty would flow to Dutch waterworks and related enterprises instead of a foreign polluter. Dutch payments would thus be somewhat less offending to those criticizing the chloride agreements as being against the polluter pays principle. Second, as noted above, Dutch investment in a new water treatment facility, which was made on the assumption of chloride reductions by upstream polluters, would be at risk if the international effort failed.

The opponents of the plan, particularly the German government, eventually gave in and the Dutch proposal was adopted. Since the second half of the 1980s, the Dutch government had lost much of its interest in chloride reductions upstream in exchange for international financing. Other types of pollution had become more important. During the many years of international stalemate waterworks and farmers had adapted to increased chloride pollution. In addition, MdPA's emissions were decreasing automatically with its economic decline since the mid-1980s. The opponents of the Dutch plan feared, therefore, that without additional incentives for the Netherlands the cooperative venture could collapse. Because they had invested so much time and effort in the endeavor, they were unwilling to abandon their investment without at least the symbolic political benefit of an agreement. They also feared that they might, for face-saving purposes, be forced to continue to find a solution to the chloride problem at a time when they were trying to get the controversial issue off the table and concentrate on more important Rhine pollution problems. In a rather strange way the Dutch bargaining power had increased because the non-agreement alternatives of the Netherlands had improved considerably.

Implementation and Effectiveness

The agreed reductions at MdPA have been implemented effectively so far. The project in the Netherlands has just started and can not yet be assessed. Since 1987, MdPA has stockpiled around 15 kg/s (equivalent to approximately 750,000 tons of salt per year) in land-based dumps. This reduction is scheduled to last until the end of 1998. The modulated reductions at MdPA have been implemented since 1992. No detailed data on the implementation of this project are available. But officials of the four countries are generally satisfied with MdPA's performance and the agreed funding has been forthcoming. Moreover, the four riparian countries seem to have complied with the chloride pollution quotas that they received under the stand-still agreements.[15] At a more general level, they have also achieved their broader environmental goal that was set in 1972: an average annual chloride concentration equal to or smaller than 200 mg/l at the German-Dutch border.

Despite formal compliance with agreed solutions, the environmental effect of financial transfers and the associated projects has been minimal. When the four governments were finally able to agree to and implement very modest reduction projects, the environmental problem had waned for reasons unrelated to the international effort. The 15 kg/s reduction by MdPA decreases the total chloride load of the Rhine only by around 4 percent. The modulated reduction removes only small quantities of salt because chloride concentrations have rarely exceeded the 200 mg/l threshold that triggers reductions. The internationally financed reductions are probably much smaller than unintentional reductions. In recent years, the production of several major chloride polluters along the Rhine, particularly potash mines, coal mines, and soda works has declined for economic reasons, and so have their chloride emissions. In addition, winters in the Rhine basin have been quite mild in the past few years, thus decreasing chloride pollution by salt that is used for the de-icing of roads and is washed into the Rhine through drainage systems. The latter source used to account for around 9 percent of the Rhine's chloride pollution.

2   Analysis

This section explains three principal outcomes that characterize the problem-solving process in the chloride case: the persistently narrow framing of the issue (problem definition); the form and use of financial transfers; and the "too little, too late" outcome of the process. The narrow framing of the issue can be explained by the characteristics of the problem, the preferences of the relevant actors, and the institutional setting. The existence of solutions based on financial transfers, as well as their form, are explained by preferences, the institutional context, and the difficulty in using alternatives to financial transfers. The long duration of the bargaining process, with its negative implications for the effectiveness of solutions to the chloride problem, has resulted from strategic behavior (not revealing true preferences in terms of costs and benefits, and attempts to capture a greater share of the benefits) and information problems (lack of information about the costs and benefits of cooperation of various domestic and national actors).

Problem Definition

The problem of chloride pollution was framed already in the 1930s and remained a specifically treated issue throughout the entire history of the chloride case. Chloride pollution emerged as the first Rhine pollution issue to a large degree because it was simple and scientifically well understood. There was one substance, a few point sources (chiefly MdPA and German coal mines), and two rather well-organized victim groups (farmers and waterworks in the Netherlands).

The persistently narrow framing of the issue can be explained as follows. In the 1960s negotiations on chloride pollution were institutionalized in a special working group of the ICPR. When other Rhine pollution problems increased in importance in the 1970s, the bargaining structure for chlorides was already well in place. As negotiations on the chloride problem reached a stalemate in the 1970s and the first half of the 1980s, the four riparian governments engaged in a strategy of "containment." In the 1960s they had still hoped that a quick and effective solution to the chloride problem might open the door for a solution to other, more complex Rhine pollution problems. From the 1970s on, they isolated the chloride negotiations from ICPR negotiations on chemical and thermal pollution to prevent negative spillovers and to ensure that the consideration of financial transfers as a solution to chloride pollution would not act as a precedent for other areas of negotiation. This strategy was facilitated by the fact that the chloride issue was, at least in the medium term, a self-solving environmental problem.

Financial Transfers
The existence of financial transfers in the chloride case, as well as their form, can be explained by the preferences of the actors, the institutional context, and the difficulty of bringing alternatives to financial transfers to bear. The persistently narrow framing of the issue was associated with a strong asymmetry of preferences (concern). To homogenize preferences, which meant to increase the demand by upstream polluters for pollution reductions, the four riparian countries resorted to financial transfers. Financial transfers in the chloride case do not simply involve an exchange between France and the Netherlands, or between Germany and the Netherlands, as one might have expected, but an exchange between all four riparian countries. This outcome can be attributed to the institutional setting and the skillful tactics of the Netherlands. As the bargaining process became institutionalized in the ICPR, initially against the opposition of the upstream countries, the Netherlands was able to maneuver Switzerland and Germany into a position where they could avoid Dutch or domestic pressure for (more expensive) reductions on their own territory by financing reductions in France.

Financial transfers may evoke problems of legitimacy because they conflict with the polluter pays principle (PPP). This principle, according to which polluters must internalize externalities at their own cost, is today widely accepted in Western countries. It should be noted, however, that the PPP was formally adopted by OECD countries only in the mid-1970s. One might argue that, had the OECD formally adopted the PPP earlier, financial transfers in the chloride case would not have been seriously considered. This proposition is hard to uphold, however, because the negotiations were pursued into the 1990s despite the unqualified endorsement of the PPP by all riparian countries of the Rhine. In a more indirect way, however, concern over the inconsistency between the PPP and financial transfers in the chloride case has probably contributed to the persistently narrow framing of the issue and the "containment" strategy noted above.

Lawsuits have been the most important alternative to financial transfers. In the second half of the 1970s, Dutch farmers and waterworks initiated lawsuits against MdPA and the French government over compensation payments for the damage caused in the Netherlands. This strategy became possible largely because the legal framework changed: victims of pollution in member countries of the European Community obtained the right to file lawsuits against polluters in any country of the community. The effect of these lawsuits on the development of the chloride case is mixed. On the one hand, uncertainties over the extent of damage made clear verdicts virtually impossible. Only in one case was a small amount of compensation paid to Dutch farmers in an out-of-court settlement. The effect of lawsuits in spurring upstream polluters into chloride reductions therefore has been small. On the other hand, prospects in the 1980s that Dutch lawsuits might succeed contributed to a reduced demand by Dutch actors for jointly financed pollution reductions. This circumstance contributed to the disarray at the international level over jointly financed chloride reduction projects.

Two other alternatives, supranational solutions and issue linkages, did not materialize. Already in the 1960s the Dutch government had proposed a majority voting rule for the ICPR. Such a rule could have been used to form a coalition to overrule France. For example, the Netherlands could have promised the German and Swiss governments that it would not seek chloride reductions in their countries if they would vote for reductions in France. But the upstream countries persistently rejected any such proposals and, because of the consensus rule in the ICPR, no change in voting rules has taken place. Given the consensus rule in the EC, carrying the issue into this institution would not have produced more progress and was, therefore, not attempted by the Netherlands. Before the Maastricht Treaty, any member state could have blocked decisions. Moreover, national regulations on chloride pollution in the EU/EC vary enormously. Any common standard would probably have been less stringent than the Netherlands desired. In addition, Switzerland is not a member of the EU/EC.

Linkage of chloride pollution to other forms of Rhine pollution would probably not have produced any progress and was never explicitly attempted. For example, the Netherlands could not trade lower chemical pollution on its part, the most important candidate for linkage, against lower chloride pollution by France. France would not benefit from the Netherlands' action. Moreover, chemical pollution is more complex and less understood than chloride pollution. Risk-averse negotiators in particular will rarely link a clearly understood issue with one where the costs and benefits of cooperation, and thus the potential gains from trade, are poorly understood.[16] Finally, for reasons noted above, the Rhine riparians had a strong interest in isolating and containing the chloride issue.

Financial transfers in the chloride case consist of decentralized payments to France and the Netherlands. The establishment of a joint fund was never considered because the amount of money involved was relatively small and was to flow to previously agreed projects. Thus a joint fund, which might have allowed for the competitive bidding by the four countries for projects to reduce chloride pollution, was not necessary. Moreover, the upstream countries would have been against a joint fund because this could have equipped the ICPR with more important functions than the upstream countries wanted. The step-by-step process and strict reciprocity used for the financing of the second phase of reductions contrast with the lump-sum payments in the first phase. This difference can be attributed to the nature of reductions (modulated instead of fixed), the larger amount of money involved, and the negative experience with the first phase of reductions.

Explaining the "Too Little, Too Late" Outcome

The enormous difficulties that the four countries encountered in trying to solve the chloride pollution problem are inconsistent with our initial expectation. One should have thought that the favorable contractual environment, the absence of capacity problems, and the narrow framing of the pollution issue would lead to a quick and effective solution. The discrepancy between this proposition and the actual outcome can be accounted for by information problems and the strategic behavior of the actors. These two factors account for the long duration of the bargaining process in terms of making it very difficult to find a stable zone of agreement. The long duration of the negotiations, in turn, led to a progressively smaller zone of agreement for reasons explained below.

The term "information problems" refers to the difficulties that the four governments faced when trying to determine their own costs and benefits of cooperation over time, as well as the costs and benefits of other actors (other governments, MdPA, local interest groups in the Alsace, Dutch waterworks and farmers). Information problems result from the limited capacity of decision makers to collect and analyze information. Strategic behavior involves attempts to hold out for more, for example by deliberately distorting information, particularly by overreporting costs and underreporting benefits of cooperation. It may also involve extortion.

The major information problems in the chloride case involve problems of determining the level of reductions that would satisfy Dutch farmers and waterworks, and determining possible levels of reductions, methods, and costs at MdPA. In France and the Netherlands the information problems stem from the fact that the actors paying and/or benefiting from reductions, and the actors who would implement reductions are not identical. The preferences of these actors have reacted differently to changes in the context of the cooperative effort. Moreover, the date when MdPA would close remained uncertain for a long time. Hence the French and Dutch governments faced serious difficulties in determining exactly how much they were willing to pay for what level and type of reductions. The instances when France failed to ratify the 1976 chloride agreement due to (arguably unanticipated) local opposition and when the Dutch government unexpectedly revised its position in 1988, due to the particularistic interests of some waterworks, highlight these information problems.

As to strategic behavior, agreement on the cost-sharing formula set an important parameter in the bargaining over which country should pay how much for chloride reductions. The negotiations thus focused more on the necessary level of reductions and on the total cost. Nevertheless, distributional bargaining continued at a more subtle level as each of the four countries continued to hold out for more.

France succeeded in lowering its reduction target from 20 kg/s to 15 kg/s and received undiminished compensation. Switzerland achieved a reduction in its financial contribution to the joint effort. Both governments had argued that soda factories on their territory had closed earlier, thus reducing their chloride emissions. Germany linked its acceptance of the 1976 agreement to the closure of MdPA's buffer basin. This linkage eliminated a groundwater contamination problem for Germany. But it undid much of the environmental effect that planned reductions by MdPA would have had, thus creating an externality affecting the Netherlands. The Netherlands obtained financing for a controversial reduction project on its territory. Most of these demands for more favorable treatment were made at times when agreements had almost been reached. At these points, the chances of obtaining concessions from the other parties were usually higher because they did not wish to spoil the political benefits of agreement.

Information problems, coupled with distributional bargaining, reinforced the tendency to engage in strategic behavior. The Dutch government sought to exaggerate the impact of the externality to pressure the upstream countries into larger reductions for less compensation. France and Germany, on the other hand, had an incentive to downplay their contribution to the Dutch chloride problem by pointing to the uncertainty over the extent to which the Dutch problem was homemade. France sought to exaggerate its marginal reduction costs by capitalizing on the uncertainty over whether reductions would be acceptable to local interest groups in the Alsace. This behavior allowed the French government to use delaying tactics and eventually obtain more money for smaller reductions. The preferences and bargaining strategies of the other countries have put heavy constraints on France's possibilities in this regard. Switzerland and Germany, which benefited only indirectly from the reduction effort, were themselves trying to capture an additional share of the joint gains by paying less or, in Germany's case, linking cooperation to the closure of MdPA's buffer basin. The latter increased Germany's benefit at the expense of the Netherlands. The outcome of this distributional bargaining given such uncertainty has been, quite predictably, smaller chloride reductions and smaller financial transfers.

The duration of the bargaining process as such had a negative impact on cooperation. The longer the bargaining went on, the more the zone of agreement shrank. Asset-specific investments in private good substitutes grew in the Netherlands, the chloride issue lost in salience because of the rise of chemical pollution, and the point in time where MdPA would close for economic reasons approached. Thus the willingness of the Netherlands to pay for reductions upstream diminished steadily. In the end, sixty years of chloride diplomacy ended with an agreement that was struck more for face-saving purposes than for environmental reasons.

3 Evaluation

In some regards, the outcome of the international chloride reduction effort can be regarded as a success. First of all, financial transfers have facilitated international environmental cooperation by homogenizing the otherwise asymmetric concern over chloride pollution. In contrast to other cases examined in this study, they have not been employed to resolve problems of asymmetric capacity or an unfavorable contractual environment. The overall goal of bringing the chloride concentration at the Dutch-German border down to 200 mg/l has been achieved. At the declaratory level, the parties share the view that, if chloride pollution ever was a problem for the ecology of the Rhine, the problem is now solved. Agreements, once ratified, have been implemented effectively. Moreover, these agreements have induced various actors to behave in ways they would not have in the absence of the exchange (for example MdPA).

In most other respects, however, the chloride reduction effort has been a failure. As noted above, the impact of these reductions on the extent of chloride pollution of the Rhine has been very small. Since the late 1980s, chloride pollution has declined somewhat. But this decline has been largely caused by factors unrelated to the operation of internationally financed measures. The small effect of international cooperation on the environment contrasts starkly with the length of negotiations and the significant amount of time, money, and other resources involved.

The above assessment raises questions of efficiency. Would different approaches to negotiating the exchanges, different types of financial transfer mechanisms, or solutions other than financial transfers have produced Pareto-superior or environmentally more significant outcomes? Because of asset-specific investments in private good substitutes in the Netherlands, the self-solving nature of the problem due to industrial restructuring, and the rise of chemical pollution, time has been a major factor. The longer it took to negotiate exchanges, the smaller the zone of agreement and the lower the environmental benefit of agreed measures became. The crucial issue therefore is whether and how the actors could have reached agreements on equal or deeper pollution cuts within a shorter time frame.[17]

Some of the information problems that contributed to strategic behavior and thus frustrated the international effort could have been alleviated by involving local interest groups and by focusing more on environmental impact than on emissions. One of the most serious problems on the French side was that local interest groups other than MdPA had very little to gain from international agreements, as they were concluded. Their most preferred outcome was the status quo, where MdPA dumped its salt waste into the Rhine. As long as there was even a small probability that MdPA's chloride reductions could increase the salination of local groundwater, these interest groups would oppose reductions. It is unclear, however, whether the direct involvement of local interest groups in the bargaining process would have led to earlier agreement. But in any event, it might have avoided the situation where France first signed the 1976 agreement but then failed to ratify it. It might also have opened up new options, such as side payments to local interest groups or an internationally financed project to reduce MdPA's emissions *and* clean up groundwater pollution in the Alsace.

The premature closure of the bargaining process and its strong focus on emissions instead of environmental impact created serious information problems. The Netherlands in particular, but also Germany and Switzerland, remained uncertain about how much they were willing to pay because the benefits to Dutch actors remained unclear. Independent investigations (perhaps by the ICPR) into the impact of chloride pollution, as well as cost-benefit analysis of adjustment measures in the Netherlands versus reductions upstream, might have avoided some of these problems.

The principal alternative in terms of assigning property rights would have been to determine, through independent investigation, the homemade share of the Netherlands' chloride pollution in those Dutch river and lake systems that are fed by the Rhine, and to define the pollution shares of the three upstream countries. On this basis, a stand-still agreement for all four countries could have been reached and the upstream countries could have paid the Netherlands for adaptation measures. This approach is somewhat reminiscent of the IJsselmeer project. An additional stand-still obligation for the Netherlands, however, would have averted some of the criticism that was raised in connection with the IJsselmeer project. This criticism focused on uncertainty over the extent to which the Netherlands was itself responsible for its chloride problem. This option was never seriously discussed, largely because the riparian countries focused much more on emissions upstream than on the environmental impact downstream. This circumstance may stem from the fact that it is generally more popular to reduce emissions than to carry out adaptation measures. It may also be attributed to premature closure of the negotiations and the excessive focus by the Dutch government and Dutch interest groups on MdPA.

Yet another option might have been to establish a joint trust fund. Let upstream polluters bid for project funding in exchange for chloride reductions, and let suppliers of environmental protection know-how and technology compete for contracts to supply conceptual and technical solutions for these projects. It is likely that with such a financial transfer mechanism MdPA would have received the most funds because its marginal reduction costs are much lower than those of other chloride polluters. It also has the know-how and technology to implement reductions. The important difference of this approach, as compared to the approach actually chosen, is that it would have provided the actors with a greater incentive to reveal their true preferences, therefore avoiding some of the information problems discussed above.

4   Conclusion

The analysis of the chloride case shows that when the demand (concern) among a group of countries for environmental protection is strongly asymmetric, the contractual environment is rather benign, and there are no capacity problems, exchanges of international funding for environmental protection can be designed and implemented. It also shows, however, that despite generally favorable conditions, information problems and strategic behavior that are associated with distributional bargaining

over exchanges can seriously hamper international cooperation and reduce joint gains. Information problems in the chloride case stem from
incomplete knowledge about the costs and benefits of cooperation, and
information asymmetries across different governments and subnational
actors. Strategic behavior has consisted of attempts by all four countries
to capture a larger share of the joint gains. In engaging in strategic behavior, the actors have usually capitalized on information problems.

Due to information problems and strategic behavior, it has taken an
extremely long time to negotiate agreements. The longer the bargaining
process went on , the smaller the zone of agreement became because the
Dutch demand for internationally financed reductions upstream decreased for the following reasons: various actors in the Netherlands had
made asset-specific investments in private good substitutes during the
long stalemate; chloride pollution decreased due to industrial restructuring upstream; it became increasingly clear that MdPA would close in the
near future; and Rhine pollution by heavy metals and pesticides became
much more important. The outcome of the sixty years of effort is characterized by "too little, too late," that is, very small if any joint gains and
very little impact on the environment.

International cooperation in the chloride case might have been more
efficient if the negotiating parties had focused more on environmental effects than on MdPA emissions. Two alternative solutions could have been
considered in this regard.

First, the four countries could have determined the homemade share of
the Dutch chloride problem and could have defined pollution quotas for
all four riparian countries. The upstream countries could then have paid
the Netherlands for various types of adaptation or reduction measures.
This solution could have been based on a joint trust fund and bidding by
companies in all four countries for contracts to supply technology and
know-how for approved projects. Second, the four countries could have
established a joint trust fund and let upstream polluters bid for project
funding in exchange for chloride reductions. This solution could also
have involved bidding by suppliers of environmental protection know-
how and technology for approved projects.

It is impossible to determine with any certainty whether these two alternatives would have been approved by the negotiating parties had they

been seriously considered. Nonetheless, given the small joint gains and enormous transaction costs of arriving at the solutions that materialized, the early closure of the bargaining process and the persistent focus on one option appear, in retrospect to have been serious handicaps.

Beyond these rather issue-specific lessons, the analysis of the chloride case also points to some wider-ranging propositions about when and why financial transfers are used in the international environmental realm. The Rhine case is, with the exception of the EU, one of the very few cases where financial transfers have been used among Western countries to facilitate environmental cooperation. The reasons for this are probably the following.

As the chloride case shows, the use of financial transfers is associated with potentially high transaction costs. Second, and related, in highly integrated and institutionalized settings, such as in Western Europe, states can employ other strategies to overcome asymmetries of concern. The most important strategy consists of simultaneous or intertemporal issue linkages. Because highly integrated settings are usually characterized by greater transparency, possibilities of legal action against polluters across boundaries and iterated interaction (generally: a favorable contractual environment), such linkages can often be based on relatively weak conditionality (diffuse reciprocity). In many cases, the transaction costs of using these strategies may be lower than the costs of using financial transfers. The analysis of the chloride case shows that financial transfers tend to be a last-resort strategy, which is employed only if other strategies, such as liability rules, issue linkages, supranational solutions, private good substitutes, and so on do not produce the desired effect. The polluter pays principle, on the other hand, does not seem to have inhibited these states from choosing the financial transfers option. Issue linkages and other alternatives to financial transfers, however, cannot resolve problems of asymmetric capacity. This may ultimately be the reason why financial transfers are, out of necessity, mostly used in the North-South and East-West context.

Notes

1. Data by Walter Jülich, IAWR. The relationship between chloride load and chloride concentration of Rhine water is shaped by the water flow. Assuming

constant chloride emissions, a higher water level of the river (e.g., after rainfall or in spring) leads to a lower chloride concentration and vice versa.

2. MdPA, German coal mines, and other chloride polluters upstream, and the national and subnational governments controlling these actors, have had the administrative and technical capacity and income to regulate emissions. As to the contractual environment, international monitoring of polluter behavior along the Rhine has been practiced effectively at rather low cost for many years. Moreover, the countries involved in the case are highly interdependent and have collaborated closely on many environmental and other issues in various international forums. This interdependence, which allows for strategies of reciprocity, creates strong incentives against reneging and thus facilitates international cooperation.

3. Transposed onto international relations, the argument by Ronald Coase holds that externalities (such as transboundary pollution) do not generate a misallocation of resources in the absence of supranational intervention if bargaining between the recipient and the producer of the externality is possible; if there are no transaction costs; and if property rights are well defined. Under these conditions, the producer and recipient of the externality will have an incentive to negotiate a mutually beneficial trade that internalizes the externality (Coase 1960).

4. Some IAWR reports quote damage cost estimates for Dutch waterworks, and farmers have come up with their own estimates in the context of lawsuits. But no independent evaluation of these figures has been undertaken and they remain contested.

5. MdPA mines sylvanite. Sylvanite contains around 27 percent potassium chloride. The latter is extracted and sold, mostly as fertilizer. The remainder of the material, consisting of 59 percent sodium chloride and 14 percent insolubles, mainly clay, has no commercial value and is disposed of (Schreiber and Dufond 1992).

6. French and MdPA reluctance to unilaterally reduce its chloride emissions has been enhanced by the fact that mining at MdPA has become increasingly uneconomical in the 1980s. The potash market in Europe has contracted. Cheaper potash is imported from republics of the former Soviet Union and other countries. And easily accessible sylvanite reserves at MdPA have decreased. MdPA is expected to cease operations by the year 2004. The principle reasons why it has not closed already are political: it is the second largest employer in the Alsace region, and the retirement packages and continuing environmental cleanup after closure would cost almost as much as the deficit that MdPA is currently running up (*Le Monde*, 5 June 1993). See also *Reuter Textline, Chemical Business News Base*, 16 September 1993, 24 June 1993, and 11 June 1993.

7. A large portion of the country consists of land that has been reclaimed from the sea and lies below sea level. As a result, saltwater intrusion from the North Sea has caused high chloride concentrations in groundwater and irrigation and drainage canals, particularly in the polders. Polders are land areas that were originally part of the North Sea but have been sealed off by dams and are drained by pumping the brackish water into the adjacent sea.

8. This and all subsequently discussed treaty texts can be found in the annual reports of the ICPR (Tätigkeitsberichte).

9. To alleviate French concerns about groundwater contamination in the Alsace, the salt waste injection could be temporarily halted if it posed a significant risk to the environment. If the injection caused damage, and if this damage was not covered by anyone else, the parties would consult on financial compensation. France had demanded full liability of the contracting parties according to the 1972 cost-sharing formula, but the other states refused. The 1976 treaty also contains a stand-still agreement. The parties agreed to report all sources of more than 1 kg/s to the ICPR and to control them. Each country received a national quota with sublimits. These quotas could be redistributed domestically, or internationally through the ICPR. The latter provision, which might have evolved into a tradable permit arrangement, has never been resorted to. The national quotas corresponded approximately to the 1976 contributions to total chloride pollution: Switzerland received 10 kg/s Cl−, France 168 kg/s, and Germany 134.9 kg/s. Chloride pollution by Dutch sources as well as smaller and diffuse sources (less than 1 kg/s) was not covered (ICPR 1974; MdPA 1988: 13). That is, the property rights of the Netherlands were defined only in terms of the transboundary influx into the Netherlands. This approach reflects the international norm that a country can pollute itself as much as it wants as long as this behavior does not affect other countries. The problem with this norm in the chloride case was that the upstream countries contested Dutch demands that upstream polluters should reduce their chloride emissions while the Netherlands could increase its homemade share of the pollution problem and benefit from upstream reductions. Against this background, it is surprising that upstream countries have not made any serious effort to define the property rights of the Netherlands in terms of homemade and imported chloride pollution and ask for a stand-still commitment by the Netherlands in regard to homemade pollution.

10. VEWIN, the organization of Dutch drinking-water companies, argued that dropping the issue would save the Dutch government 100 million guilders, but that drinking-water companies would then have to invest around 300 million guilders to comply with public health requirements of 120 mg/l sodium and 150 mg/l chlorides.

11. By exchanging calcium ions for natrium ions, the calcium concentration of the water was to be reduced. Without chloride (natrium-chloride) reductions upstream, the investment would have been at risk because the purified water would have contained too much natrium (Dieperink 1992: 21–22).

12. Downgrading the chloride problem to a drinking-water issue was also possible because the Dutch ministry of agriculture had long before 1988 withdrawn its support for the negotiations. It considered that the high chloride concentration of the Rhine might have been a problem for some sectors of Dutch agriculture. But it concluded that the problem had in the meantime been resolved through adaptation measures in the Netherlands, such as improved irrigation and pollution control methods. Henceforth, the chloride issue remained in the hands of the Dutch ministry of transportation and public works.

13. MdPA can retain between 42 and 56 kg/s, in addition to the 15 kg/s it already stockpiles, when working at normal capacity. It can thus reduce its emissions by up to 71 kg/s if necessary; that means, more than the fixed 60 kg/s reduction originally envisaged. On the other hand, the modulated reduction will reduce the salt load of the Rhine and the annual average chloride concentration much less than the previously planned 60 kg/s reduction at a fixed rate.

14. MdPA will start the modulated reductions one year after all parties have paid the first of their yearly contributions to the French government. Subsequent payments must be made in full each year by 31 January. France and MdPA are not bound by their obligation until all parties have paid their share, unless money from previous years is left over and can be used. Stockpiling may also be discontinued if at any point the money deposited with the French government is insufficient to cover the expenses in accordance with the fixed prices for investments and units of stockpiled and discharged salt waste. The payment schedule is as follows: 1991, FF 90 million; 1992, FF 38m; 1993, FF 27m; 1994, FF 73m; 1995, FF 36m; 1996, FF 36m; 1997, FF 50m; 1998, FF 50m.

15. MdPA, for example, has stated that its emissions have remained below the permitted level since 1988. Data by Henry Schreiber, MdPA, 1993.

16. See Sebenius 1983.

17. Of course, there might be a trade-off between transaction costs, in terms of the costs of negotiating agreements, and transaction costs, in terms of implementing these agreements (hastily negotiated agreements might be incompletely implemented or might be more unstable). However, given that the contractual environment in the chloride case has been rather benign and implementation of reductions, once agreed, has not been a significant problem, quicker agreements might not have been more unstable or less effective in the chloride case.

8

Nuclear Safety in Eastern Europe and the Former Soviet Union

Barbara Connolly and Martin List

On 26 April 1986, at block 4 of its Chernobyl nuclear power plant, the Ukraine was rocked by the largest accident so far in the history of the civilian use of nuclear power. The accident, blamed on operator error during an attempt to test the reactor's response to freak power surges, led to sudden and uncontrollable increases in power that destroyed the reactor along with the building it was housed in and caused a raging fire that defied control for several days and released radioactive material into the environment. Official estimates put the human costs at thirty-one casualties of plant staff and emergency crew working to contain the accident (Commission of the European Communities 1989: 3). More than 60,000 people had to be evacuated from the area originally, and since then more than 200,000 people have lost their homes due to long-term radiation exposure (Herttrich, Janke, and Kelm 1994: 89). Damages were not limited to the Ukraine. Persistent northeast winds spread the radioactive cloud across Western Europe and Scandinavia, contaminating agriculture and endangering human health. Indeed, because Soviet authorities at first tried to conceal the accident, the West first detected its occurrence after measurements of air radioactivity levels in Sweden indicated that an accident must have occurred somewhere. The long-term health consequences and the ensuing death toll caused by released radiation will continue to be studied and contested for years to come. More than any other single event, the disaster at Chernobyl brought the issue of the safety of nuclear power to new heights of international political concern.

Chernobyl was not some freak accident. Subsequent revelations regarding the condition and operation of Soviet-built nuclear reactors throughout Eastern Europe and the former Soviet Union suggest that this was an accident waiting to happen, and indeed, that more such accidents are possible unless urgent action is taken. This chapter examines the international response to the global risk posed by the failing safety of nuclear reactors in Eastern Europe and the former Soviet Union. We will show that the huge risk emanating from continued operation of some of these reactors, although a liability to civilian populations in countries throughout Europe and to the continued political viability of nuclear power in the West, also created tremendous opportunities for the Western nuclear power industry, providing welcome relief from stagnant markets at home in most countries. The nuclear safety issue has given rise to a fierce political battle between advocates of drastic measures to ensure that nuclear power is run safely in the East, or not at all, and commercial interests seeking to capture potentially huge profits by promoting the continued operation and expansion of nuclear power in Eastern Europe and the former Soviet Union.

Pressed into action by public concern that they do *something* about the problem, Western governments have contributed substantial sums of assistance to alleviate the nuclear safety risks in the East. We argue that existing patterns of Western assistance to the East, no matter how justified by the urgency of the nuclear risk, seem likely to lead to unintended long-term consequences in prolonging a still dangerous situation that may ultimately outweigh the short-term successes this assistance has been able to achieve. In particular, we argue that the assistance effort has been very much captured by the particularistic interests of the nuclear power industry. It would not be overstating the point to suggest that many of the Western governments underwriting nuclear safety assistance have succeeded more in subsidizing their own nuclear companies than in genuinely resolving Eastern safety issues. The Eastern governments, for a variety of reasons, also exhibit a clear pronuclear bias that strongly colors the international bargaining over solutions to the nuclear safety dilemma.

The nuclear lobby's capture of this issue at the very early stages when the problem was first being defined has certain implications that do not bode well for improving nuclear safety. First, it led Western providers of assistance to jump into an effort to conduct short-term upgrades on some of the most dangerous nuclear power plants in the region without long-range planning, supposedly so these reactors can operate at lesser risk for a few years while alternate energy supplies are established, and then be closed down. Western governments, however, have failed to secure adequate guarantees or elaborate reliable strategies to ensure that Eastern governments will indeed shut down these reactors after the interim period, and in fact, the plant improvements may have the unintended effect of prolonging the existence of the very plants that Western governments most wanted to see shut down.

Second, the intense competition among nuclear suppliers for contracts in the East, both for the larger retrofitting operations at existing nuclear plants and for contracts to complete new nuclear reactors, has created a failure of collective action among Western donors. Torn between their desire to solve the safety problems in the East and political pressures to support ailing domestic nuclear industries, several Western governments have bowed to commercial interests and agreed to provide financial and technical assistance for nuclear projects in Eastern Europe and the former Soviet Union without conditionality directed at lasting solutions to the safety problems. As long as some major donors are willing to provide unconditional aid, the international community loses its principal source of leverage in bargaining for commitments from Eastern governments to shut down the most dangerous nuclear reactors.

Third, the generally pronuclear orientation of the problem-definition phase in nuclear safety assistance has resulted in the neglect of other, potentially cheaper and more effective solutions to the nuclear safety problem. Although it is widely recognized that the Eastern economies are much more energy intensive than their Western counterparts and thus offer enormous potential for profitable improvements in energy efficiency, most of the nuclear safety debate has turned on the question of how to increase energy supply rather than how to reduce demand. Some of the countries running Soviet-built nuclear power plants even have enough excess energy capacity to export electricity to their Eastern and Western neighbors. The narrow focus of the donor community on the supply side of the energy sector to the neglect of the demand side is likely to make solving the nuclear safety problem much more expensive than it needs to be, and may even result in a pattern whereby electricity consumed in Western Europe is imported from nuclear reactors in Eastern Europe that are inferior to Western standards of design and operation.

1   Description

To see how and why the Western nuclear safety assistance effort has landed in this unsettling state of affairs, we first take a look back at how the most important Western and Eastern actors understood the issue of nuclear safety in the East and what sort of solutions they resolved to apply. The most obvious dimension of the problem exposed by the Chernobyl accident was the environmental risk posed by the continued operations of some fifty-nine remaining Soviet-designed nuclear reactors in Eastern Europe and the former Soviet Union. Although Chernobyl had occurred three years earlier, the 1989 revolutions across Eastern Europe created the first real opportunities for Western governments to address their concerns about the safety of Soviet-built nuclear plants in an institutionalized fashion. The new openness led to surprising new revelations about how widespread the nuclear safety problem really was. Starting in late 1990, the International Atomic Energy Agency (IAEA) launched a series of fact-finding missions to Eastern Europe and the former Soviet Union, initially focusing its efforts on inspections of the oldest generation of pressurized-water reactors (PWRs) in the region. The conclusions of the IAEA inspection teams startled the international community and prodded Western governments to make nuclear safety a priority within their assistance efforts to the East.

There are two types of Soviet-designed reactors: RBMK and VVER. These reactors were designed in multiple generations and vary considerably in safety level. The RBMK design, not used in any Western nuclear reactors, was chosen by the Soviets primarily for military purposes, because it facilitates access to weapons-grade nuclear material. For this same reason, no RBMK reactors were exported outside the Soviet Union. Following that country's breakup, however, three states now claim RBMK reactors: Russia, Ukraine, and Lithuania. Several of the design features of the RBMK, notably the absence of containment to prevent the release of radioactive materials should an accident occur, the use of flammable graphite as a moderator, and finally what experts call a positive void coefficient that creates the possibility of a runaway chain reaction if the coolant evaporates, have led many—although not all—Western experts to conclude that RMBKs can never be brought up to adequate safety levels, even through extensive upgrading.

The Soviet VVER design was the focus of initial IAEA and other Western assessments. VVERs are pressurized-water reactors, a design type that is commonly used in the West. The Soviets built three generations of VVERs. The oldest of these, the VVER-440/230, was built prior to 1974. The IAEA's report on these reactors stated that "All of these units lack safety features basic to other PWRs. The weaknesses include limited emergency core cooling capability, insufficient redundancy and separation of safety equipment, deficient instrumentation and control systems, insufficient fire protection, and the lack of a containment to enclose the reactor systems" (IAEA 1992: 1). Experts consider this generation of VVERs, along with the RBMKs, to be the most dangerous nuclear reactors in operation in the region, and generally recommend that they be shut down as soon as feasible. The second generation, VVER-440/213, was built between 1974 and 1980 and has remedied many, although not all, of the design deficiencies of the earlier model. Experts deem the 213 models upgradeable to Western safety standards. The most modern generation of Soviet pressurized-water reactors is the VVER-1000. The IAEA states that this design "is similar to that of non-Soviet plants in operation worldwide and includes a full containment structure. However, some concerns related to design and operational problems remain, even for the more advanced 1000 MW units, mainly about core safety and instrumentation and control" (IAEA 1992: 6). Most experts believe that redesigns of the reactor core and instrumentation and control systems will enable this reactor type to approximate Western safety levels.

Besides the design flaws, Eastern nuclear power plants (NPPs) are plagued by a host of operational safety problems that carry across all reactor types. For example, the IAEA found that plant organizational structures were often overly bureaucratic, with unclear and fragmented lines of authority for plant performance. Operating procedures were found to be inadequate and there were either no adequate written procedures for emergency and abnormal operations or else plant management did not enforce the use of existing procedures. IAEA inspectors determined that control rooms were too meagerly staffed to safely manage emergency conditions. Finally, in several cases the material condition of plant equipment had been so poorly maintained as to jeopardize the function of essential safety mechanisms (IAEA 1992: 40–45 and Hoensch 1993: 130–133). Practitioners speak of this cluster of operational failings as an absence of "safety culture" in the East.

Most of these problems existed well before the West knew much about them. The safety problems were compounded, however, by the pullout of Russian technicians from their former Soviet satellites starting in 1989. All of the Eastern countries, with some degree of variation, depended on the Soviet Union for the supply of nuclear technology and fuel, as well as for the handling of spent fuel.[1] Some of these countries were left in particularly dire straits when Russian crews left. At the Kozloduy plant in Bulgaria, for example, the remaining local personnel was left without adequate operating manuals or experience. Similarly, the blueprints for the Lithuanian NPP at Ignalina were available only in Moscow. Both technically and in terms of operating skills these "orphan countries," as one EC official called them, were thus in need of outside assistance.[2]

Competing Definitions of the Problem

The design flaws of the Soviet-built nuclear reactors, coupled with serious deficiencies in safety culture, undeniably heighten the risk of another nuclear accident of Chernobyl proportions in the East. As they are, these reactors impose serious environmental risks to all of Europe. This is the central element driving concern among Western publics, which translates into political pressure on Western governments to provide nuclear safety assistance to Eastern Europe and the former Soviet Union. It is also the only element on which virtually all Western donors can agree. At that point the consensus breaks down.

What is most interesting—and problematic—about the nuclear safety issue is that there is not one but rather several competing notions of how the problem should be understood among the most important actors. The specter of another Chernobyl in the East puts the political viability of nuclear power in jeopardy in the West as well. Nuclear energy already has a serious image problem; to preserve its long-term prospects, the industry needs to prove to concerned publics that nuclear energy is safe. Improving the safety of operational nuclear reactors in Eastern Europe and the former Soviet Union is thus a task that many nuclear companies and utility owners perceive as part of their own long-term self interest.

The nuclear industry enjoys staunch support from the government in many countries where nuclear energy figures prominently in electricity generation—for example, France, Belgium, Germany, and the United Kingdom. These governments themselves are greatly concerned with preserving the nuclear option at home. Both nuclear industry actors and the pronuclear governments thus find themselves in a rather awkward bind: pressing public concern pushes them to do *something* to prove that they are responding to the dangers of nuclear power in the East, but at the same time they need to respond in such a way as to avoid any further appearance that nuclear power might be unsafe. In addition, governments in almost all the major nuclear states encounter political pressure from the nuclear companies themselves, who have been severely hurt by stagnating markets for their wares at home. In the United States for example, the last time a nuclear plant was ordered without subsequently being canceled was 1974. In Western Europe, every nuclear country except France now has a moratorium on new construction of nuclear plants, either official, as in Britain, or de facto, as in Germany and Spain. Even in France, new orders have slowed down dramatically (*Economist*, Nov. 21, 1992: 21). For Western nuclear companies, Eastern Europe offers the promise of new and expanding markets. Domestic nuclear industries have been clamoring for government support in their bid to capture a piece of the action in the East.

Other Western governments see the problem differently. Governments that have undertaken political commitments to dismantle their own nuclear programs, or never to develop one, are more inclined to interpret the safety problems in the East as yet another nail in the coffin of nuclear power. Austria is the most vocal spokesman for this position. Addressing the international nuclear establishment at a 1986 IAEA meeting, Austria's foreign minister preached, "For us the lessons from Chernobyl are clear. The Faustian bargain of nuclear energy has been lost. It is time to leave the path pursued in the use of nuclear energy in the past, to develop new alternative and clean sources of energy supply and, during the transition period, devote all efforts to ensure maximum safety. This is the price to pay to enable life to continue on this planet" (Flavin 1987: 64). For those states that have renounced nuclear power, although they are fewer in number and less influential than the major nuclear powers, the nuclear safety problem is best understood as an imperative to dismantle nuclear facilities in the East on the way to environmentally safe, sustainable development.

The position of Eastern governments could not provide a starker contrast. These governments, along with their utility operators, do not generally share the West's concern about the safety of the Soviet-built nuclear reactors. Although foreign assistance programs have begun to make inroads in raising concern in some countries, the reality is that Eastern governments and their mostly state-owned utilities are much more concerned with securing cheap energy supply, and for some states an export potential as well, than they are with the safety deficiencies of existing nuclear reactors. Nowhere is this disdain for Western safety concerns more pronounced than in Russia, the only Eastern country aside from the Czech Republic with its own nuclear supply industry and the design source of all NPPs in the East except for those in Romania.[3] The nuclear establishment continues to hold an influential position in Russian politics.[4] The fact that members of the Russian nuclear establishment remain psychologically attached to the technology they have developed on their own increases their resistance to recognizing flaws in the Soviet reactor designs. Indeed the Russian nuclear industry has much to lose, in terms of reputation as well as nuclear export potential, from Western assistance programs that operate under the assumption that Soviet-designed reactors are inherently less safe than Western designs. Although nuclear energy currently comprises a relatively small part of electricity production—about 12 percent—policymakers are officially committed to continuing rapid development of Russia's nuclear program, including construction of new nuclear reactors, to maintain technological competence and for military reasons (European Parliament 1993: 14).

Outside of Russia, the other Eastern governments and utilities also downplay nuclear safety problems and instead emphasize a variety of energy and economic constraints. Some of these countries experience very severe energy constraints and are therefore caught between the nuclear risk and an unacceptable social and economic situation. For example, the two RBMK reactors at Lithuania's Ignalina NPP furnish at least 50 percent of the country's electricity, some of which represents excess capacity. Bulgaria's Kozloduy nuclear plant supplies more than 31 percent of the country's electricity, and 50 percent of Hungary's electricity comes from nuclear energy. Armenia faces a dire energy situation, because of the disruption of its imports of oil, gas, and electricity by hostile neighbors and regional conflict. Lacking any indigenous energy resources except hydroelectric power, and observing how its inability to meet domestic energy demand has only aggravated the country's economic and social crises, the government may well succumb to pressure to reopen the country's one nuclear plant at Armyanskaya, closed in 1989 because of seismic risks (European Parliament 1993: 12–14). For all of these countries, the choice for the near term is to rely on nuclear power or to impose untenable electricity rationing.

Other Eastern countries have a somewhat more flexible energy situation. Nuclear energy accounts for 25 percent of electricity production in Slovakia, 24 percent in Ukraine, and 20 percent in the Czech Republic (European Parliament 1993: 12–14). A key policy objective for these countries, as well as for those that rely more heavily on nuclear power, however, is to reduce their characteristic dependence on energy imports from Russia. Reorienting toward new import supply sources will require long lead times to build the appropriate infrastructure, such as pipelines for oil and gas. Critical shortages of foreign exchange encourage all of these countries to rely as heavily as possible on their very limited indigenous energy supplies. Several of these countries have substantial reserves of domestic lignite, but they are reluctant to further exploit these resources because of the environmental consequences of burning poor-quality lignite, including very high emissions of sulfur dioxide, nitrogen oxides, and ash. Diversification and security of supply concerns thus increase the attractiveness of nuclear power.

For all of these countries, existing nuclear plants provide a cheap source of electricity, given that the capital costs have long since been paid. For a few of them, including Lithuania, Ukraine, and in the near future Slovakia as well, nuclear plants provide the capacity to export electricity surpluses to neighboring countries in exchange for hard currency. Given the objectives of diversification and security of supply, least-cost supply, and export potential, nuclear energy is the favored choice of policymakers throughout the region. All of these factors conspire to make governments in Eastern Europe and the former Soviet Union bent on keeping their nuclear reactors in operation through the end of their natural design lifetimes.

Safety Assistance: Something for Everyone
With these very different views about why and for whom nuclear safety in Eastern Europe and the former Soviet Union poses a problem, not to mention whether it is even a problem at all, it comes as little surprise that the dominant actors in this issue have pushed different solutions. At their 1992 Munich Summit, the G-7 countries agreed, although with differing enthusiasm and emphases of motives, that the West would need to provide financial assistance to tackle the problem. Led by Germany and France, the G-7 called for Western donors to mobilize $700 million in grant funds for emergency repairs to the most dangerous nuclear plants in the East. Convinced that the safety problems demanded an urgent and coordinated Western response, the countries called for a five-point program of action including operational safety improvements, near-term technical improvements based on safety assessment of each reactor, improvement of the nuclear safety regulatory systems in each country, examination of the potential for upgrading the more modern nuclear plants, and examination of the scope for replacing less-safe plants by the development of alternative energy sources and the more efficient use of energy (Commission of the European Communities 1993: 9). What the 1992 Munich Summit accomplished was to place nuclear safety in the East high on the international political agenda. The call for action, however, masked considerable disagreement among Western donors about what should be done.

Launching a high-profile aid program, it seemed, was good for everyone. Donor governments would thereby give the impression to concerned publics that they were indeed taking the safety problems seriously. With grant money on the table, recipient governments in Eastern Europe and the former Soviet Union were getting something for nothing and could hardly complain. The Western assistance program came as particularly good news to business. In fact, nuclear suppliers such as Westinghouse (based in the United States), Germany's Siemens-KWU, the Swedish-

Swiss-German multinational Asea Brown Boveri, and French state-owned Framatome had heavily lobbied their governments to provide extensive nuclear safety assistance to the East. Besides protecting the nuclear industry from yet another debilitating accident, the assistance program would deliver welcome contracts for nuclear suppliers. Moreover, the fact that Western governments were paying for the emergency retrofitting work removed the financial risk of getting involved in the East. Most important, these companies saw the short-term retrofitting contracts as a way to get a foot in the door for much more profitable ventures in the future, including contracts for major retrofitting work on the most modern nuclear plants and for completing the large numbers of partially built nuclear reactors in the region. Official government assistance provides a politically acceptable way for governments to subsidize their domestic nuclear industry, helping them to gain competitive advantage in the new Eastern markets.

The main lines of disagreement dividing the Western donors were over how to handle existing nuclear reactors in the East over the medium to long term. Some Western governments have argued vehemently that the most dangerous nuclear reactors in Eastern Europe and the former Soviet Union (that is, the VVER 440/230s and the RBMKs) must be permanently shut down as quickly as possible. Germany and Austria both represent this position. Following the IAEA's June 1991 inspection of the Kozloduy NPP in Bulgaria, home to four VVER 440/230 reactors, the German Federal Minister of Environment and Nuclear Safety Klaus Töpfer called for a crash international program to finance replacement power to allow the old reactors to be shut permanently. (*Nucleonics Week*, July 11, 1991: 1). In fact, the Germans had implemented this same solution at home, by shutting down the VVER-440/230 reactors at Greifswald which they had inherited through reunification, citing the impossibility of retrofitting to reach adequate Western safety standards in any economical fashion. Töpfer has vowed that no German aid would be given for power sector projects unless recipient states made firm commitments to shut down unsafe RBMKs and the oldest generation of VVER reactors.

Austria has taken an even more extreme position, opposing any and all nuclear power projects in its Eastern neighbors. The Austrian government, prodded by intense NGO lobbying, has urged the Slovak government to shut down its Bohunice NPP, which also houses two of the oldest generation VVER reactors. The government tried to mobilize international political pressure by sending Austrian experts as part of an international inspection of the plant, but even though the Austrian delegation recommended that the plant should be closed, the other three expert delegations from Siemens, the IAEA, and Westinghouse concluded that the two VVER-440/230 units could reasonably be retrofitted for operations at least until 1995 (interview with IAEA official, 1 June 1992, Vienna). The Austrian government has also strongly protested completion of work on the Czech Republic's Temelin NPP, site of the most recent generation of Soviet pressurized-water reactors, the VVER-1000. In this case, the U.S. firm Westinghouse won a contract worth some $317 million to redesign Temelin's fuel assemblies and to install a new instrumentation and control system, which many experts believe will be sufficient to bring this new reactor up to Western safety standards upon its completion. For Westinghouse, the contract represents a major breakthrough in the markets of Eastern Europe, as it will likely be able to parlay its expertise for more contracts on the many unfinished VVER-1000 reactors in the region. The Austrian government has done its best to break up the deal, including sending a high-level delegation to Washington in February 1994 to protest the U.S. Export-Import Bank's loan guarantee on the financing for this project (*New York Times*, Feb. 23, 1994: A5).

The drawback of the German and Austrian positions, and others like them, is that success in resolving the nuclear safety problems hinges on obtaining credible commitments from Eastern governments to shut down their most dangerous nuclear plants. Without such commitments, the most serious nuclear safety problems remain essentially untouched. Demanding closure of the most dangerous nuclear plants, however, requires that donor governments be willing to present a united front to Eastern aid recipients. Divisions within donor governments about whether they can realistically achieve closure of these reactors, or about whether commercial opportunities for their own nuclear industry outweigh the potential safety gains of concerted action in imposing conditionality, undermine the chances that Eastern governments will concede to any Western demands for plant closure.

Several Western governments have taken a more pragmatic approach, noting that energy constraints imply that many of the most dangerous reactors will likely continue in operation for several years. For these countries, the goal of assistance is to increase safety levels in the interim period until alternate capacity can be arranged. Among the advocates of this position are France, the United Kingdom, and Sweden. For example, the director of the French nuclear regulatory agency DSIN, Michel Laverie, has stressed a more long-term approach to resolving the nuclear safety issue that would focus both on short-term safety analyses and on longer-term actions that would create the material conditions necessary so that older plants could be shut down. Laverie argues that the best way to get the older reactors shut down is to help Eastern utilities start up their VVER-1000 reactors, some of which are already operational while many others are partially constructed, under good conditions (*Nucleonics Week*, Oct. 24, 1991: 9). Not coincidentally, the French position of accepting that we will have to live with existing nuclear plants in the East for a long time to come and focusing on bringing the newest reactors into operation also offers the best business opportunities in the East for the stated-owned French utility, Electricité de France (EdF), and its nuclear supplier, Framatome. It also presages a further, though perhaps safer, expansion of nuclear power in Eastern Europe and the former Soviet Union.

The Swedes have taken a similarly pragmatic position in their bilateral assistance efforts, concentrating on adopting Lithuania's nearby Ignalina plant. The Ignalina NPP houses two RMBK reactor units and produces at least half of Lithuania's electricity. Because of this high level of dependence, the Swedes quickly concluded that immediate shutdown of this plant would be out of the question, and thus proceeded from the assumption that these two reactors would remain operational for at least ten more years (Swedish Nuclear Inspectorate 1992). Figuring that somewhat higher safety levels would be preferable to the status quo, Sweden therefore set out to improve safety at the nuclear plant for the interim period. The Swedes granted assistance to strengthen Lithuanian regulatory authorities, to train plant operators in nondestructive testing of materials, and to develop technical improvements in the reactor itself, mainly relating to improved fire protection.

Although small in volume—about $13 million[5]—(Swedish Nuclear Inspectorate 1993: 5) Swedish assistance to Lithuania assumes greater significance as a precedent. Prior to the Swedish work at Ignalina, the Western donor community had generally steered away from safety retrofitting work at RBMK reactors. Based on safety analyses conducted on these reactors and on recognition of their critical design weaknesses, most Western donors had insisted that the RBMKs simply could never be brought up to Western safety standards and therefore should not be touched except to shut them down. Now, however, the Swedish work at Ignalina has catalyzed a $36.6 million grant from the European Bank for Reconstruction and Development's multilateral nuclear safety account, earmarked for hardware improvements at the Ignalina RBMKs. By accepting that they would have to live indefinitely with existing RBMK reactors in Lithuania and then setting out unilaterally to conduct what short-term safety improvements were possible at the reactors, Sweden may have in fact ensured that these reactors would remain in operation for a long time to come, when other options were possible. They might, for example, have directed their bilateral assistance toward capturing large potential improvements in energy efficiency that could have reduced the Lithuanian dependency on nuclear power. They might have worked on providing an attractive financial package that would enable the Lithuanians to construct alternate plant capacity. Instead, the seemingly pragmatic Swedish effort placed the energy-efficiency and supply issues on the back burner and gave a (probably unintended) signal that perhaps it is possible to live with these RBMK reactors after all.

In keeping with their lower concern for the safety risks posed by their own nuclear reactors, the general response of Eastern governments and utilities has been to accept whatever Western assistance is offered on grant terms, but to resist donor attempts to attach any conditionality to that aid. Some governments have gone farther than others in their resistance to perceived Western "interference" in their nuclear programs. In Russia and Ukraine, for example, the European Union was unable to provide on-site assistance for the first two years of its Technical Assistance to the Commonwealth of Independent States (TACIS) nuclear safety program because of the unwillingness of Russian and Ukrainian officials to supply commission officials with important construction and operational plans to allow the preparation of safety studies and the conduct of safety work on nuclear plants (*Official Journal of the European Communities* 1993: 183). The Russian attitude, contemptuous of the commercial interests driving Western assistance programs, was to expect EU member states to simply hand over the requisite grant money and let the Russians do upgrading work themselves. Consequently, the EU limited its first two years of assistance to funding generic safety studies of Soviet-designed reactors as well as training programs and support for nuclear safety authorities.

Other Eastern governments have been more cooperative but still strongly resistant to any Western attempts to dictate energy policy in the region. For example, Bulgaria holds the dubious honor of possessing the first NPP (Kozloduy) for which the IAEA ever recommended immediate shutdown until extensive safety retrofitting operations could be conducted. Bulgaria has welcomed extensive on-site assistance from multilateral and bilateral sources, and the Bulgarian nuclear regulatory authority has proved very receptive to upholding much higher safety standards than were in practice when the IAEA first inspected the Kozloduy NPP in 1991. In fact, Bulgaria has received more nuclear safety assistance than any other East European country, due to the extreme urgency of the safety situation at Kozloduy. Despite accepting large amounts of Western grants, the Bulgarian government still insists that it cannot and will not close the four VVER-440/230 reactors at Kozloduy until alternative supply capacity has been created, with responsibility for securing adequate financing falling largely on Western shoulders. In sum, the dominant Eastern response is to accept whatever nuclear safety assistance the West is willing to provide unconditionally and in grant form, but otherwise to proceed with business as usual in relying on existing nuclear plants as a cornerstone of domestic energy policy.

Western Assistance Commitments: Who and How Much?
Within the bounds of this very limited consensus over how to handle nuclear safety problems, Western governments had committed about $874 million in grants for nuclear safety assistance to Eastern Europe and the former Soviet Union by the end of 1993[6] (*Nucleonics Week* Dec. 9, 1993:8; Commission of the European Communities 1993: 10). These re-

sources are being channeled to Eastern governments through bilateral aid programs and multilateral funds, including the European Union's PHARE and TACIS programs and the EBRD's nuclear safety account. Virtually all of this money, except for the nuclear safety account's grants, has gone for "software" or technical assistance, such as safety studies, training, preinvestment studies, and regulatory or legislative assistance. Additional forms of assistance not noted in these monetary totals include technical assistance provided by the IAEA under extrabudgetary programs and twinning programs run by a consortium of Western nuclear plant operators, whereby operators from Eastern and Western plants exchange experiences and expertise.

As the only institution with a prior mandate in nuclear safety as well as the relevant expertise and the necessary geographical reach, the IAEA emerged as a central actor in the nuclear safety issue following the Chernobyl accident. In fact, IAEA Director-General Hans Blix headed the first international on-site inspection of Chernobyl on 7–8 May 1986. IAEA experts, however, did not gain access to Eastern NPPs to assess their operational safety and design features until 1990. Since then the agency has launched or completed safety studies of each of the four main Soviet designs, paid for by voluntary contributions from IAEA member states. The IAEA is the recognized industry expert on technical issues and the principal forum in which research takes place on whether the safety flaws of Eastern NPPs can be remedied, and what technical and operational measures would need to be taken to compensate for prior design flaws.

In terms of monetary contributions, the European Union is far and away the dominant actor in the nuclear safety field. By the end of 1993, the EU had committed $560 million to nuclear safety in Eastern Europe and the former Soviet Union, nearly two-thirds the total amount of Western assistance. The EU thereby set the early standard for how assistance would be provided. The Union channels nuclear safety assistance to Eastern governments through its PHARE (for Eastern Europe) and TACIS (for the former Soviet Union) programs, which have the much broader missions of providing grant assistance for overall political and economic restructuring in the region. Funds committed under PHARE and TACIS overwhelmingly took the form of technical assistance, which many critics characterize as blatant subsidies to consultants from member states' nuclear industries. The Union does not attach political conditionality to its provision of grants; it takes the position that determination of national energy policy remains the responsibility of each recipient government.

On-site assistance to improve operational safety at nuclear plants has dominated PHARE programming in Eastern Europe. For example, PHARE has supported extensive on-site assistance to the Kozloduy NPP from 1990 to 1993, which included engineering studies to identify and solve safety problems at units 1–4 (the VVER 440/230 models at the site), a program to solve urgent housekeeping problems and improve maintenance at the plant, cooperation with the Bugey NPP in France whereby French operators came to Kozloduy to share their expertise, strengthening of Bulgaria's regulatory body, and a study of the Bulgarian electricity supply system (IAEA 1992: 13). PHARE has also paid for studies and on-site assistance for the Bohunice NPP in Slovakia, safety studies of the Paks NPP in Hungary and the Dukovany NPP in the Czech Republic, supply of safety equipment and technical expertise to the Ignalina NPP in Lithuania, and a regional program of assistance to nuclear safety authorities (Commission of the European Communities 1993: 1–8).

Limited by the reluctance of Soviet authorities to allow Western European access to their nuclear stations, most TACIS funds through the end of 1992 went for generic safety studies of the main reactor designs. The program has also included training programs and support for safety authorities. In 1992, for the first time, TACIS funded on-site assistance to Russia and Ukraine that was directed at both human resources and plant safety equipment. For example, on-site action in Russia involved providing technical assistance to develop operating procedures, to carry out nondestructive inspections, to improve fire protection and training, as well as supplying basic safety equipment. On-site assistance in Ukraine took the form of staff training, review and improvement of operational procedures, setting up better preventive measures against accidents such as fire or power failures, and assessment of possible deficiencies of plant components. TACIS also paid for provision of basic safety equipment (Commission of the European Communities 1993: 1–8).

On examining the EU's nuclear safety assistance, three striking characteristics are evident: the domination of the program by professional (nuclear) organizations, the overwhelming preponderance of technical assistance, and the absence of explicit conditionality in the provision of assistance. Because of the technical nature of the nuclear safety problems, EU experts take more of an administrative than a technical role, and instead contract out to specific (always Western) professional organizations for program monitoring, evaluation, and preparation of technical specifications for projects, which perhaps lends weight to the impression that the program is more about subsidies to the Western nuclear industry than about safety. Member state operators and regulatory authorities are heavily involved in developing on-site actions. Although understandable in view of the technical complexity of the issue, what this means is that decisions about the direction of EU assistance are heavily influenced by nuclear industry experts, as opposed to generalists with a wider view of energy choices and a less-specific political agenda.

The preponderance of technical assistance derives primarily from the financial means available to PHARE and TACIS. Because member states provide these programs with limited amounts of grant funds, they are unable to take on the vastly more capital-intensive tasks of funding extensive provision or retrofitting plant safety equipment. Instead, the EU programs focus much more heavily on developing human resources, which makes use of the concentration of technical expertise among EU nuclear plant operators and regulators and also requires less capital. In fairness, then, the EU's central contribution should be an improvement in the safety culture among plant operators and regulators in the East, but the programs cannot be expected to make a substantial impact on the major issues of hardware design.

The third striking characteristic of the EU programs is that the Commission makes no explicit attempt to leverage the grant assistance it provides in exchange for commitments by Eastern governments to close down the most dangerous NPPs. Two possible explanations come to mind. First, PHARE and TACIS lack adequate financial inducements to pursue such a conditionality strategy. Given that Eastern governments simply are not as concerned about nuclear safety as Western governments, Western threats to withhold short-term safety assistance are neither credible nor significantly costly to decision makers in the East. In order to make compelling demands for nuclear plant closures in the East, EU member states would have to link those demands to commodities that

Eastern governments value highly—for example, access to EU markets or financing for other energy sector projects that would generate significant economic benefits, such as building new capacity.

Divisions among member states about how highly they value the potential safety benefits of concerted demands for plant closures in the East, as opposed to the political and commercial benefits of subsidizing their own nuclear companies' quest for competitive advantage in Eastern markets, constitute a second and deeper reason for the EU's apparent disinterest in strategies of conditionality. Although no guarantees exist that even unified EU demands for closure of the most dangerous NPPs in exchange for further grant-based assistance would prove successful, it is certain that once the Commission disburses grants to Eastern governments for short-term improvements in existing NPPs, those governments have less incentive than ever to defer to Western wishes about how to deal with those reactors in the longer term.

At a sum total of $198 million in firm commitments by mid-1993,[7] bilateral aid programs provide the second largest source of nuclear safety assistance. The main feature of bilateral assistance is that it is uncoordinated by design, despite the existence of the G-24 Nuclear Safety Coordination Unit, which ostensibly coordinates strategies among Western donors. This allows donors leeway to pursue quite different solutions, from the German vision of facilitating closure of the oldest reactors as quickly as possible, to the French and Swedish vision of accepting their indefinite operation and aiming for safety improvements on the margin until recipient governments restructure their energy policy.

Even more important politically, the lack of coordination allows each bilateral donor to guarantee the receipt of domestic commercial benefits in exchange for its aid contribution. In the protracted political debate over whether another multilateral fund (which became the EBRD nuclear safety account) should be created to mobilize increased Western assistance for nuclear safety in order to address the critical but capital-intensive problems of replacing or renovating plant safety equipment, the United States and Japan led a campaign of resistance, largely because they preferred to continue their direct bilateral aid programs which gave them more control over how the money was spent and more leverage in winning business contracts (*New York Times*, Jan. 29, 1993: A2). Even after the creation of the multilateral fund, these countries made only nominal contributions while maintaining and strengthening their bilateral programs. Lack of coordination and the underlying commercial battle of national industries virtually ensures the failure of any bilateral attempts to attach policy conditionality to nuclear safety assistance. Where one bilateral donor tries to impose particular conditions on its aid program, another donor will be willing to provide the same assistance without conditions in exchange for the promise of commercial returns, either now or in the future.

Because of their generally small size and short time horizons, bilateral programs are limited in what they can accomplish in the absence of coordination. For these reasons, most bilateral assistance falls under the category of "technical assistance," including studies, training, seminars, and visits and exchanges by experts. Like EU assistance, then, we might expect that bilateral donors would make their strongest marks on the human resources or "safety culture" side of the nuclear safety problem.

A relative newcomer to the nuclear safety assistance community, the EBRD's nuclear safety account became operational in April 1993 with initial pledges of 115 million European currency units (ECU) (nearly $140 million) from thirteen donors. Donors created the account as an autonomous entity within the EBRD, specifically to manage Western grant assistance to Eastern Europe and the former Soviet Union for nuclear safety. The nuclear safety account stands out in two respects. First, in contrast to the rest of nuclear safety assistance, which is dominated by "software" or technical assistance, the EBRD account specifically targets hardware problems. It is the only multilateral source of grant funds for the purchase of essential safety equipment.

Second, the nuclear safety account is also the only forum that has attempted to apply concerted policies of conditionality. To date the account has allocated only two grants: 24 million ECU (about $27 million) to Bulgaria's Kozloduy NPP for the purchase of safety equipment including fire protection devices, vessel inspection equipment, safety valves, electrical components, a new emergency feedwater system to cool the reactor, and safety parameter display systems for the control rooms (EBRD 1993), and 33 million ECU (about $36 million) to the Lithuanian government for a similar range of equipment to improve safety levels at the Ignalina RBMK reactors. Before allocating either of these grants, the holders of the nuclear safety account insisted on conditional commitments from the recipient governments to close down the nuclear power plants of greatest concern. Bulgaria therefore agreed to close the four oldest reactors at the Kozloduy NPP: units 1 and 2 by 1997, and units 3 and 4 by 1998 *provided that* replacement capacity had been made available by that time. (*Nucleonics Week*, June 24, 1993: 13.) The Lithuanian government, in exchange for its grant, agreed that it would shut both RBMK reactors at Ignalina around 2004 (unit 1) and 2010 (unit 2) *if* alternate capacity can be installed (*Nucleonics Week*, Feb. 17, 1994: 1).

What gives the nuclear safety account bargaining leverage its position within the EBRD? If these countries renege on their commitments to shut down nuclear plants when energy constraints allow, they are likely to suffer reputation costs that will impair further access to EBRD financing, particularly for energy sector projects, in a situation where the EBRD promises to be a leading source of capital to the region. Ongoing EBRD negotiations with Slovakia, which is seeking financing for completion of a new NPP at Mochovce, make explicit use of this bargaining power. The EBRD and most other involved parties in the West are conditioning financing for Mochovce on a commitment from the Slovak government to close the only two VVER-440/230 reactors in the country, Bohunice units 1 and 2, by around 1995, in contravention of the Slovak preference to backfit these units for longer operation (*Nucleonics Week*, Nov. 11, 1993: 1). This sort of conditionality might work, but only if the Slovaks cannot find unconditional sources of finance for the project, such as bilateral export credits from an eager supplier country. In the most recent round of negotiations on the proposed loan, the Slovak government requested postponement of a final EBRD decision shortly after announcing that the Mochovce completion could also be accomplished, more cheaply even, with the help of Russian loans and Czech (Skoda) technology.

Despite this rather large array of nuclear safety assistance programs and the fact that nuclear safety has drawn much more interest from Western donors than any other environmental issue, as measured by financial commitments, it is still the case that financial needs in the East overwhelmingly exceed available assistance funds. An influential study conducted recently by the World Bank, International Energy Agency, and the EBRD calculated the costs of shutting down all VVER-440/230 and RBMK reactors in Eastern Europe and the Soviet Union, replacing them with alternative energy supplies to make closure politically feasible, and conducting the necessary safety upgrades on the remaining nuclear plants to allow them to operate to the end of their design lifetimes. This study estimated the total investment costs only of this program to range from $21 to $23 billion over the period 1993–2000, depending on how quickly the nuclear plants are shut down (World Bank/IEA/EBRD 1993: 6). Incremental operating costs (that is, the cost of replacing nuclear-generated electricity with imports of gas, oil, or coal) would also increase significantly as nuclear plants are shut down more rapidly—an unattractive prospect for most of these countries which are already burdened by severe balance-of-payments difficulties. Although certainly Western governments cannot and should not shoulder the entire cost of such a strategy for closure of the most dangerous nuclear plants in the East, even the World Bank/IEA/EBRD study concedes that narrow economic rationality will cause Eastern governments to favor a strategy for continuing operations of their nuclear plants to the end of their design lifetimes, unless the West can provide sufficient financial or political inducements to alter their preferences. Existing Western financial commitments fall short by orders of magnitude.

Implementation: A Failure of Long-Term Planning?

In tracing the flow of nuclear safety assistance to the East from these various programs, we find no evidence that allocated funds have been in any way misused or diverted from their intended purpose. To be sure, PHARE, TACIS, and the nuclear safety account have received some criticism for their alleged slowness in disbursing assistance, although this is a generalized phenomenon in multilateral aid programs. Most of the serious questions to be raised about the nuclear safety effort, however, can be traced back to the initial stages in which the problem was defined and potential solutions proposed, or more precisely to the absence of a consensual, long-term strategy among Western donors for dealing with existing nuclear power plants in the East which has enabled companies working in the nuclear sector to dominate strategic planning of the assistance effort.[8]

In the implementation stage of the nuclear safety programs, only one major problem stands out, which is the real prospect that attempted policies of conditionality, applied primarily by the EBRD nuclear safety account and by some bilateral donors, will fail to achieve their aim of nuclear plant closures. The *Economist*, editorializing about the nuclear safety account grant for short-term upgrades at Kozloduy in exchange for the Bulgarian government's conditional commitments to shut four reactors at the plant several years later, expressed the dilemma most clearly: "Once the four unsafe units have been made a bit safer, Bulgaria may wonder why they should be shut down, especially as they will be producing extremely cheap electricity for an economy that may, by the late 1990s, have begun to recover" (*Economist*, June 26, 1993: 58). Why indeed, unless the Western investment community is sufficiently unified to condition Bulgaria's future access to bilateral and multilateral sources of finance on their reputation as a cooperative recipient of nuclear safety assistance.

The commercial battle among Western nuclear companies exhibits enormous potential to erode any unified donor front that the nuclear safety account has been able to erect. One example suffices to make the point. The French utility Electricité de France, according to its chairman Gilles Ménage, has joined forces with the nuclear safety account by threatening the Bulgarian government that it would pull out of its upgrading work at Kozloduy unless the government made a reciprocal commitment to a long-term safety and energy program including closure of Kozloduy's oldest reactors. Ménage accused the "American presence," apparently Westinghouse, of being a "disruptive factor" at the site, because it gives the Bulgarians "an excuse to get out of the requirements that we are trying to impose" on the Bulgarian National Electricity Company. Another EdF official claimed Westinghouse was courting Bulgaria, trying to gain competitive advantage by not demanding commitments for plant closure in exchange for aid. Said a senior EdF official, "The Bulgarians won't hear of closing [units 1 and 2]. They say that someone else is waiting to help them if we leave" (*Nucleonics Week*, Dec. 9, 1993: 10).

Subsidies at the Expense of Safety

If we consider whether the various actors involved in the nuclear safety issue have made progress in achieving their preferred solutions, one fact stands out: so far, the entire nuclear safety assistance effort has produced not one single closure of the very plants that sparked this expensive program of emergency aid. Even at Chernobyl, site of the accident that sparked international concern in the first place, two RBMKs of the same type as the one that was destroyed continue to operate.[9] Of course real energy and economic constraints made immediate closures impractical for most of these reactors. Continuing political obstacles, however, suggest that the status quo is more than temporary. Western actors who have lobbied for closure of the most dangerous reactors as quickly as possible, and therefore have justified not providing interim safety assistance to those reactors, confront the continued operation of these plants and their own inability so far to induce significant changes in the nuclear energy policies of Eastern governments. The more resigned approach of accepting continued operations of the most dangerous reactors, but then providing assistance for what safety upgrades are possible to at least reduce the risks of their operation, raises the somewhat different but still disturbing specter that these limited upgrades will unintendedly prolong the lifetimes of the most dangerous plants.

The real winners have been the Western nuclear companies and, to a lesser extent, the Eastern governments. Because the least common denominator of the various actors' preferred answers to the nuclear safety problem was a crash program of grant assistance, Western business concerns have essentially walked away with a highly profitable subsidy program, in addition to their improved prospects for new markets. Eastern governments, because they did not care enough about safety risks at their nuclear reactors to devote already scarce foreign exchange resources to unprofitable (in the narrow economic sense) safety upgrades, will nonetheless enjoy the modest safety improvements that Western assistance programs have been able to deliver, at very little cost and with no major changes thus far in their existing energy policy.

2   Analysis

This and the following section will examine several more specific hypotheses about why the nuclear safety problem has been defined and approached in the ways described above and will suggest where missed opportunities occurred or remain open for more effective solutions. In addition, we will call attention to the limited but not insignificant successes of the assistance effort in order to suggest possible lessons about what makes financial assistance for environmental problems more or less effective.

Industry Capture of the Problem Definition Phase

From the very beginning of the assistance effort for nuclear safety in the East, the nuclear lobby clearly dominated the process of conceptualizing and fashioning responses to the issue. As evidence, consider the conspicuous absence of one group of actors who might have played a key role: electricity consumers and purveyors of energy-efficiency technologies. It is a well-known fact that the economies of Eastern Europe and the former Soviet Union are much more energy-intensive than the West European average, and that potential economic gains through increases in energy efficiency abound.[10] If we know this fact, it appears strange that the nuclear safety problem was framed almost exclusively as a problem of energy supply, in a framework that basically assumed the necessity of achieving constant or increasing electricity production. An alternative understanding of the problem was possible. With their very large potential for savings in energy efficiency and temporarily repressed demands on energy production induced by recession, many (although admittedly not all) Eastern countries possessed a unique window of opportunity to institute demand-side energy savings programs that would in effect leave them with excess generation capacity, and thereby ease energy and economic constraints on retiring the more dangerous nuclear plants.

Why did nuclear supply interests so heavily dominate the framing of the issue? At least four factors contributed to this outcome. The reason that the issue was framed as a supply-side rather than demand-side energy problem appears most straightforward. The utilities and plant manufacturers that supply and operate nuclear plants, and generally nonnuclear energy technologies as well, are very large, well-organized, often state-owned and politically influential. By contrast, consumers of electricity are not at all organized politically in Eastern Europe and the former Soviet Union. Companies that sell energy-saving technology are much smaller than the supply industries, and consequently, carry much less political weight. Although progressive legislation can give utilities a profit incentive to pursue energy savings rather than always expanding supply, Eastern utilities are still trapped in the paradigm of supply-side planning and show little concern for saving energy. For Western industry, which lobbied heavily for the existence of aid programs, the potential profits in the East lie in building capital-intensive new energy capacity that would generate increased revenue streams, rather than in the much smaller and more diffuse energy-savings projects from which revenues would ultimately accrue to the Eastern consumer.

Once the nuclear safety issue was framed as a supply-side problem, why did the nuclear lobby manage to dominate over suppliers of other kinds of plants, such as fossil fuel–fired plants? In fact the situation is somewhat more complex. The same Western companies and utilities that supply and run nuclear plants usually deal in nonnuclear supply options as well, and certainly these actors have enjoyed opportunities in the East to build and retrofit coal-fired power stations as well as nuclear. The nuclear branches of these companies, however, clearly needed the business opportunities in the East to improve their profitability, to hold onto jobs for highly skilled workers, and to prove the long-term viability of nuclear energy in international markets. In addition, the nuclear establishment held an advantage within the Western donor community, because the major donors are all states with significant nuclear power programs at home. Some of the smaller donor countries have taken the nonnuclear route at home, but no powerful donor constituency exists to frame the nuclear safety problem in the East as an opportunity (or obligation) to phase out nuclear power.

The highly technical nature of the nuclear safety issue necessarily required high levels of dependence on expert judgments and dictated that nuclear engineers conduct evaluations of the reactors in Eastern Europe and the former Soviet Union and assess possible solutions. Not surprisingly, these organizations defined the problem technocratically rather than politically, recommending the desirability of a "technical fix." The political tightrope many Western governments were forced to walk by the combination of their own great reliance on nuclear power and the political pressure to take action regarding dangerous reactors in the East only reinforced the attractiveness of defining the problem as a narrowly technical one. By portraying the nuclear safety problems in the East as solvable through financial and technical assistance, these governments sent attentive publics the message that these reactors were indeed problematic enough to require Western assistance, but yet not so dangerous as to present a generalizable lesson that nuclear power might inherently involve unacceptable safety risks.

One more factor contributed to the very early separation of the nuclear safety issue from other aspects of energy policy and international assistance to the energy sector in the East: the institutional division of authority within Western governments and international organizations that traditionally separates those ministries or directorates dealing with nuclear energy from other energy and environment issues, because of the military implications of nuclear power. Although rooted in domestic politics, these divisions are replicated in foreign assistance programs. The tendency to see nuclear safety issues as separate from other aspects of energy policy implies a real loss of synergy among related subjects of assistance that might be addressed more cheaply as a cluster of overlapping problems than as entirely separate issues.

Technical Fixes and the Slippery Slope

It was not a difficult jump from understanding the nuclear safety problem as a primarily technical issue to deciding to undertake a substantial program of financial and technical assistance, particularly given that the crash aid program served the interests of all of the major actors. Western governments got to show they were acting responsibly in the face of a major environmental risk, and to win political points from the Western companies whose business benefited from government funding of the assistance program. The nuclear companies received subsidized introductions, at low political risk, to new and potentially very profitable electricity customers in the East. Eastern governments received free safety upgrades, with virtually no conditions attached; this was a program they could go along with even though nuclear safety was a much lower priority for them. How could taxpayers object to such a high-profile effort to address their safety concerns?

The real problem in the assistance effort is that although the main actors could all agree on the desirability of Western aid, they absolutely did not agree on how to balance commercial versus safety priorities in the assistance program, nor did they agree on the medium-to-long-term implications for what should be done about existing nuclear power plants in the East, even where safety was the primary issue. Economic competition among the major nuclear companies, as exemplified in the dispute between EdF and Westinghouse at Kozloduy, different levels of exposure to the environmental risk of a nuclear disaster in the East due to geographical location, and lower concern about safety in the East all contribute to deep divisions among the main actors about how to resolve the nuclear issues. These divisions, and the consequent inability of the donor community to collectively impose strong conditionality on Eastern governments, only open the door further to special interest lobbying by nuclear suppliers.

Political divisions among donor countries about the best longer-term solutions to the nuclear safety problem were probably inevitable. The "muddling through" style of the subsequent assistance effort, shaped by incremental decision making without long-term planning, surely was *not* inevitable. Western donors started providing nuclear safety assistance in the absence of any medium-to-long-term energy strategy. They expected that the recipient countries would establish these plans themselves, which they have not, in part because of unusually high levels of uncertainty about future energy demand, attributable to the unprecedented ongoing structural change in their economies. In addition, the EU's PHARE and TACIS programs, which provide the largest share of nuclear safety assistance, were set up as emergency assistance measures with considerable uncertainty about their duration. Seeking to address quickly the most pressing needs for Western assistance, the Commission allocated little time for longer-term planning in the initial phases of this emergency assistance program, which encouraged a bias toward incremental quick fixes. Two or three years into the assistance program, donors began to confront the need for energy planning more systematically, but meanwhile short-term solutions to the nuclear safety issue already dominated the array of possible responses. With the complicity of Western donors, the governments in the East were able to put off or avoid difficult political choices among competing energy objectives—safety, economy, and security of supply—and instead continue incrementally along existing energy pathways.

The absence of a medium-to-long-term strategy among Western donors for dealing with existing nuclear plants in the East has introduced a very serious risk that we call the "slippery slope" problem, which refers to the danger that incremental actions, designed to reduce safety risks at existing NPPs in the East for a specified interim period, may unintentionally prolong the operational lifetimes of the most dangerous reactors. Recall the *Economist*'s question about why the Bulgarians would want to shut Kozloduy down once the plant has been made safer and is producing cheap electricity. From the start, Eastern governments and utilities preferred to continue operations of existing nuclear reactors, while conceding the necessity of some retrofitting for longer-term operations. Conflicting preferences among Western donors and Eastern recipients potentially create a rather sinister dynamic: the more resources donors allocate for incremental risk reduction at operating NPPs, the worse the problem is likely to become that Eastern plant owners will resist closing these plants. Once grants have already been disbursed and implemented for safety improvements, recipients surely have less incentive to comply with earlier conditional commitments for plant closure. Even within the donor countries, political support for providing adequate assistance to shut down older nuclear reactors and construct alternative energy capacity may seriously erode as their investments in making existing reactors safer increase. After all, competing political demands such as recession in the donor countries and humanitarian and military crises elsewhere in the world place sharp limits on available assistance funds.

Small donors can have a large impact in legitimizing technical quick fixes to the nuclear safety problem and thereby starting the trip down the slippery slope. For example, Sweden's "go-it-alone" bilateral assistance to the Ignalina RBMK reactors, partly inspired by exasperation with what they perceived as very slow-moving multilateral efforts, lends legitimacy to the idea that it is technically feasible to upgrade RBMK reactors to what the Swedes believe to be an adequate safety level for the medium term. The Swedish example also incurs the risk that certain RBMKs, once upgraded, may continue to operate much longer than parts of the donor community had hoped. From one perspective, all the Swedes have done is to acknowledge that the Lithuanian government was strongly committed to continuing operations of their RBMK reactors for many years to come; and rather than allow the reactors in their own back yard to expose Sweden to continuing high safety risks, they have set out to make the reactors as safe as possible. Interpreting the precedential consequences of the Swedish program, however, we note that the Swedes have sown further division within the donor community about how to deal with RBMK reactors. As long as important donors are willing to tolerate their continued operation, it becomes virtually impossible for the rest of the donor community to impose strong conditionality (i.e., shutdown at a given point in time) on their nuclear safety assistance to countries with RBMKs. The point is generalizable: given that donors differ in their view of the risk presented by NPPs in the East and the most appropriate (and realistic) solutions to the risk problem, certain donors are able to undermine the solutions of others by pursuing their own bilateral strategies.

The existence of a number of bilateral agreements between Eastern and Western utilities for the import/export of electricity invites further concern about the potential for bilateral strategies to destroy any possibility that Western governmental assistance might be leveraged for plant closures in the East. Agreements for export of electricity from East to West have already been signed between Austria and Ukraine,[11] Finland and Russia,[12] and Hungary and Italy. Western officials are reluctant to speak on the record about these politically sensitive agreements. Although electricity experts point out that these agreements involve only small-scale exchanges of electricity, because of the very limited physical connections between the Eastern and Western electricity grids and that the electricity in question may not be generated by nuclear plants, the larger point is that electricity exchanges of this kind only encourage Eastern utilities with Western neighbors to maintain overcapacities in order to earn hard currency through electricity exports. Financial incentives thereby provide one more reason why Eastern utilities and their governments will resist pressure to retire NPPs before the end of their design lifetimes.

Further agreements of this type are highly likely. Lithuania already exports electricity to its East European neighbors, which accounted for about 15 percent of Lithuania's total exports in 1992 (World Bank/IEA/EBRD 1993: 19). Surplus capacity generated by the RBMK reactors at Ignalina makes these profitable exports possible. Lithuania's electricity export market may soon expand westward. Sweden's industry and commerce minister recently commented that "Sweden could eventually import electricity from the Ignalina and Leningrad nuclear stations," as part of a plan to create a common energy market among Baltic countries (*Nucleonics Week*, March 24, 1994: 13). Advanced negotiations between the Slovak Energy Company (SEP) and a French-German consortium for a contract to complete Slovakia's Mochovce NPP envisage repayment of Western loans for the project through electricity exports to Slovakia's Western neighbors, including Austria and Germany, once the plant starts working around 1998 (*Aegis* 1994: 2; *Petroleum Economist* 1993: 9).

The real danger of these existing and potential electricity exchange agreements is that Western governments are taking an official position on nuclear safety in the East that is completely undermined by secretive agreements among Western and Eastern utilities. By importing electricity from Eastern Europe, Western countries that officially express strong concern about deficient safety at Eastern NPPs in effect prop up the nuclear sector in Eastern Europe. Environmentalists understandably object to these electricity exchanges on the grounds that they encourage Eastern states to build or maintain much more electricity generating capacity than they need, and that they allow some Western states to circumvent domestic opposition to nuclear power by importing electricity generated by nuclear plants in the East, where safety standards are lower and the environmental risks consequently higher.

Capacity Building and Conditionality Strategies

Although donors of nuclear safety assistance have fallen short in resolving their own differences and therefore also in formulating a longer-term strategy for dealing with nuclear power in the East, they have achieved smaller successes by increasing concern about safety among critical constituencies in the East—a strategy that may ultimately prove more successful and more cost effective at resolving the nuclear safety issues than bribing or coercing Eastern governments into closing reactors that they would rather continue operating. Donors take two different tacks. First, they follow a strategy of rewarding Eastern countries that demonstrate sympathy for Western safety concerns with further assistance, while limiting involvement with countries that resist Western interference in their nuclear programs. PHARE, the nuclear safety account, and bilateral donors all gave substantial amounts of financial support to the Bulgarian government, rewarding its openness to extensive on-site assistance from international experts, and because of the nuclear regulatory body's willingness to withhold licenses for restart from reactors that had been shut down for upgrading—even in the face of electricity shortages—until international approval of the licenses was given. By contrast, TACIS restricted its assistance to Russia and Ukraine mostly to generic safety studies of reactor types when those countries refused to allow Commission officials the access they requested to information about their nuclear sectors. Officials at the EBRD nuclear safety account have expressed clear preferences for dealing only with countries that will accept their conditionality policies. It is therefore likely that the account will provide more funding to East European countries than to Russia or Ukraine, both of which vehemently object to any Western attempt to decide energy policy for them.

Second, Western donors have focused attention on strengthening capacity among nuclear plant regulators, operators, and managers. This strategy seeks to increase concern about nuclear safety and to improve the ability of small but critical groups of experts to implement and maintain higher levels of safety. Success depends not on winning over the whole of Eastern governments, nor on influencing public opinion in the East, but rather on ensuring that the people who license, operate, and manage NPPs understand the critical importance of running them safely. To be sure, transformation of safety culture in the East is a long-term project. Nonetheless, one of the central objects and the more important successes of Western assistance has been to create and financially support independent nuclear safety inspectorates, and to instill a more Western safety culture in plant operators and managers. This effort is already well under way in Eastern Europe. Regulatory concerns linger in the former Soviet Union, where bilateral donors have emphasized the importance of devoting more attention to the regulatory infrastructure.

These small successes are overshadowed by donor countries' failure to generate assistance funds on anywhere near the scale of the Eastern countries' financial need if they were in fact to retire older reactors or even to perform major retrofits as required on the more modern reactors. It is something of a puzzle that nuclear safety has not received higher levels of funding, given the strong self-interest in Western donor countries of reducing the transboundary risk of a nuclear accident. Several explanations are possible. Other priorities may simply have had more important claims on the resources of possible donors: for example, domestic recession, the war in Bosnia, and concerns about nuclear weapons proliferation stemming from the dissolution of the Soviet military machine. Certain donors, who for political or geographical reasons were more able to discount safety risks of the NPPs in the East (for example, France, the United Kingdom, the United States, and Japan), seem to have fashioned their assistance effort more as a subsidy program to help their own domestic nuclear industry capture new business opportunities than as a safety program; existing levels of funding may be more than adequate for this purpose.

Even for those donors whose primary concern was increasing safety, it may have been clear that a permanent solution to the nuclear problem, which would entail replacing the electricity capacity from the most dangerous reactors with new power plants and conducting major retrofits on the rest of the NPPs, was well beyond the reach of any international assistance effort. Unwilling to shoulder disproportionate burdens of assistance while other Western nations shirked their responsibility, actively concerned nations such as Germany, Denmark, and Sweden appear to have settled for the public relations benefit of a more limited nuclear safety campaign. Indeed, without the existence of stronger mechanisms for holding Eastern governments to their political commitments to shut down the oldest reactors after disbursement of substantial assistance funds, Western governments have understandably displayed reluctance to throw good money after bad.

The latter failure of Western donors to implement strong policies of conditionality is one of the major missed opportunities of nuclear safety assistance, although recent institutional developments suggest hope for greater progress on this front. As noted earlier, most nuclear safety assistance (PHARE, TACIS, and bilateral funds) has been disbursed without explicit policy conditionality. This situation may well draw donors into a longer-term commitment than they had intended. As long as the more dangerous Eastern NPPs continue to operate, it will be politically difficult for Western governments to withhold assistance altogether, after having justified existing programs on the claim that these reactors posed an unacceptable risk to the West. Worse yet, those very assistance programs may only prolong the lifetimes of the dangerous NPPs.

Earlier we noted the major reasons why it is so difficult for the donors to enforce conditionality. These included the nonsimultaneous nature of the commitments involved, whereby donors provide grants for safety assistance immediately in exchange for the commitments of Eastern governments to shut down NPPs at some imprecisely defined point in the future. The other major obstacle to conditionality policies is a collective action problem among donor states. The fact that Western nuclear suppliers are engaged in stiff competition with each other over contracts for safety retrofitting and nuclear capacity-building work, both now and in the future when more lucrative commercial contracts will be involved, undermines the resolve of Western governments to stick to common conditionality policies. At issue is whether Western governments will be willing to block loans from commercial or multilateral investment banks or withhold government loan guarantees for new nuclear construction or major retrofitting projects if the recipient country has reneged on earlier commitments to shut down its older reactors, given the likelihood that some other government may offer more favorable financial terms to give its own suppliers an advantage in securing the contracts.

The creation of the EBRD's nuclear safety account in the spring of 1993—an important innovation in the institutional framework through which donors channel nuclear safety assistance—offers some hope of progress on the conditionality front. The major advantage of the account is that its institutionalized connection with the long-term capital lending operations of the EBRD proper provides valuable leverage for contracting negotiations with Eastern governments. The EBRD will certainly play a central role in providing much-needed loans for energy projects in Eastern Europe and the former Soviet Union over the next decade or more, which puts the nuclear safety account in a better position than PHARE, TACIS, or individual bilateral donors to apply issue linkage in order to improve the enforceability of policy conditions it chooses to impose on its grant assistance. The first two grants disbursed by this multilateral fund—to the Bulgarian and Lithuanian governments—included commitments by the recipient governments to indicative dates for closure of units 1–4 at

Kozloduy, all first-generation VVER reactors, and both RBMK units at Ignalina, provided alternate generating capacity had been established by those dates.

In addition, the nuclear safety account represents the only existing forum that draws all Western donors, including nonmembers of the EU, into a common assistance framework. This institutional innovation makes the account better equipped than other arenas to resolve collective action problems among donors in their adherence to common conditionality policies. Although it cannot change the underlying competitive pressures among donor states, the nuclear safety account at a minimum provides a source of political pressure on donors to place priority on safety results over commercial gains. With capital totaling only $140 million so far and limited participation of important donors including the United States and Japan, however, the nuclear safety account is not adequately funded to single-handedly redirect assistance strategies.

The potential contractual benefits of this multilateral fund came only at the cost of considerable delay, as donor states squabbled over the appropriate institutional locus of assistance. The protracted political struggle over whether to have a new multilateral fund at all turned on the issues of how to spread the burden of financing nuclear safety assistance, when safety aspects were far more important to some donors than to others, how to resolve contractual problems both with Eastern recipients (securing plant closures) and among donor states (sticking to common policies of conditionality), and how to divide the commercial benefits of the assistance program among donors. Existing EU and bilateral programs guaranteed that the retrofitting work on Eastern NPPs would be conducted by contractors from the donor states (but did not adequately resolve German concerns that safety had not been significantly improved in the East, especially given that no NPPs at all had been shut down) and that the burden of assistance should be spread among more donors, particularly including the United States and Japan. Existing programs also did not satisfy French desires to provide more substantial subsidies to EdF and Framatome in a politically acceptable way. The United States and Japan opposed the idea of a multilateral fund, no doubt partly out of reluctance to get pulled deeper into a long-term and high-cost engagement in the East, but also because of their fear that a multilateral fund dominated by the Europeans would shortchange them when it came to commercial returns on their investment, which they could better control through bilateral aid programs.

In the end, the creation of the EBRD nuclear safety account represented a German and French victory in the institutional struggle, although they have not achieved the desired level of U.S. or Japanese participation in the fund. For the Germans, the nuclear safety account delivers substantial improvements in the contractual environment over uncoordinated bilateral or EU assistance, and hence better guarantees of an increased safety return on their large investment. For the French, the nuclear safety account's provision that contracts will be awarded back to donor states in rough proportion to their contributions allows them to support their enormous and suffering nuclear power sector both politically and financially.

It is noteworthy, although perhaps not surprising, that across the entire nuclear safety assistance effort, donors states set the agenda as far as which projects will be funded, in what amount, and through which institutional arrangements. Look at EU assistance; even the very existence of a nuclear safety program, in contrast to the other areas PHARE and TACIS support, was imposed on recipient countries by the EU donors. All of the various institutional arrangements for nuclear safety assistance ensure the delivery of commercial benefits primarily to Western donors, *not* to recipients. Perhaps the lack of Eastern governments' control over programs, projects, and institutional arrangements is inherent in the fact that Western states were footing the bill. In this case, however, two additional factors exacerbated this tendency. Western donors cared much more about the nuclear safety issue to begin with, which naturally meant they would seize the agenda. In addition, nuclear safety is one area where the requisite safety technologies and expertise really had to come from the West, because Russian technologies and regulatory, managerial, and operating practices were the source of the problem, and the other Eastern countries lacked indigenous expertise in the nuclear field.

Unresolved Problems

Stepping back to survey the outcomes of nuclear safety assistance as a whole, we draw the following conclusions. First, Western assistance programs, despite measured improvements in operational safety at some sites, appear to have given Eastern NPPs a push down the slippery slope of prolonged operation. Western donors have implemented interim solutions, supposedly to raise safety levels at the more dangerous plants until plans can be made to permit their permanent closure in the near future, without actually planning a long-term strategy for closure. Because of this, their quick-fix solutions risk becoming permanent solutions.

All of the solutions to the nuclear safety problems that have been pursued thus far are essentially pronuclear—that is, they all involve extensive continued reliance on nuclear power in Eastern Europe. The pronuclear orientation is not surprising, given domination by professional nuclear organizations during the stages in which the nuclear issue was defined and potential solutions evaluated. The generally pronuclear orientation of the major donors and their political preferences to support domestic nuclear industries also contributed heavily to this outcome.

Donors have left key contracting problems in the provision of assistance unresolved. It appears likely that some or all recipient countries will eventually defect from the few conditional commitments they have made so far to shut down particular reactors that cannot feasibly be brought to Western safety levels. The EBRD nuclear safety account represents an improvement in the contractual environment, but it commands neither adequate funding nor wide enough political support from all the major donors to single-handedly resolve the problems of imposing and enforcing conditionality on nuclear safety assistance. We point to the underlying competition among donors over lucrative contracts to provide new energy capacity as the principal explanation of persistent collective action problems among donor states in attempting or enforcing stricter conditionality.

Finally, donors have done little about providing alternative electricity capacity to facilitate retirement of Eastern NPPs or making the capacity in question unnecessary through demand-side energy management programs. If the West ignores this issue, interim solutions for emergency retrofitting will certainly turn into permanent ones, and all the assistance for improved nuclear safety will have only prolonged the lifetime of the most dangerous Eastern reactors. Contributing to donors' apparent shortsightedness is the sheer enormity of the bill for replacement capacity, estimated at $21 billion in investment costs alone just for closing the highest risk (RBMK and VVER-440/230) reactors in the region by the mid-1990s or shortly thereafter (World Bank/IEA/EBRD 1993: 6). Such sums will necessarily come from commercial and multilateral loans, rather than official grant assistance. The fact that nuclear safety programs remain administratively separate from other energy and environment concerns has also impeded joint resolution of these linked issues, as has the failure of Eastern countries to devise medium-to-long-term energy plans of their own. Energy efficiency programs have excited little self-interest among Western donors, although they could bring significant economic and environmental benefits to Eastern Europe and the former Soviet Union.

In spite of these constraints, one fact is critical. On the dictates of narrow economic rationality combined with their lower concern about safety aspects of nuclear power, Eastern governments left to their own devices will prefer to continue operation of all NPPs to the end of their design lifetimes. To succeed at all in overcoming nuclear safety problems in the East, Western countries will either have to infuse Eastern states with far greater concern about these issues themselves or make it economically worthwhile for Eastern governments to retire substantial nuclear capacity. As of now they have achieved neither condition.

3   Evaluation

This final section will draw attention to some of the missed opportunities of nuclear safety assistance to suggest, from a normative perspective, mistakes that decision makers who are concerned about environmental benefits from foreign assistance should try to avoid in future issues, or perhaps remedy in the nuclear case. Also highlighted in this section are positive achievements in the nuclear safety field that might merit replication in another case.

Premature Closure of Problem Definition
The first missed opportunity consisted in premature closure of the initial phase in which a coalition dominated by professional nuclear organizations defined the problem of nuclear safety in the East and proposed viable solutions. The highly technical issues involved contributed to the closure of this phase, in which the dominant actors clearly overlooked a

much wider possible set of definitions of the problem and complementary solutions. Once the nuclear safety problem had been defined as a "technical" issue and contracted over to professional organizations, it was no surprise that the privileged response became technical fixes for the plants and training for operators, managers, and regulators, as opposed to an open debate within the energy field over whether any individual Eastern country needed its nuclear capacity at all, or an open political debate over whether citizens of either Western or Eastern countries were willing to tolerate the environmental risks of maintaining or expanding nuclear power if the capacity was in fact needed. The lesson, perhaps familiar, is that it is of crucial importance to keep early evaluations even of a highly technical environmental issue open to the broadest possible set of actors, lest proposed solutions correspond more closely to the interests of the particular groups that capture the problem definition phase than to the actual environmental problem.

Building Knowledgeable Concern

Another lesson from the nuclear safety case, which in this case points to more successful efforts by international donors, is that it is possible to achieve significant environmental benefits by building administrative, legal, and managerial capacity in recipient countries even when technological capacity lags behind. Nuclear reactor safety depends as much on operational safety (i.e., "safety culture" among operators, regulators, and managers) as it does on reactor technology. Reactors with identifiable technological weaknesses in their safety systems can be run much more safely when operators use established routines and procedures that compensate for those weaknesses. Improving safety culture by training operators, giving them hands-on experience with how Western operators work, training managers and regulators, and enhancing their human and financial resources all constitute very cost-effective ways to improve the operational safety of a nuclear power plant. The bulk of Western assistance for nuclear safety has concentrated on these human factors,[13] and significant successes have been achieved at some sites. For example, international assistance to the Kozloduy NPP in Bulgaria stimulated surprisingly large improvements in plant maintenance and operations, as well as strengthened the Bulgarian nuclear regulatory commission by augmenting regulators salaries to retain skilled personnel. So dramatic were the improvements that this plant for which the IAEA had recommended immediate closure in 1991 managed to reopen unit 2 in December 1992 and unit 1 at the end of the following year with an internationally approved license for short-term operation (*NUSAC News,* Dec. 7, 1993: 7).

Despite the real benefits of building administrative, legal, and managerial capacity as far as improved operational safety of NPPs in the East, a more permanent solution to the nuclear safety problems cannot ignore the need to address shortfalls in technological capacity. Eastern governments will probably refuse to shut down any nuclear reactors—no matter how dangerous—until they can meet energy demand with remaining (or newly constructed) capacity. Eastern utilities have little incentive to pay for construction of new capacity themselves, when they can already generate electricity cheaply by using existing nuclear plants; and they have yet to show a genuine interest in demand-side management programs. Even the most modern (VVER 1000) set of reactors in the East, which very few Western governments even talk about closing, will require significant and costly safety improvements—principally for new fuel assemblies and instrumentation and control systems—before they can operate at safety levels equivalent to Western standards.

Missed Opportunities for Conditionality

We have stressed that the failure of the major donors in the nuclear safety problem to agree on an appropriate set of solutions imposes serious collective action problems that limit the degree to which the donor community will ever be able to successfully apply common conditions requiring plant closures to their assistance. Even so, donors missed a second major opportunity by not seeking to apply less-demanding but potentially more successful strategies of conditionality that they might all have been able to agree on.

Environmental NGOs have criticized EBRD energy policy for playing lip service to "least-cost" energy planning, but in practice relying on traditional planning for supply only, rather than genuinely considering possible demand-side options as part of a comprehensive least-cost strategy. Of course, energy planning in Eastern Europe and the Soviet Union has always turned on options for increasing supply capacity, with little respect for the demand side. Even since the revolutions, energy planning in Eastern Europe remains a protracted debate over building fossil fuel versus nuclear plants. Current EBRD lending in the region has gone mostly to the supply side, especially for oil and gas.[14] The International Institute for Energy Conservation has suggested that analysis of efficiency potential should become a central part of prefunding evaluations for energy projects, and that the EBRD should itself conduct detailed cost assessments of supply *and* demand resources, rather than leaving the latter to the potential loan recipient (Boyle 1993).

How do these concerns relate to conditionality? Donors who value environmental improvements, both in Eastern states and internationally, should consider attaching conditions to grants and loans for all energy projects that would require Eastern recipients to evaluate and exploit the potential for improvements in energy efficiency, before projects for new capacity or for major retrofitting work on NPPs are even considered. This strategy has three significant political benefits over previously attempted forms of conditionality. First, it would be extremely difficult for major donors to oppose conditionality fostering efficiency gains, particularly given that the Environmental Action Plan for Central and Eastern Europe (endorsed by all major donors and recipients of environmental assistance as well as the EU, EBRD, and World Bank) stressed that improvements in energy efficiency were one of the key "win-win" reform policies for the region, promising both environmental and economic benefits for the countries in transition.[15] Second, recipient countries would have little reason to oppose such conditionality, because both the economic and environmental benefits of improved energy efficiency will accrue primarily to Eastern consumers and utilities, rather than to Western companies. Third, the capital requirements for investments in energy efficiency would likely be much lower than for building energy capacity to completely replace nuclear plants; and because such investments will bring quick economic returns, Eastern utilities or governments should be able to gather most of the necessary capital themselves.

Another lesson deriving from the nuclear case, which points to a positive feature of the assistance effort, is that environmental assistance is more effective when it entails a careful balance of participation by foreign and local experts. Exactly what this balance should be depends on issue-specific conditions. We have noted that for nuclear safety issues, it was quite appropriate that Western organizations should dominate the provision of safety technologies and expert advice, whereas other issues might fare better to solicit greater contributions from recipient country experts. One aspect of nuclear safety assistance that appears to have been particularly effective is the institutionalized matching of foreign and local experts. For example, the World Association of Nuclear Operators (WANO), a voluntary, international, intra-industry group comprised of nuclear plant operators, directs very successful "twinning" operations whereby Eastern and Western operators of nuclear plants exchange experiences and expertise. WANO also created its own incident reporting system so that operators can learn from mistakes made at other NPPs. Western utilities even fund these programs themselves, in the interests of protecting their own nuclear assets from further political damages from perceptions of mismanagement in the East, as well as keeping operational safety of NPPs an intra-industry affair. Western nuclear regulatory officials were the major actors involved in building independent, adequately funded, and well-staffed regulatory bodies in the East. This kind of one-on-one matching of experts has proved itself to be a highly effective, low-cost means of building capacity, and one that takes advantage of the high concentrations of human expertise but low willingness to provide large amounts of capital among donor countries.

Strategic Capacity Building

A final and related lesson from the nuclear safety case is that the choice of where to build capacity in the provision of environmental assistance— within an environment ministry, the financial sector, private companies, regulatory systems—is both highly political and genuinely consequential for environmental outcomes. The greatest success of the nuclear safety assistance effort has been in capacity building, and specifically in instilling a more Western appreciation for safety culture along with the resources to uphold higher standards among operators and regulators of NPPs. This is a highly selective strategy for capacity building, focused on building knowledgeable concern about nuclear safety among those who are in a position to decide which plants will be licensed and how they will be run.[16] The great advantage of this strategy is that it is a relatively low-cost way for Western donors to increase Eastern concern about the nuclear safety problem by creating and strengthening a small but key constituency for safety within Eastern countries, one that might eventually exert decisive power over the safety issue.

4 Conclusion

Given that reducing the risk of another major nuclear accident in Eastern Europe or the former Soviet Union is an important and ongoing issue on the international environmental agenda, we use this concluding section to reiterate the essential characteristics of the existing nuclear safety assistance effort and to point to useful directions for key actors to move in to produce further progress in this area. The nuclear safety issue is shaped significantly by the fact that its major actors pursue competing rather than complementary solutions, a situation resulting from divergent priorities and from the economic competition among nuclear suppliers that underscores the whole issue. Continued pursuit of competing solutions leads to failures of collective action that might well undermine donors' ability to reach any solution at all to the safety dimensions of the nuclear issue.

It is a real and serious possibility that Western nuclear safety assistance may lead to a slippery slope problem, whereby aid intended to make NPPs slightly safer for a short period of operation until Eastern states can implement plans for shutdown has the unintended effect of prolonging the operations of the more dangerous NPPs, which leaves concerned Western states facing perhaps a slightly lower safety risk from Eastern NPPs, but over a longer period of time. The slippery slope effect derives from failures on the donor side to formulate a long-term strategy for Western action that would necessarily involve comprehensive energy planning in concert with Eastern governments and the formulation of implementable plans for closures of the oldest NPPs. The latter condition requires that donors confront and work around prior failures to formulate common policies of conditionality that would facilitate resolution of the nuclear safety problems. Exacerbating the slippery slope effect are the "go-it-alone" operations of certain bilateral donors, the deliberate lack of coordination among bilateral assistance programs for commercial benefit, and agreements signed for the import of electricity produced in Eastern Europe by Western utilities, which provide incentives for Eastern states to maintain excess capacity generated by NPPs.

Existing disagreements about appropriate solutions to the nuclear safety problem both within the donor community and between donors and recipients have enabled the nuclear lobby to dominate the agenda of assistance activity. As a consequence, the safety problem was defined very early as a narrow, technical problem about energy supply, and even more narrowly, about how to remedy the technical and operational deficiencies of nuclear plants. Bringing home contracts to nuclear suppliers in the donor states has been a central concern in all existing arenas of nuclear safety assistance. Actors on all sides have largely overlooked the much broader set of energy options (and attendant political trade-offs) that might have provided feasible solutions to the nuclear safety dilemma, some of them perhaps at significantly less cost to Western donors. The assistance program looks more like an ingenious way to revive nuclear industries in the West than like a serious program to secure Europe against further risk of radioactive contamination from another nuclear disaster like Chernobyl.

Symptomatic of this shortsighted, narrowly defined understanding of the problem is the fact that Western donors (along with Eastern recipients) have conducted no systematic consideration of the long-term requirements for permanent solutions to the nuclear issue. Specifically, neither side has formulated an implementable strategy either for building (and financing) replacement electricity capacity so that the oldest NPPs can in fact be shut down after temporary operations for a short period of time, or for instituting demand-side management strategies that would reduce the need of Eastern countries for the capacity produced by these NPPs, and thereby facilitate their closure. Donors cannot rely solely on Eastern governments to conduct this long-range energy planning exercise themselves, because the Eastern side has exhibited much less concern about nuclear safety issues and little understanding or appreciation of the potential economic and environmental gains of demand-side management.

Several roads might lead away from the current quagmire of the assistance effort. Rather than throwing good money after bad, Western donors might be better advised to step back to the conceptual phase of the nuclear safety programs and to begin thinking about nuclear safety less in narrow, technical terms and more in terms of broad political choices about energy and environment. Such reevaluation could open the door to a new and broader set of solutions that might include cheaper, more effective, or more implementable solutions. As part of this exercise, donors must explicitly acknowledge their inherent disagreements on appropriate objectives within the nuclear issue in order to gain an understanding of what they can realistically accomplish within the constraints of their disagreement. The object is for donors to avoid undermining any solution at all to the safety problem by continually attempting to implement competing solutions.

More specifically, donors need to formulate concrete, long-term plans to build (or obviate the need for) replacement capacity for the NPPs which they deem unworthy of retrofitting, and for financing safety retrofitting on the more modern reactors. It is not necessarily the case that the West will have to shoulder the entire bill for these energy sector transformations, but it will certainly have to provide substantial financial and political inducements to force Eastern governments out of status quo energy planning. Donors need to consider ways to devise and apply more creative conditionality policies on which all the major donors could agree and that would induce Eastern states to undertake more comprehensive energy planning, including evaluating and implementing projects to exploit the huge potential gains in energy efficiency.

Finally, donors should continue and build on existing assistance policies which build capacity in a discriminating way, in order to create influential constituencies within Eastern governments and utilities with the incentive and ability to lobby for higher standards of nuclear safety. Attempts to imbue Eastern operators, regulators, and managers of NPP with Western understandings of safety culture are unambiguously productive ways both to reduce safety risk levels at Eastern NPPs in the short term and to increase knowledgeable Eastern concern about nuclear safety. If successful, these efforts can be expected to increase the willingness of Eastern governments to contribute a greater share of financial and political resources to permanent solutions for the safety risks emanating from their nuclear reactors.

Notes

1. The Czech Republic is an exception: it has its own nuclear supply industry, Skoda, which was responsible for manufacturing some of the Soviet designs. The Czechs built one Skoda-designed reactor on their own in the mid-1970s, which was shut down not long after because of fatal design flaws. The Czech Republic continues to hold interests in developing and exporting nuclear technologies.

2. The situation in Hungary is much brighter. Its Paks NPP has an excellent record of operation; and Soviet, French, Finnish, WANO (World Association of Nuclear Operators), and IAEA experts have all conducted inspections and pronounced the reactor and its operators essentially up to Western standards.

3. Five Canadian-designed CANDU reactors are under construction in Romania with Canadian financial backing.

4. The nuclear establishment is so influential, in fact, that Alexei Yablokow, environmental advisor of Russian president Boris Yeltsin, said in a recent interview: "Our [nuclear] inspectors are working quite well, despite the fact that they earn only a quarter of the salary a worker in a nuclear plant gets. The major problem is that the Ministries of Defense and Atomic Energy do not permit inspection of their nuclear plants—in spite of explicit presidential orders" (*Die Zeit* 1993: 33).

5. The exact figure is 100 million Swedish Crowns.

6. This figure includes $560 million from the European Union's PHARE and TACIS programs, $116 million in the EBRD's nuclear safety account, not including the EU's contribution which came out of the PHARE and TACIS budget, and $198 million in bilateral contributions.

7. This figure excludes PHARE and TACIS funding, which is often counted as part of G-24 (bilateral) assistance.

8. Indeed, the Court of Auditors of the European Community made this very criticism of the PHARE and TACIS nuclear safety programs, claiming that the Commission "had still to equip itself with a real policy for interventions in the field in question" (*Official Journal of the European Communities* 1993: 183).

9. The Ukrainian government has reversed itself more than once on commitments to close the Chernobyl NPP. Most recently, the Ukrainian environmental minister Jurij Kostenko was quoted in the German weekly *Die Zeit* as saying that Chernobyl's block 2, currently off-line, will be permanently closed in 1996, while blocks 1 and 3 will be closed in 1997 and 1999, respectively (*Die Zeit,* July 7, 1995: 39). Negotiations with an international consortium of companies led by Asea Brown Boveri to build a nonnuclear power plant are under way, which increases the likelihood of Ukraine sticking to its current course of action.

10. According to the *Economist,* Ukraine, for example, "wastes much of the energy it gets. In 1992 the country used almost as much energy as Britain, whose GDP was 12 times as large. . . . A recent study of ten large industrial enterprises

concluded that quick, cheap measures such as insulating hot-water pipes and switching off idle machinery could cut their energy consumption by 20–30%" (*Economist*, 12 March 1994: 55). The energy intensity of Eastern economies varies considerably, but for all countries the potential gains from improved energy efficiency are enormous.

11. The Austrian utility Österreichische Elektrizitätswirtschaft and the Ukrainian Interenergo signed a fifteen-year contract for the import of up to 800 million KWh annually of electricity (FOE 1992: 15). This agreement runs completely contrary to the official position of the Austrian government, which strongly opposes any operation of nuclear power in neighboring Eastern states.

12. Finland has agreed to import electricity from the Leningrad area. However, Finnish authorities say this electricity is *not* generated by the Sosnovyi Bor NPP (Hiltunen 1994: 27).

13. Donors' concentration on human factors was not fully intentional; it stems from a number of factors, such as the small size of bilateral assistance programs as well as PHARE and TACIS country budgets which made it impossible to take on more capital-intensive hardware projects, donors' desire to start the flow of assistance while technological needs for individual reactor types were still being determined, and some donors' contentions that certain reactors could in no case be brought to adequate Western safety standards and therefore should not be touched.

14. Only a very small proportion of the Baltic states emergency funding and the Polish heat supply and conservation loan was for demand-side management (Boyle 1993).

15. At the same time, donors may possess less self-interest in efficiency improvements in Eastern Europe and the Soviet Union, which would diminish the market for new power plants manufactured by Western companies.

16. In the United States, the nuclear safety issue was managed with a much broader transformation of political processes, whereby the transparency of and access to the licensing process of NPPs was greatly increased, allowing domestic NGOs to play a decisive role as a watchdog of safety, despite their apparent disadvantages in dealing with such a highly technical issue.

9

Organizational Inertia and Environmental Assistance to Eastern Europe

Barbara Connolly, Tamar Gutner, and Hildegard Bedarff

The fall of the Iron Curtain in 1989 revealed in Eastern Europe[1] the most polluted countries in Europe. Since then, environmental protection has emerged as a new and relatively prominent issue on the agenda of East-West cooperation, at least rhetorically. Within the larger effort to support democratic reform and economic restructuring in Eastern Europe, Western governments have mobilized significant levels of financial assistance to help these Eastern states combat environmental degradation and begin the process of developing and adopting new environmental policies.

This chapter examines the largest donors of environmental assistance to Eastern Europe: the World Bank, the European Union (EU), the European Bank for Reconstruction and Development, (EBRD) and bilateral donors. The East Europe case distinguishes itself from many of the other cases in this book in that donors have nested environmental issues within a much broader network of assistance programs. The institutions responsible for most of the environmental assistance to Eastern Europe did not fashion their efforts as a direct response to one particular environmental problem, or even as an explicit answer to a large array of environmental problems afflicting the region. Rather, these institutions tended to tack on the additional goal of environmental protection to a broader, largely preexisting set of economic and political development objectives—such as privatization, market liberalization, and the expansion of public participation in decision making.

Even though these institutions became involved in Eastern Europe at the same time and confronted the same broad set of environmental problems, each has selected distinctly different solutions and mechanisms to achieve them. What accounts for this variation? This chapter argues that the donor institutions, confronted by new or additional environmental objectives, tend to apply "garbage can"–style a familiar set of preferred solutions, constrained by each institution's specific interests and available set of financial transfer mechanisms (Cohen, March, and Olsen 1972). Although the existence of organizational inertia may not be surprising, it has had important and systematic consequences on the effectiveness of environmental assistance to the region. Organizational inertia results in the use of financial mechanisms that do not always fit the specific environmental problem under consideration; in practices that may return more economic benefits to donor countries than they leave environmental benefits to the recipients; and in institutions that are hampered in their attempt to fulfill environmental mandates by their own set of financial tools and procedures. On an aggregate level, environmental assistance to Eastern Europe is much less than the sum of its parts. "Garbage can"–style assistance has resulted in the neglect of certain priority environmental problems (e.g., local air pollution, which poses the greatest threat to human health in the region), duplication in numerous assistance efforts, and rampant and even deliberate failures to coordinate assistance programs in related areas.

Suppose the donor institutions had set out instead to tailor their assistance programs and mechanisms specifically for environmental purposes, aiming to achieve the greatest possible environmental benefits in Eastern Europe within existing resource constraints. In this hypothetical scenario, donors and recipients might forge collective agreements on the region's most pressing environmental problems and then parcel out each of those areas to individual institutions according to a system of specialization by comparative advantage. They would create, adapt, or combine financial mechanisms to fit the problem and increase East Europeans' ability to absorb the assistance. Although we stress that the policy process is largely driven by "garbage can" decision making instead of attempts to optimize environmental performance, we also argue that donors can and should narrow the gap between current practice and this optimized model if they care to enhance the effectiveness of their environmental assistance. Perhaps the best evidence that progress toward enhanced environmental effectiveness is possible are the repeated attempts by donors and recipients to increase coordination and to set common priorities for assistance. We argue that this process shows signs of improving but not ending the current organizational inertia.

Like most of the other cases in this book, environmental assistance to Eastern Europe shows a strong pattern of donors setting the agenda of which environmental problems will receive financial assistance and how acceptable solutions will be defined. What stands out in this chapter from other chapters, however, is the argument that in addition to various kinds of capacity within recipient countries, institutional capacity of the donors can be an important factor in determining effectiveness in the implementation of financial transfers once problems and solutions have been identified. Each of the donor institutions manages certain kinds of problems quite well, but in other cases their organizational goals and existing financial transfer mechanisms may stand in the way of achieving environmental objectives. Hence, improving the effectiveness of environmental assistance may require capacity building within donor institutions as well, in the sense of enhancing their adaptability and flexibility to tailor old mechanisms to new problems.

Another argument in this chapter that stands out from those in many of the others is about how to evaluate effectiveness. While it makes intuitive sense to evaluate the effectiveness of a financial transfer in terms of how well the solution eventually implemented matches a particular environmental problem, this criterion misses additional factors which gain importance when we consider that in the real world, many donor agencies are involved in providing financial transfers for many problems, all at the same time. In this aggregate sense, the effectiveness of environmental assistance hinges also on how well donors and recipients coordinate their aid efforts, whether the solutions advanced by various aid efforts in related project areas are complementary or contradictory, whether priorities are established so that aid flows target the most important problems, and whether synergies develop among different aid efforts. This chapter explores impediments to such aggregate effectiveness as well as suggestions on how it might be enhanced.

The remainder of the chapter is structured as follows. Section 1 describes the environmental situation in Eastern Europe and discusses the shortcomings of the aggregate picture of environmental assistance. Section 2 analyzes the strategies of the main donors and how, together, they contribute to this outcome. Section 3 provides a normative evaluation of the assistance effort and describes the nascent efforts of the "Environment for Europe" process to redress some of its weaknesses. Section 4 summarizes the argument and suggests implications for policy and research.

1   The Aggregate Picture of Environmental Assistance

The degradation of the environment in Eastern Europe has been widely publicized since the fall of the Iron Curtain. Numerous articles have described dramatic environmental damage, including the death of vast stretches of rivers, whose waters are unfit for human drinking or industrial use; the irreversibly damaged forests such as those in northern Bohemia; and the high degree of birth defects and retarded development in some of the more seriously polluted areas (Ackermann 1991; Fisher 1992; French 1991; DeBardeleben 1991; Moldan and Schnoor 1992). Information on environmental issues was limited before the regime changes due to tight government control over technical information, and a limited amount of research being conducted on ecological issues (Kabala 1993: 50). Often data collected on environmental conditions and issues were declared state secrets and suppressed (Andrews 1993: 12).

Shock over the degree of environmental degradation led to some early exaggerations of sweeping environmental disaster in the region, when in fact the worst of the damage is fairly localized. We now know, for example, that average levels of exposure to most major pollutants, such as sulfur dioxide and particulates, are not significantly higher in Eastern Europe than in the rest of Europe (World Bank 1992d: 2). While the overall state of the environment in Eastern Europe does not match some of the media's apocalyptic descriptions, however, certain smaller areas come close.

In fact, the environmental damage in Eastern Europe is mostly concentrated in a limited number of heavily populated industrial regions, or "hot spots," such as the Black Triangle region (incorporating Upper Silesia, northern Bohemia, and southeastern Germany) and Copsa Mica in Romania. In these areas, environmental damage is truly dramatic. For example, lead and cadmium levels in the soil of the heavily industrialized Upper Silesian towns of Olkusz and Slawkow exceed previously recorded measurements anywhere in the world, making food contamination a formidable problem (World Bank 1992d: 2). In northern Bohemia, the concentration of lignite-fired power stations and industrial plants has resulted in sulfur dioxide emissions that are around twenty times the national average, and maximum limits are exceeded 128 days a year (Alcamo 1992: 124). These "hot spots," where human health is clearly jeopardized by the extreme environmental degradation, are prime targets for environmental expenditures.

More generally, massive air pollution is undoubtedly the worst pollution problem in Eastern Europe from a human health perspective. It is caused largely by the burning of plentiful domestic lignite, a soft brown coal characterized by much higher sulfur and ash content and lower heating efficiency than hard coal, which is much less available in the region. The largest sources of particulate air pollution, one of the worst health threats in the region, are low stack emissions from domestic heating and small- and medium-size enterprises, which rely on low-quality lignite to a much greater extent than in Western Europe. Air pollution is also a transboundary problem; indeed, Eastern Europe has the distinction of producing over 66 percent of Europe's sulfur dioxide emissions, but only 33 percent of its GDP (Alcamo 1992: 235).

Water pollution is another serious problem in the region. While to date there is no thorough comparative data that measures the extent of the problem throughout the region, available data is still grim. In Poland, the proportion of surface water unfit for human consumption has been put as high as 95 percent (Schreiber 1990). Former Czechoslovak government statistics for 1992 show that around 50 percent of the country's drinking water was below government purity standards, and that 30 percent of the country's river water was not even capable of sustaining fish (Moldan and Schnoor 1992). Among the main causes of water pollution in the region are inadequate drinking water processing systems and sewage treatment, unmonitored municipal and industrial discharges, contamination from agricultural runoff, contamination from industrial air emissions, and saline discharges from coal mining operations. Finally, in the case of both water and air pollution, it is clear that monitoring is either inadequate or nonexistent, and technology is out of date. Data on toxic waste, biodiversity loss, and soil pollution also provide evidence of serious environmental degradation in the region.

Environmental Assistance Programs

In the wake of the regime changes in Eastern Europe, environmental issues were high on the agendas of both donors and recipients, because
environmental groups were among the key domestic East bloc actors
pushing to remove the communist regimes. The region's massive restructuring needs triggered the creation of numerous Western assistance
programs.

It is difficult to provide a concise breakdown of donor environmental
assistance. Available data are highly distorted by the lack of any common
definition among donors of what is or is not "environmental assistance,"
and likely underestimate the total. In addition, with the exception of the
World Bank, there have been no concerted attempts to calculate environmental components of projects and loans in other sectors (such as energy
or transport), especially since the environmental benefits built into other
projects are difficult to quantify. Individual institutions also appear to
have trouble agreeing on in-house definitions. The World Bank's 1994
annual report, for example, lists only one loan ($18 million) in the envi-
ronmental sector for Eastern Europe for the period 1990–1994. Yet another publication it produces adds four other projects to that category,
which it labels as projects with "primary environmental objectives," and
points out another four projects which have "environmental components" (World Bank 1994e). This brings the total to $788 million. Still
other Bank documents include certain structural adjustment loans which
contain energy or environmental components, and they push the total up
by another $1 billion (World Bank 1994b).

Available data, although difficult to compare, show that the World
Bank has been the largest donor to environmental projects and assistance,
followed by the EU's PHARE program,[2] bilateral donors, and the EBRD.
An important qualification is that while the World Bank and EBRD rely
primarily on lending, PHARE and the bilateral donors provide mostly
grants. The World Bank has approved at least $788 million in loans for
projects with environmental objectives and components through fiscal
1994 (World Bank 1994e). The EBRD, in turn, passed its first two "environmental infrastructure" loans in July 1994, for around 46 million ECU;
while it does not highlight the environmental components of its energy
projects in its official documents, it does note that it has spent over 24

million ECU on grant-funded technical cooperation projects with "significant environmental objectives" (EBRD 1994b: 11–14).

The European Union ranks as the largest grant-based donor of environmental assistance. Through PHARE, the EU committed 337 million ECU (about $381 million) to its environment programming in Eastern Europe from 1990 to 1994.[3] Excluding the EU's programs, total contributions for environmental projects and programs from bilateral donors, collectively known as the G-24,[4] amounted to 468 million ECU from 1990 to 1993,[5] according to estimates by the G-24 Coordination Unit[6] (G-24 Coordination Unit 1994b). Over the same time period, Denmark was the second largest bilateral donor (behind the EU), with 127 million ECU ($143.5 million), and the United States was third, with $91.4 million (G-24 Coordination Unit 1994b; U.S. General Accounting Office 1994: 42). Bilateral totals, however, are tricky to compare; as is the case with the MDBs, bilateral donors have different ways of defining environmental assistance. While some use a narrow definition of environmental assistance, others employ a more expansive definition which includes energy projects and thereby inflates the total.[7]

On the horizon is a potentially important new source of funding: the U.S.-Japan Joint Initiative, a $1 billion fund to provide low-interest loans to environmental projects in Eastern Europe. The Initiative, announced in May 1994, is funded by the Japanese government, but will be run by a joint committee of U.S. and Japanese officials. Disagreements between the United States and Japan, and within different U.S. agencies over how and to whom the fund should lend money have delayed its initiation.

Through the end of 1993, Poland was the leading single recipient of environmental assistance in Eastern Europe, having received 29.3 percent of G-24 commitments. The Czech Republic and Slovakia combined[8] were the second largest recipients, with 13.6 percent of commitments, while Hungary was not far behind with about 10.5 percent. One-third of assistance commitments went to regional programming (G-24 Coordination Unit 1994b).

Viewed against estimates that the cost of cleaning up Eastern Europe runs into the tens of billions of dollars, foreign assistance is only a drop in the bucket. The Environmental Action Programme, for example, notes that international assistance for Poland and Hungary through 1993 amounted to only slightly more than 5 percent of national environmental expenditures from all public and private sources. The document projects that countries in the region can expect to meet more than 90 percent of the costs of environmental expenditures out of domestic resources (Environmental Action Programme for Central and Eastern Europe 1993: I-3).

Despite the fact that total Western assistance is low relative to the region's needs, Western governments and international institutions are playing an influential role in environmental reform and clean-up in the region (Gutner 1993). From the donors' perspective, the degree of "concern" in tackling Eastern Europe's environmental problems reflects their interest in reducing transboundary pollution, their desire to provide humanitarian assistance, pressure from their domestic environmental communities, and also their interest in promoting their own exports. In Eastern Europe, by contrast, the strong domestic constituencies pushing for environmental reform in the wake of the political transformation have largely disappeared, since the overwhelming problems related to economic restructuring have pushed the environment out of the limelight. As a result, East European governments are most interested in environmental assistance to augment their own overstretched budgets; therefore they vastly prefer assistance in the form of grants rather than loans. This has meant that support and pressure from the international donor community have become increasingly important in pushing environmental reform as a regional priority.

Solutions Looking for Problems

A central argument of this chapter is that donor institutions prefer selectively to define and work on particular kinds of problems that fit into their established agendas and practices. What difference does this tendency make? Among the important consequences are that donors usually set priorities for assistance projects, determine funding levels, and receive many of the benefits; that resources made available may not match the most important environmental problems; and that donor strategies toward capacity building vary, as a function of the established practices of donor organizations.

Donors Drive the Project Agenda   With few exceptions, donors set the agenda for areas that will receive environmental assistance, either through an explicit selectivity bias or by imposing strong project or policy conditionality. PHARE exerts influence on recipient decisions through its in-house presence in recipient country environmental ministries and insists on some degree of economic and transboundary environmental benefits for EU donors. Financial and policy conditionality imposed by the EBRD and World Bank, in turn, effectively defines which kinds of projects will be eligible for assistance. The majority of bilateral donors also tend to fund those projects that fit well with their own established priorities, which run from transboundary effects to areas where donors have considerable expertise or environmental goods to export.

Recipients exhibit a strong tendency to accept whatever assistance is offered in the form of grants, regardless of whether the proposed project conforms to weakly established domestic priorities (Smith and Coppendale 1993: 34). Recipients exert more control over project selection where loans are concerned by turning down projects. In practice, this results in the rejection of environmental projects in favor of economic investments. Despite recipients' ability to decline loans, few assistance programs are truly "demand-(recipient) driven."

Because donor institutions are able to dominate the selection of assistance priorities, donors also capture a disproportionate share of the economic benefits of assistance. The most common tactic for bringing home commercial benefits to the donor state is the routine practice of donor institutions to hire short-term Western consultants for project preparation, feasibility studies, and expert advice. One such consultant from Great Britain, PA Consulting Group, was hired by the EC Commission to assess major donor environmental projects. The ensuing report stressed the prevalence of recipients' criticism that all donors in the region fail to make adequate use of local experts, who understand local conditions and who usually charge a fraction of Western consultancy fees (Smith and Coppendale 1993: 37). This practice diverts attention from cultivating a local consulting sector and indeed becomes a drain on recipient country capacity as well, because it forces environment ministries to spend large amounts of their time and energy supporting what they call "consultant tourism." In addition, the donors have been criticized for failing to give recipient countries a large enough role in managing projects, which would be a means of building sorely needed local management capacity. Besides the economic benefits from technical assistance, firms in donor states often benefit from sales of major equipment, financed with the assistance of MDBs loan, and perhaps cushioned by foreign assistance grants as well.

The low levels of foreign assistance relative to environmental needs are also a direct result of donors setting the project agenda. The environment, after all, is only a secondary concern for MDBs, and many environmental projects cannot meet prudent lending criteria within these institutions. Many costly environmental problems do not incur transboundary effects or provide attractive export opportunities, and hence are of less interest to bilateral donors.

Donors' biases have caused them to systematically overlook certain kinds of environmental problems or geographical regions, with potentially very serious consequences. Evaluations of the coverage of environmental assistance programs have uncovered substantive gaps in important areas such as air pollution from distributed sources, including vehicles or low stack emissions. This gap is not trivial; indeed, local air pollution is potentially the most serious short-to-medium term environmental problem in Eastern Europe in terms of its consequences for human health (Environmental Action Programme for Central and Eastern Europe 1993: I 6–8, II 6–11).

Understanding the particular biases of donor institutions makes such substantive gaps predictable. PHARE and the bilateral donors will be less inclined to intervene in local air pollution problems where donors would not share prospective environmental benefits.[9] Additionally, PHARE lacks access to adequate financial resources to run hardware projects, which means it could establish inventories and studies of local pollution problems, but not underwrite the investments to remedy those problems. For the MDBs, local air pollution projects are less attractive because they require smaller-size loans than the banks prefer along with substantial commitments to institution-building, in return for relatively small environmental benefits per project. By contrast, investing in improved efficiency and pollution controls on the large power plants in the region promises very significant benefits with much less preparation work and enables the banks to disburse large sums of assistance more quickly. Such projects are also much more likely to enjoy international political support, because of their transboundary environmental benefits.

Geographical gaps in the coverage of environmental assistance to Eastern Europe also follow predictable patterns. Poland, the Czech Republic, Slovakia, and Hungary have received far more assistance than other countries in the region in part because they were the first to initiate broad reforms and have stronger institutional capacity compared with their neighbors. In addition, it is surely no coincidence that the political, economic, and environmental importance of these countries to their Western neighbors is higher than for countries farther to the East. Again, these gaps are consequential with respect to environmental improvements. Consider the example of the Danube River: it is likely that the bulk of environmental assistance for mitigation of water pollution will go to upstream countries which are better creditors and have more developed institutional capacity to absorb assistance, such as Hungary or Slovakia. From an environmental perspective, however, downstream countries Bulgaria and Romania, less favored by Western donors, suffer far greater environmental costs from Danube pollution and have less national means to counteract those damages.

Resources May Not Match the Problems Variation in the sources of environmental expenditures influences the total amount of resources available, the efficiency with which those resources are spent, and the types of projects that receive funding. First consider grant-based assistance, where taxpayers in the donor country ultimately bear the burden of environmental spending. One of the reasons that bilateral aid budgets exist at all is because of the political constituency they find in domestic firms by tying aid to the supply of goods or services from the donor country—a practice which can significantly reduce the cost-efficiency of the project. EU assistance also carries the restriction that goods and services be procured only from member countries (either on donor or recipient side). Although these patterns bring donors criticism for serving the needs of their own consultants more than those of the recipients, it is important to recognize that the sum total of resources available for environmental assistance is not altogether fixed; the amount available expands to the extent that there is a political constituency for aid programs within the donor and recipient countries. Indeed, recipients may face a choice between environmental remedies that meet the political and economic needs of suppliers in the donor countries versus no remedies at all.

The choice of grant- and loan-based assistance implies very different types of projects. The EU's PHARE program defines its primary mission as institution-building, which is no coincidence—relying as it does on small grants, PHARE lacks the financial tools to implement environmental projects requiring significant capital. PHARE and the bilaterals both receive heavy criticism for conducting too many feasibility studies, many of which fail to provide information that results in public and private investments. However, studies and training are what these donors are best equipped to do. Larger environmental investments will necessarily come dominantly through loan-financing, which means that the place to lay blame for studies that do not turn into concrete projects is usually within MDBs or at the interface between the grant-based donors and the MDBs, in their failure to communicate about or coordinate objectives and procedures so as to ensure that project preparation work takes into account or alleviates banks' constraints for implementable projects.

Because the World Bank and EBRD have access to much larger amounts of capital, but primarily on a loan basis, they are able to tackle much larger projects such as retrofitting power plants or building waste water treatment plants. On the other hand, because the loans must eventually be paid back, the banks can only extend financing to projects with guaranteed means of generating a financial return. Recipients generally resist financing projects such as institution building on a loan basis, because such projects usually would not generate the financial resources to repay the loan. They also resist accepting loans for projects not on the government's priority list. The Czech Republic and Hungary, for example, have been particularly averse to taking on additional government debt for environmental expenditures. Loan assistance is often more of an obstacle than an opportunity for environmental projects.

The point is not so much that PHARE should give loans, or the World Bank should offer grants, but rather that inflexibility in the financial toolset used by the various donors and insufficient efforts to combine their different financial arsenals to improve the absorption of assistance are responsible for the persistent gaps in coverage identified above. More creative efforts to fine-tune the financial mechanisms of each institution and to blend their respective mechanisms by collaborating on projects (which has always occurred to some extent through co-financing arrangements), would improve the fit of available financial assistance with environmental needs.

Different Approaches to Capacity Building: A Strategic Choice   The various donors take strikingly different approaches to where they target their capacity-building efforts, from the level of the central government down to small private investors, which has important effects on the outcome of their assistance. For example, PHARE and the World Bank's focus on lending to or through central government ministries plays an important role in augmenting the central government's financial and technical resources. Yet there are differences in what they seek to accomplish. The World Bank, for example, emphasizes the use of proper economic incentives to implement environmental reform and protection. PHARE, in turn, works to build strong national environmental regulation backed up by strong environment ministries to police the environmental impacts of economic activity. The EBRD takes a different approach altogether to capacity building, prioritizing the growth of private industry. Although that growth is a laudable objective for economic development, it is unclear that private industry will ever be a strong advocate for the environment independent of government inducements. Bilateral donors take a much more eclectic approach to capacity building, but may still have important effects in particular cases, such as building monitoring networks for municipalities or national governments, or providing funding for environmental NGOs in Eastern Europe which might not otherwise be able to survive.

It may be that capacity building in all of these areas is necessary; however, whether the particular institution's approach to capacity building will yield environmental benefits depends heavily on the context of the environmental problem in question. These institutions do not greatly alter their targets for capacity building to meet divergent circumstances. Which institution a recipient selects for a particular environmental project is thus an important strategic choice.

The East Europe case suggests the importance of carefully tailored strategies of capacity building, to ensure beneficial environmental effects at the lowest possible cost. The much-criticized tendency of the major donors to under-use local (recipient) expertise has important (and sometimes legitimate) reasons. For example, in the earlier stages of the assistance program, East European officials were not at all familiar with the normal practices of the EU or MDBs relating to project appraisal procedures, bidding, and contracts. Significant participation of Western consultants was desirable in order to facilitate rapid disbursement and to transfer know-how to recipients on how to prepare, propose, and implement projects with international assistance. Effective capacity building depends on carefully calibrating the balance of outside and domestic experts to account for the level of local expertise on technical issues, the need for transfer of know-how, technology, or management skills from the donor, the familiarity of recipients with standard operating procedures of donor institutions, and the donor's degree of willingness to finance the environmental transfer under untied conditions.

The following section looks more closely at the interests, strategies, and financial mechanisms favored by the MDBs and bilateral donors, which lie underneath this aggregate picture. This analysis will highlight the particular stamp each of these donors imprint on environmental assistance. For purposes of comparison, we group the two MDBs together because of the common constraints incurred by their lending-based approach. Subsequently, we compare PHARE and bilateral donors, who share reliance on a grant-based approach.

2   Structural Biases: Explaining the Characteristic Stamp of Each Donor Institution

As described above, the major donor institutions in Eastern Europe follow distinctively different organizational practices. The emphasis of the multilateral development banks on loans profoundly affects their strategies, in comparison to the grant-based programs, PHARE, and the bilateral donors.

Multilateral Banks: The EBRD and the World Bank
The multilateral development banks (MDBs) share the overarching priority of promoting and financing economic development and growth,

mainly through loans. Even though the environment is not a centerpiece of the banks' operations, both banks have specific environmental strategies and activities. Yet the fact that these environmental mandates are embedded within broader objectives, and the fact that loans are the main form of financial transfer has created clear structural problems that make it difficult for the banks to carry out their environmental policy goals. The outcome has been only a handful of loans with explicit environmental goals in Eastern Europe. While the banks' traditional strategies and financial mechanisms make it difficult for them to tackle the region's environmental problems directly through specific environmental projects, they have sought other ways to address these problems. These include the incorporation of environmental components into other types of loans (e.g., the World Bank's structural adjustment loan to Poland includes the decontrol of coal prices, which is likely to reduce demand for coal); the banks' requirement of environmental impact assessments for projects that have a significant impact on the environment, and technical assistance projects that help to increase recipient country capacity in developing environmental policies.[10]

The greatest obstacle to MDB environmental activity in East Central Europe is the difficulty of producing loans that are attractive to both the recipient countries and to the banks. Environmental loans face three obstacles. First, in terms of public sector loans, environmental lending has not been a priority for recipient countries, which are focusing their attention on other sectors and are not eager to increase their debt. While the banks may suggest "environmental" projects to the countries, they cannot force the finance ministries to agree to them. Indeed, there are many examples where there is demand from environmental ministries for environmental projects, which the finance ministries essentially veto.

Second, a major problem facing the banks is the difficulty in finding "bankable" environmental projects to offer governments or private sector actors (nor is it common for them to receive such project ideas from these actors). An ideal bankable project is one that generates a revenue stream (preferably in foreign exchange); has a strong local contribution; has a sound institutional framework, and sufficient guarantee that the borrower will be able to repay the loan. Many environmental projects are relatively less "bankable" because they do not generate short-term financial return and rarely earn foreign currency needed to repay MDB loans denominated in hard currency.

Finally, bank project staff feel pressure to get projects out as quickly as possible, and therefore tend to make use of existing lending expertise (for example the World Bank's vast experience with big energy projects). Often environmental projects are more time consuming to develop than projects in other sectors because of the need for intensive institution-building to make the project viable; in the environmental sector, the banks are dealing with particularly weak ministries, institutional structures, and sets of problems that were not addressed in the past.

Despite facing similar constraints, the bank's lending operations differ in a number of important ways, reflecting the banks' different institutional interests, and differences in the types of loans they offer, to whom, and for what purposes. There are also important differences in the arsenal of non-loan tools the banks have used to try to carry out their environmental strategies. Clearly, the World Bank has had a bigger arsenal of tools (e.g., options provided through structural adjustment, sectoral restructuring, action plans, and so on) than the EBRD to use in Eastern Europe. These options have given the World Bank a broader latitude to influence environmental development in East Central Europe than the EBRD, which despite early hopes has not turned out to be a leader in "green" banking or environmental reform.[11]

The World Bank[12] is the biggest single creditor for the developing world, now lending over $20 billion annually. With its expert staff of over 8,000 economists, engineers and other professionals, the Bank is also a major source of policy advice and technical assistance for borrowers. Unlike the other major donors, the World Bank has a history of involvement with Eastern Europe.[13] The bulk of World Bank lending goes to governments in the form of project- or policy-based lending, including structural adjustment lending, which is conditioned on macroeconomic policy reform. Bank staff prepare country strategy papers that set the Bank's priorities for lending (Rich 1994:85). Loans are made following an intensive policy dialogue with recipients and must be guaranteed by recipient governments. In Eastern Europe and the former Soviet Union, the top three areas of lending between 1990 and 1994 have been infrastructure and urban development ($3.1 billion); energy ($2.3 billion), and industry and finance ($1.9 billion) (World Bank 1994a).

Since a major reorganization in 1987,[14] the World Bank has a large environmental staff, now totalling around 200. Other outcomes of the restructuring included increased lending for environmentally beneficial projects; and addressing environmental issues in recipient countries by helping them prepare National Environmental Action Plans, which establish environmental priorities and offer policy solutions to them (World Bank 1991c: 1–5; Rich 1990: 307). Additional measures undertaken since then include the creation of a new vice presidency for Environmentally Sustainable Development, a set of environmental assessment procedures, and the placement of environmental coordinators inside each country department to offer advice on environmental issues while identifying environmental projects. The Bank's main environmental objectives are to: help borrowers with environmental priority setting, institution building, and program implementation; address the environmental side-effects of Bank projects; better integrate environmental protection and economic efficiency; and address global environmental issues, mainly through the GEF (World Bank 1993a: 2–3).

In Eastern Europe, the Bank advocates a hierarchy of four measures for its environmental work: first, market reforms with energy price liberalization as a central step toward an environmentally sustainable economy; second, economic and industrial restructuring; third, specific environmental policies including effective regulations and institutions; and fourth, targeted environmental expenditures in sectors where other measures are not effective (World Bank 1992b: 52). Within this framework, the Bank has made only modest progress in specific environmental loans; it has been much more active in trying to incorporate environmental components into its energy-related lending, particularly in helping countries liberalize energy prices. Because energy prices in the region have been highly subsidized, price liberalization produces environmental benefits by sharply reducing energy demand.

Indeed, the vast majority of World Bank loans with environmental objectives or components are in the energy sector. This emphasis on energy lending supports the hypothesis that the Bank does in its environmental lending what it has always done. Typically, energy lending consists of large scale supply-side projects that are mostly cofinanced. They are compatible with the Bank's emphasis on lending for large-scale projects developed with relatively few staff hours. These projects are aided by the fact

that the energy industry usually constitutes an economically strong domestic partner that can provide the technical and political skills necessary
to prepare and undertake such large projects.

These energy loans get mixed reviews. On the one hand, they appeal
to recipients because of the priority accorded to increasing energy supply
in most of Eastern Europe. And it is certainly true that wasteful and inefficient energy generation, distribution, and consumption are major causes
of the region's environmental problems. Notwithstanding the number of
reforms the Bank has undertaken in its environmental policies, however,
NGOs remain highly critical of the Bank's environmental record,[15] and
particularly its weakness in supporting demand-side management and
end-use efficiency in its energy projects (Environmental Defense Fund,
NRDC 1994).[16]

The 1994 annual report lists an $18 million environmental manage-
ment loan as the one "environmental" project in the region between 1990
and 1994. This loan, approved in 1990, was aimed at strengthening the
ministry of environment, and addressing some of the country's most urgent environmental problems (e.g., air pollution management in the
Katowice-Krakow area; river water management). The project has generally been seen as successful in carrying out its objectives, and has served
as a prototype for a $110 million loan to Russia, passed by the Bank
board in November 1994.

An additional three World Bank loans are classified as having "primarily environmental objectives" (World Bank 1994e). These are all energy-
sector loans, and include a $38.4 million district heating rehabilitation
loan to Estonia; a $246 million Czech Republic power and environmental
improvement loan; and a $340 million Polish heat supply restructuring
and conservation loan. Loans with environmental "components" include:
an energy and environment loan to Hungary; a petroleum sector rehabilitation loan for Romania; and two non-energy loans—a Bulgarian water
companies restructuring loan, and a water supply loan to Albania. The
Bank's structural adjustment loans to countries in the region also include
components to decontrol energy prices (World Bank 1994b). Among the
projects in the pipeline are a loan to Slovenia focusing on reducing air
pollution in towns and cities; and a geothermal loan and waste water
treatment loan for Poland.

Outside of its traditional lending activities, the Bank, along with the EBRD, has been active in a number of multidonor regional programs for the Black Sea, Baltic Sea, and Danube River. The World Bank has been particularly visible in its role as the author of the Environmental Action Programme for Central and Eastern Europe (EAP), making use of its expertise in writing national environmental action plans in a major effort to establish environmental spending priorities for the entire region.

In contrast with the World Bank, the EBRD is a small, regional bank, still in the process of defining its role. The EBRD is the world's newest MDB, created in May 1990 with around $13 billion in capital as a European initiative specifically to help foster Eastern Europe's transition toward market economies through cultivation of the private sector and promotion of democratic institutions. Part of the political bargain which brought the EBRD into existence gave the bank the mission of building the private sector through project-based lending.[17] The EBRD has sought to fulfill its mission by emphasizing projects in privatization and economic restructuring, by promoting foreign investment (in part by investing in projects with foreign companies), and by focusing on the promotion of small and medium-sized enterprises. While it is able to lend money to central governments, its focus is on assistance to the nascent private sector. It thus sees itself having a "bottom-up," more decentralized approach, versus the World Bank's "top-down," public lending approach.

The EBRD's overall strategy differs from the World Bank's in three important respects. First, being project driven, the EBRD does not nest its lending within the broad type of macro policy dialogue the World Bank establishes with borrowers. Thus, EBRD staff do not have the same incentive as World Bank staff to make sure a project fits in with a broader strategy (environmental or economic) for the country. Second, the EBRD's emphasis on private sector lending gives it the flexibility of working without government guarantees in the majority of its lending, and also provides it with a number of ways in which it can tailor its financial mechanisms to projects: the Bank can provide loans, take minority equity stakes in projects, guarantee bond issues, and so on. To date, however, this increased flexibility does not appear to have given the EBRD noticeable advantage in environmental lending. Finally, the EBRD limits its commercial financing to 35 percent of total project costs, and its minimum loan to 5 million ECU. This means that the Bank has found itself cofinancing big projects with groups of large Western companies; indeed, foreign joint ventures accounted for almost half of the Bank's private sector investments (EBRD 1994a: 13). The result of this strategy is that projects are often driven by demand from foreign investors, a factor that has not produced many environmental projects. By the end of 1994, the Bank's three major sectors of lending were: finance/business (1.6 billion ECU); transport (1.1 billion ECU); and telecommunications (796 million ECU).

The EBRD is unique in the sense that it is the first MDB with an environmental mandate built into its articles of agreement. Article 2 (1) (vii) commits the bank "to promote in the full range of its activities environmentally sound and sustainable development" (EBRD 1990). Over the first two years of its existence, the EBRD developed a set of environmental "policy priorities." In its desire to play a "leadership role in the environmental recovery of central and Eastern Europe," the Bank outlined its goals to: (1) help countries develop their environmental policy; (2) promote the use of market-based instruments in environmental management; (3) encourage development of environmental goods and service industries; (4) adopt "adequate environmental assessment" procedures in all of its activities (EBRD 1992a).

To date, the EBRD has made minimal progress in carrying out the first three goals of its ambitious mandate, and has ended up emphasizing the fourth. Like the World Bank, the EBRD is also interested in pursuing some stand-alone environmental projects, and has sought to do so by focusing on municipal government environmental infrastructure, but project development has proceeded slowly. The idea is that economic decentralization in Eastern Europe will give increasing responsibility to local governments, which will have growing power to affect the region's transition to market economies (EBRD 1992b).

A number of municipal environmental projects are in the pipeline, but by late 1994 only two had made it past the approval stage. The first is a loan of about 23 million ECU ($28.5 million), supplemented by grants from PHARE and Finland, to the Tallinn Water and Sewage Municipal Enterprise. This loan aims to rehabilitate and expand the city's wastewater treatment plant, while restructuring the operator into a self-financing, self-managed utility (EBRD 1994b: 6). The second is a 22.6 million ECU ($28 million) loan to rehabilitate water services in five Romanian cities. Other planned projects, including two Polish wastewater projects, were stymied by a lack of commitment from the local government. Bank staff have pointed out that municipal governments are often not eager to bring private sector actors into their work. Other observers have added that in some cases, the local governments were unwilling to borrow the amount of money recommended by the EBRD.

Ultimately, the EBRD's ability to carry out its environmental mandate has been thwarted by the common difficulties MDBs have in making environmental loans attractive, as well as a general lack of incentive within the Bank to emphasize environmental projects. The EBRD is under pressure to prove its worth, given a very low disbursement rate, and criticism of the management practices of its first president, Jacques Attali. Of the 2.1 billion ECU in financing approved in 1993, for example, only 409 million ECU was disbursed. News that in 1992 the Bank spent twice as much on outfitting its offices as it disbursed in loans did not help the Bank's reputation, ultimately resulting in Attali's resignation and his replacement by former IMF chief Jacques de Larosiere (Preston and Burns 1993). The Bank is project-driven, and staff are under extreme pressure to get loans approved and disbursed quickly, to the neglect of environmental projects. Disbursements are often slowed by factors beyond the Bank's control; for example, as a brand new organization, it takes time before it can operate smoothly; also, it is difficult to provide funds to newly born private sectors while meeting prudent lending criteria.

As a way of dampening criticism over its dearth of environmental projects, the EBRD makes a point of emphasizing its attempts to address environmental issues in design of all projects through its environmental procedures. These contributions, however, are difficult to quantify. For example, a loan to complete a section of a Hungarian highway was designed to have minimal impact on natural habitats and human settlements, and includes mitigation measures such as forest belts to help reduce noise and air pollution to nearby communities.

Outside of traditional lending activities, the Bank has also undertaken some technical cooperation projects that may help increase environmental concern and awareness in the private sector and affected public. For example, it sponsored a research project on environmental standards and the implications of impending harmonization of those standards with Western Europe as a way to assist local and foreign investors in making investment decisions. The project also analyzed institutional changes that are needed to improve environmental monitoring and enforcement. In addition, it initiated a research project to help Eastern European countries adopt better public participation policies, responding to longstanding NGO pressure for the banks to better involve the affected public in their project decisions. (EBRD 1993).

Ultimately, both banks do what they do best as MDBs, applying preferred solutions to new problems, while constrained by the limitations of a lending-based approach. Variation in the banks' institutional interests and tools is also evident. The result has been an emphasis by the World Bank on its traditional strength in energy lending, as well as its involvement in macro agenda-setting exercises. The EBRD, in turn, looks more like a merchant bank, with an even greater gap between its environmental policy goals and its actions.

Grant-Based Assistance: PHARE and the Bilateral Donors

PHARE and the bilaterals, the major grant-giving donors, face very different but still serious constraints in translating their assistance into positive effects on the environment in Eastern Europe. For each of them, an identifiable set of political interests and financial transfer mechanisms cause them to favor distinctive environmental strategies. Both the EU's PHARE program and bilateral environmental aid rely almost exclusively on relatively small sums of grant assistance. As a consequence, both lean toward small technical assistance projects, such as sectoral and feasibility studies, training, pilot projects, technical expertise, legislative assistance, and small grants for the purchase of essential equipment. Limitations on what can be accomplished with relatively small sums imply that the role of these donors is at best catalytic of other efforts; often, underlying political motives and resultant procedures reduce the catalytic effect of PHARE and bilateral grants.

The PHARE program began at the July 1989 Paris Summit, where the G-7 countries decided to provide aid to consolidate democratic reforms and market economies in Poland and Hungary. The European Union later decided to set up an integrated EU program for that purpose, which became PHARE. Over the next three years, the program was expanded to include Bulgaria, Czechoslovakia, the GDR,[18] Romania, the three Baltic states, Albania, and Slovenia.

Less altruistic purposes lay behind the PHARE agenda as well. EU members supported reform efforts in part due to fear that political and economic instability in its Eastern neighbors might spill over into the West. At the same time, Poland, Hungary, and Czechoslovakia's requests for eventual membership in the Union and for trade liberalization in "sensitive sectors" to benefit East European exporters met strong resistance from EU member states. Unwilling until 1993 even to discuss specific conditions under which these states might eventually be allowed to join the Union, member states opted for an aid strategy preparing East European countries for eventual EU membership while holding them off for as long as possible.

The EU's environmental programming does reflect a certain inclination to remake recipients governments in the EU's own image. For example, one of PHARE's defining characteristics is its relatively strong emphasis on command-and-control mechanisms. Much of its institution-building effort is aimed at strengthening national environmental ministries and augmenting their available resources for legislation, monitoring, and enforcement, whereas other donors tend to extend their capacity-building efforts to a more diverse set of government and private actors, or focus more on capturing environmental benefits by strengthening market institutions and incentives instead of strengthening government. Since eventual accession to the EU is a likelihood for several PHARE countries,[19] the program focuses on promoting policy reform and environmental legislation to harmonize recipient and EU standards. The PHARE program also takes a markedly top-down approach, channeling assistance to national government ministries, rather than to local governments, for instance. This preference for implementing top-down, command-and-control solutions mimics the EU's own Commission-dominated, bureaucratic approach to environmental policymaking.

Bilateral assistance, often referred to as G-24 assistance, has much less coherent goals than the EU program. Bilateral aid features some of the most blatant examples of "solutions chasing problems," where donor institutions with their own agendas seek out collections of problems that allow them to export favored goods or services, to achieve domestic political gains by mitigating transboundary environmental problems in neighboring countries, or to focus on particular sectors in which they have substantial experience at home. For example, the Connolly/List chapter in this volume demonstrates how bilateral donors have seized on nuclear safety problems in Eastern Europe as an opportunity to channel politically legitimate subsidies to their domestic nuclear industries. The dominance of donor political priorities tends to work against building institutional capacity among recipient countries to develop and enact their own priorities.

PHARE, by contrast, fancies itself a demand-driven program. The EU Council of Ministers sets indicative budgets for each country. Within that set amount, recipient countries are expected to designate funding priorities and propose projects to the Commission. Hence the selection of projects and funding levels for the environment sector reflects the relative priority of those activities within the recipient country, which has led to the relative decline of environmental support as a percentage of the total PHARE budget from 20.5 percent in 1990 to 10.7 percent in 1991 to 6.3 percent in 1992 (PHARE Information Service 1993). Cumulatively over the period 1990–1994, environmental assistance made up 10 percent of the total PHARE budget (PHARE 1995: 9). The steady drop over time reflects the fact that economic needs have overwhelmed initial enthusiasm for environmental expenditures in Eastern Europe, a fact which has haunted MDB environmental lending as well. In practice, PHARE's demand-driven approach has been compromised by the difficulties East European governments initially had in establishing national priorities for foreign assistance projects expressed as concrete projects (Melzer 1993: 6–8). As a result, PHARE officials often stepped in to help the ministries define their priorities.

PHARE had at its disposal in 1993 an overall budget for all sectors of 1.1 billion ECU. The program's reliance on grant aid gives it the advantage (in contrast to the MDBs) of being able to support projects that will not yield commercial returns and face fewer obstacles to disbursing assistance, but the drawback of having the limited means most suited to support smaller-scale technical assistance projects. For example, PHARE's total environment sector budget for Poland amounted to 22 million ECU in 1990, 35 million ECU the following year, 18 million ECU in 1992, zero in 1993, and 14 million ECU in 1994. (PHARE 1995: 21). At least 60 percent of the funds allocated to any given project must go for technical assistance. Recipient countries complain about the heavy dominance of technical assistance within PHARE, claiming that they already know what the problems are and just need help to encourage investment and funds to purchase equipment and hire their own consultants (*Le Monde,* 10 April 1993).

Those demands face obstacles in the fact that PHARE assistance to Eastern Europe is subject to complex bidding procedures, which among other objectives ensure that member states get something in return for their donations, in the form of contracts for supplies and services. Only EU or recipient country companies are eligible to participate in the bidding for consultancy jobs or procurement of supplies. Lengthy bidding procedures have led to many recipient complaints that PHARE is overly bureaucratic, with long delays in disbursement exacerbated by staff shortages both within PHARE and on the recipient end. Typical delays between when consultants start work on a project and when the process begins are twelve to eighteen months, with program implementation lasting three to five years. In the future, PHARE hopes to speed up this process by using a fully decentralized implementation process begun in 1994, utilizing local Project Management Units.

By contrast, bilateral assistance is much less bureaucratic. Britain's Know-How Fund, for example, typically sends in consultants eight to twelve weeks after receiving a request (*Economist,* 10 April 1993). The advantages of speed are often offset by a loss of control for recipient countries, however. As a general rule, bilateral donors do not allocate a set amount of assistance to particular countries, let alone the environmental sector, and so recipients have to compete with each other for assistance. Recipient countries often feel compelled to accept what is offered them, because of the scarcity of grant aid and their weak capacity to define their own priorities for environmental assistance.

Potentially at least, bilateral assistance can employ a mix of instruments: offering financial assistance, which comes as loans and credits, in addition to the grant-based technical assistance. In practice, however, bilateral environmental aid has been disbursed almost exclusively as grants, which says something about the utility or attractiveness of loan mechanisms for environmental purposes. In fact, of the total G-24 contribution to environmental assistance (including PHARE) from the first quarter of 1990 through the second quarter of 1992, over 95 percent of the 501 million ECU disbursed were grants (Fennesz 1993: 198). In other sectors, bilateral donors have been able to make much more use of credits. For example, from 1990 to 1992, as a proportion of overall bilateral assistance totals, official trade credits amounted to as much as 70 percent of assistance to the CSFR and 42 percent of assistance to Hungary and Poland (Institute for East-West Studies 1992: 23). Recipients' unwillingness to borrow and a scarcity of required local counterpart funding have proved decisive obstacles to absorption of bilateral loans for environmental purposes, creating logjams of projects which have reached implementation stage but cannot proceed further (Melzer 1993: 18–19).

Given the array of institutional interests and available financial transfer mechanisms we have laid out, we can now make sense of why PHARE and bilateral assistance tend to gravitate toward distinctive kinds of solutions to environmental projects. The preferred solutions of the PHARE program tend to be institutional strengthening of national environment ministries and regulatory frameworks, regional cooperation projects, pan-European monitoring and information collection systems, and sectoral and feasibility studies of specific environmental issues with an emphasis on harmonization of policy and standards, linked to equipment supply, action plans, and pilot projects. Bilateral donors prefer less systematic collections of technical assistance projects, such as feasibility studies and training that pave the way for donor exports of technologies or expertise, or for improvements in transboundary environmental problems.

Ideally, PHARE's goal is to foster the recipient government's capacity to decide for itself on an overall environmental strategy, with priorities and projects—in other words, institution building at the level of national government. Central to this effort are PHARE's program implementation units (PIUs), which provide concentrated assistance to help Eastern governments decide upon their priorities. The PIUs were set up as offices within the recipient countries' national environmental ministries and staffed by ministry officials seconded to PHARE along with a small team of EU-sponsored consultants hired by PHARE. Some East European governments, such as Hungary and Czechoslovakia, have given these offices a great deal of responsibility for managing and coordinating environmental assistance projects (Gutner 1993).[20] PIUs provide for the direct exchange of know-how between external experts familiar with the ways and procedures of the European Commission, and recipient environment ministries familiar with local conditions and needs but not with Western bureaucracies. These offices also provide the Commission with an advantageous source of influence directly inside the region's environment ministries.

As another means of strengthening national environment ministries, PHARE provides financial and technical assistance to help these ministries set up self-sustaining funds for environmental investment and improvements, drawing their major financing from charges and fines levied on pollution and emissions. These national Environment Funds are then recycled to provide financial incentives such as grants, soft loans, or interest subsidies for investments in environmental projects. PHARE provides support for environmental funds in Poland, Hungary, Slovakia, and the Czech Republic, and plans in the future to fully integrate its environmental assistance into the funds (Regional Environmental Center 1993). In these cases, PHARE augments domestic resources, perhaps sustaining a higher level of concern for environmental issues than these countries would be able to accomplish alone. Because of the region's high budget deficits and weak financial sectors, earmarked environmental funds mixing command-and-control regulations with market-based incentives are expected to provide an important source of finance in the future. Already, such national and regional funds accounted for over 40 percent of environmental investments in Poland in 1991 (Environmental Action Programme for Central and Eastern Europe 1993: VII–6).

In one important way, the PHARE program falls short of its promise for building capacity, given its penchant for routinely using Western consultants rather than local expertise to carry out its projects. To be fair, short lead times for PHARE programs and the experience of working with countries who had not before received aid from the EU made it more efficient, especially early on, to use Western consultants already knowledgeable about EU procedures. In addition, managerial expertise, though

not technical expertise, has been a weak point in East European governments. Still, the recipients complain. "PHARE is sometimes driven more
by the needs of consultants than by the needs of the country," claimed
the secretary-general of Hungary's Foreign Trade Ministry, Endre Juhasz,
who argued that many feasibility studies have been conducted that will
never be implemented (*Economist,* 10 April 1993).

PHARE does better at stand-alone environmental projects than at integrating environmental strategies into other industrial restructuring activities—a consequence of the program's strategic choice to target
national environmental ministries for assistance, as opposed to strengthening environmental moles in multiple sectors. Between 1990 and 1994,
PHARE financed several hundred environmental projects in Eastern
Europe, mostly studies, training, small demonstration projects, or preinvestment work. Despite the fact that PHARE consultants train recipient
country staff in environmental impact assessment procedures, it is noteworthy that until 1994, PHARE's programs or projects in other sectors
were not subject to environmental impact assessment (Chabrzyk 1993: 7).

PHARE is the strongest advocate of regional programming within the
assistance community, reflecting the EU's distinctive position that regional
environmental cooperation is a means toward broader political cooperation and a necessary corollary of regional trade integration. PHARE de-
votes 15 percent of its budget to financial and administrative support for
several major regional programs, including ones for the Danube River
Basin, the Black Triangle, the Baltic Sea and the Black Sea. However,
PHARE can only carry a regional project through its early stages, such as
creating a coordination body, formulating strategies and master plans,
and setting up pollution-monitoring systems; beyond that, coordination
with other donors is essential. For example, in the Black Triangle project
aimed at reducing air pollution in the coal-producing areas of eastern
Germany, Poland, and the Czech Republic, PHARE has allocated funds
to create an organizational structure for negotiations among the three
countries and the EU, to build a joint air pollution monitoring system,
and to draw up an urgent action plan and a master plan for rehabilitation
of the region. For capital investments in "hardware" projects such as improving the operating efficiency of power plants in the region, or retrofitting them with dust filters and desulfurization equipment, however, the
countries will have to look primarily outside PHARE.

PHARE has undergone some adaptation over the last five years to improve its performance and combat limitations of its small grant mechanism. The major changes include a shift from annual to multi-annual programming as of 1993, in order to allow PHARE to conduct longer-range planning and concentrate on larger-scale programs, and a shift from a "shopping list approach" which featured large numbers of discrete, unrelated projects in the first year, to a more integrated program approach in subsequent years (Commission of the European Communities 1992). For the next five years, PHARE has pledged to devote more funds to capital projects to combat the most urgent environmental problems, especially wastewater treatment, reducing air pollution from industrial enterprises, and preserving biodiversity. The program will also continue its support for long-term, recipient-based environmental financing strategies such as the revolving national environment funds (PHARE 1995).

A new cross-border cooperation facility created in 1994 will build on PHARE's record in regional programming. The 150 million ECU facility will provide funding for multi-country problems affecting both an EU member state and a PHARE neighbor, making use of EU structural funds. For example, the facility is preparing master plans for hazardous waste storage and management in Latvia, Estonia, and Denmark. Like earlier regional programming and institution-building efforts, this new initiative aims to catalyze the harmonization of environmental laws and standards between the EU and its Eastern neighbors.

Because of heavy reliance on limited grant resources, bilateral assistance tends like PHARE toward a preponderance of training, institution building, and feasibility studies. However, the preinvestment studies of the two groups lead in somewhat different directions. PHARE, moving toward harmonization of European environmental standards, is financing more basic stocktaking studies such as inventories of environmental problems and pollutant sources in the Danube River Basin and the Black Triangle, as well as the pan-European report on the present state of environmental information called the "State of the Environment in Europe." In contrast, bilateral G-24 aid appears more investment-oriented, serving commercial motives of donors. The targets of bilateral assistance span the entire range of actors, from national environment ministries to utilities, local governments, or the private sector, which makes effective coordination of bilateral assistance very difficult for the recipient countries and exacerbates the donor-driven nature of bilateral programs. To date these bilateral programs have not yet resulted in any significant level of environmental investments (Melzer 1993: 21–22), although the new $1 billion U.S.-Japan Initiative will change the landscape in this area.

The high proportion of bilateral aid tied to the purchase of donor goods and services, such as the use of consultants from the donor country to perform feasibility studies for future investment projects or the purchase of particular technologies sold by a donor country firm, provides evidence for the claim that donor priorities drive bilateral aid programs. Aid tying creates political constituencies for bilateral aid programs (Cassen 1986), but the practice also inflates costs and reduces efficiency by constricting competition among different suppliers and technologies.

Donor interests also exert a selectivity bias toward projects promising environmental or commercial benefits to the donor. For example, Austria focuses its bilateral assistance on nearby Hungary, the Czech Republic, Slovakia, Poland, and Slovenia, and funds mainly projects which reduce transboundary air or water pollution into Austria. Finland also directs its efforts at transboundary issues affecting it, by providing assistance for reducing airborne and waterborne pollution in the Baltic Sea region. France focuses on river basin management, drawing on French expertise in this sector. Japanese bilateral aid to Eastern Europe has concentrated on municipal waste management and flue gas desulfurization, prime areas for the export of Japanese technology and expertise (Smith and Coppendale 1993). Such selectivity biases, though not necessarily detrimental to the recipient country, imply that projects selected often do not represent the most important environmental expenditures in Eastern Europe, at a time when limited assistance resources make prioritization critical. Additionally, selectivity biases result in the proliferation of discrete, disconnected projects, rather than complementary efforts in priority areas.

Not all bilateral assistance follows the above model—for example, the U.S. and Denmark both take a more regional approach. In 1989, U.S. assistance funded a few isolated projects, notably the Regional Environment Center in Budapest,[21] a facility designed to expand public participation in environmental decision making by providing small grants to environmental NGOs, coordinating task forces to address issues of regional interest, and acting as an information clearinghouse. After that year, the U.S. effort shifted toward institution building on a regional basis. USAID aims to promote policy and legal reform in the environmental sector, to help the region's governments provide efficient and effective environmental services, and to improve the private sector's efficiency at producing goods, preventing pollution, and developing business and financial skills related to the environment. To accomplish these aims, USAID relies on a combination of training courses, technical assistance projects, and demonstration projects (U.S. General Accounting Office 1994: 38–41).

Probably the most important thing to recognize about PHARE and bilateral assistance is that the effectiveness of these efforts, because of their relatively small scale, hinges on their ability to clarify priorities for environmental spending, strengthen recipient institutions for the absorption of assistance, and connect with or catalyze additional environmental investment or assistance efforts. These objectives require a greater level of coordination among donors and recipients than that realized so far.

3 Evaluation: The Need for Coordination and Prioritization

Because the major donor institutions select the set of environmental problems that fit their prior agendas and existing transfer mechanisms, efforts to coordinate and prioritize aid projects have been a struggle. Cofinancing is one important strategic element in enhancing donor coordination as well as recipient absorption. Because of its restriction to lending only 35 percent of private sector projects, the EBRD co-finances projects with the World Bank, as well as with commercial banks, export credit agencies, and bilateral institutions. A considerable number of PHARE's feasibility studies are provided upon request by the MDBs, to complement their assessment of lending opportunities. It is commonplace, however, for feasibility studies prepared by PHARE or bilateral donors not to meet MDB standards of "implementability," because those studies are written without explicit consideration of bank procedures or the prerequisites of a successful, bankable project.

Regional cooperation projects have also provided opportunities for both donor and recipient institutions to collectively arrive at more effective solutions to transboundary problems, such as air or water pollution. However, the record of these projects is mixed. While the Environmental Action Program, discussed below, has achieved some success as an exercise in agenda setting and improving coordination of assistance, the Black Triangle and Danube River Basin programs have run into major delays in administrative start-up and disagreement on priorities for intervention among affected states.

Formal efforts at coordination aside, donors have been criticized within the region for not coordinating their projects with other donors. There are many examples of cases where more than one donor was requested to fund the same project. In one case, a donor country and a private investor from that country discovered that they were each planning to set up industrial waste treatment plants within miles of each other (Smith and Coppendale 1993: 39). In addition, there is no central register of feasibility studies, even in the recipient countries, nor is there an effective common database of environmental projects underway. Bilateral donors in particular are often deliberately not open about their assistance efforts. Recipients also have certain incentives to conceal duplication of donor efforts in order to maximize assistance resources. The G-24 office at the European Commission has produced a database to facilitate coordination, but it is far from comprehensive.

Related to the failures of coordination, attempts to establish priorities for assistance have also fallen short over the last five years. The EBRD and PHARE tended to rely on recipient countries to establish their own priorities for funding, but these countries have not lived up to the task. Many bilateral donors have intervened on the basis of their own priorities, which frequently do not correspond with the most important environmental needs in recipient countries. The World Bank has been a leader in trying to establish priorities for environmental expenditures in Eastern Europe, but there is some element of imposing its own policy template on the region in this exercise as well. The real danger in failing to prioritize assistance projects is that the willingness of Western governments to provide aid may be exhausted well before the most important needs in the region have been addressed.

Donors and recipients have recognized some of the weaknesses of the assistance effort and have sought to strengthen institutionalized attempts at coordination and prioritization of environmental expenditures. Early attempts to encourage greater cooperation among donors were overcome by organizational inertia. When the PHARE program was initially set up in 1989, donors envisioned that it would coordinate assistance by all the G-24 countries to East European governments in a handful of "high priority" sectors, including environmental protection. However, this initiative quickly dissipated, given different assistance agendas among the G-24 states, many of which use bilateral assistance to promote their own commercial interests. PHARE became the EU's own vehicle for assistance to the region, and G-24 coordination in the area of environment ended up as little more than periodic meetings and a weak project database.

Portraying pan-European environmental activities as the natural route to a more politically integrated Europe, former Czechoslovak federal environment minister Josef Vavrousek lobbied in 1990 and 1991 for a permanent council of European environment ministers. The proposed council would consist of regular conferences of all European environment ministers, responsible for setting action plans and targets for environmental revival, formulating strategies and generating funding to implement those plans, coordinating pan-European monitoring and information systems, and unifying European environmental legislation (Environment for Europe 1991: 17–23). This campaign was brushed aside by donors who were not eager to create new institutions. Vavrousek did, however, succeed in initiating what is now referred to as the Environment for Europe process, which are biannual meetings of environmental ministers from East and West, as well as key multilateral institutions, which seek to strengthen attempts at donor coordination and prioritization of environmental expenditures. We argue that this process shows signs of improving but not ending the current of organizational inertia.

Perhaps the most important outcome of the Environment for Europe process to date has been the Environmental Action Programme for Central and Eastern Europe (EAP), an ambitious document that seeks to identify priority environmental problems in the region and offers ways of solving them.[22] The document, written by the World Bank with input from a variety of donor and recipient actors, was endorsed by a group of Eastern and Western environmental ministers and officials from international organizations at the Environment for Europe Conference in Lucerne, in April 1993.[23]

The EAP argues that the incredibly high costs of environmental restructuring and the reality that Western aid can only cover a minute portion of those costs requires donors and recipients to identify the highest priority environmental problems and to look for cost effective solutions. More specifically, the EAP argues that the major criterion for identifying the most important environmental concerns in the region should be the level of damage to human health. The EAP argues that the highest priority should be air pollution caused by particulates; other areas it recommends addressing include lead found in air and soil, caused by transport and lead and zinc smelters; nitrates found in water; and contaminated drinking water, resulting from poor disposal of hazardous and nuclear waste. The problems identified as highest priority issues are thus primarily local, and not transboundary problems.

In terms of policy recommendations, the EAP presents four key criteria for policy prioritization: (1) It argues that policymakers should support "win-win" economic reform policies that also have environmental benefits, such as market reform that results in higher energy prices; (2) it argues that environmental policies should be targeted and should establish a framework of institutions and incentives to discourage emissions of pollutants and biodiversity loss in a cost-effective manner; (3) it argues that environmental spending should go to projects with the highest benefit-cost ratios; and (4) it argues that "modest" expenditures should be set aside for programs that have a long lead time from start to finish but still have high benefit-cost ratios (Environmental Action Programme for Central and Eastern Europe 1993: II–26).

The EAP also discusses a wide range of specific policy options, in order to show policymakers the pros and cons of different strategies. For example, it compares command-and-control policies with market-based incentives. In general it emphasizes market approaches, but is cognizant of some of the difficulties in implementing them in the region. With respect to strengthening domestic institutions, the EAP discusses the importance of increasing interministerial coordination; shifting more responsibilities for environmental management to local authorities; and improving the "functional capacity" of environmental ministries, through greater training, and the adoption of basic environmental management techniques such as setting goals, targets, setting up sectoral studies, and so on. Chapter 5, in turn, contains a very specific list of suggested short- and long-term expenditure priorities.[24] Finally, the EAP is explicit in its desire to be seen as a process, a guide for project development, a basis for "consensus and partnership" among Eastern and Western countries and multilateral institutions.

It is too early to evaluate the EAP. Governments in the region were highly disappointed that the effort brought no additional donor funding targeted toward specific projects. Critics also argue that the EAP is "only a program," whose measures may have been endorsed by ministers but will unlikely be adopted by many (Baumgartl 1994: 148). However, the Environment for Europe process is significant for two reasons. First, the EAP is the first major attempt to set environmental priorities for the region as a whole. Second, the Lucerne meeting produced a potentially important institutional innovation, the Project Preparation Committee (PPC). Its creators devised the PPC as a matchmaker to bring together donors on project funding in order to help make the EAP "implementable." The PPC's goal is to act as a clearinghouse for projects between multilateral and bilateral donors that meet the EAP's criteria. Such blended-projects are of interest to all involved actors and have the potential to increase East European abilities to absorb available environmental assistance. Recipients have more incentive to accept an MDB loan if it is sweetened by a bilateral grant component. MDBs, in turn, are happy if they can sell their loan-based projects more easily when they are supplemented by grant money. Finally, such projects are also beneficial for bilateral donors, as a way of diffusing criticism that they fund many studies leading nowhere. Under the PPC, such studies can be directly linked to future loan-driven investments. The search for these combination projects also encourages cooperation and improved sharing of information among all involved actors.

The PPC itself includes Austria, Denmark, Finland, France, Germany, Netherlands, Norway, Sweden, Switzerland, the United Kingdom, the United States, EBRD, EIB (European Investment Bank), World Bank, and the Nordic Environment Finance Corporation. Donors have hired four people, two situated at the World Bank and two at the EBRD, to organize potential PPC projects and match donors with each other. Donors are beginning to take the PPC seriously. At a September 1994 meeting of the PPC members in Norway, donors circulated lists of projects for which they sought cofinancing from other donors, and found a generally healthy response from each other. For example, of the thirty-five projects on the World Bank's list, twenty-four received firm commitments of involvement from other donors. Bilateral donors on their own are also seeking to demonstrate that their aid is more aligned with the EAP's recommendations. For example, the new U.S.-Japan Joint Initiative, will fund projects that are "in accord" with the EAP's priorities.[25] Donors hope to present a sizeable list of PPC projects under implementation at the next Environment for Europe conference in Sofia in October 1995.

The PPC may be an important way to catalyze environmental assistance to the region by providing donors a channel for cooperation that will result in projects which are attractive to recipients. Yet some important issues remain unresolved. First, since the PPC is essentially a club of donors, recipients are not directly involved in the process. It is unclear how their position outside the club will impact their desire to accept the assistance offered. It is also unclear to what extent the EAP's priorities are in harmony with individual recipient country priorities, and what happens when those priorities are far apart. Second, because donors clearly do not want to create a new bureaucratic organization around the PPC, the PPC will remain a clearinghouse with a tiny staff; such a small group may have difficulty catalyzing the overall donor community. Indeed, the PPC itself is a watered-down version of a more institutionalized Project Preparation Facility initially proposed as part of the EAP. In addition to undertaking the tasks done by the PPC, the proposed Facility would have instituted feasibility studies, and assessed projects proposed by Eastern European countries (Environmental Action Programme for Central and Eastern Europe 1993: VII–4). Finally, it is too early to know whether or not the PPC projects will consist of new projects that would not otherwise exist, or whether the PPC vehicle will provide a way for donors to get new help carrying out projects they might have otherwise done on their own.

The history of development assistance certainly suggests that improving donor coordination will be difficult at best; in this volume, Fairman and Ross emphasize that the continuing difficulty in achieving coordination over the last twenty-five years reflects real underlying conflicts of interest among donors and recipients. Bilateral donors, for example, stand to lose major environmental and export benefits from a technocratic approach that directs donor aid to areas of recipient priority. The major multilateral donors in Eastern Europe clearly have divergent strategies for how best to improve the environmental situation in the region, which can impede coordination. The fact that donors insisted on weakening the original vision of the PPC reflects an understandable tendency by donor institutions to resist oversight mechanisms that can infringe on their autonomy.

We are convinced, however, that an institutionalized process for broad-based examination of the priority environmental needs of Eastern Europe, connected to explicit consideration of how existing donor institutions and financial mechanisms can best divide or combine efforts to cover those areas, lays a reasonable foundation for enhanced effectiveness of environmental assistance. Prodding donor institutions to move beyond business-as-usual practices and experiment with new approaches or innovative financial mechanisms can be a formidable task, and the PPC is not as strongly outfitted for this goal as it might have been. Still, the Environment for Europe process and the EAP and PPC that resulted from it do push against the current of organizational inertia in the following respects.

Common priority-setting efforts with participation both from the donor side and from East European recipients undercuts the tendency of donor institutions to run piecemeal projects divorced from a clear strategy for environmental improvement in the region and enhances recipient ability to impose a more coherent guideline on project selection. The Environment for Europe process greatly increased information flows and awareness about what activities the various donors had planned or are underway in the region, reducing the likelihood of duplication. The process created an ongoing, high-level dialogue focused on priority problems and potential solutions between donor institutions and East European recipients, which enhances the possibility that recipient needs will become a more important driver of the activities of the donors. The problem-driven orientation of the EAP is also conducive to continuing discussion about which financial and policy mechanisms are more or less appropriate to solve particular problems, and consequently which donor institutions are best equipped to address those issues, or how individual institutions might innovate within their preferred solutions or available financial mechanisms to better address a given problem. Finally, the PPC encourages formal efforts at cofinancing, both by increasing the speed and ease with which donors can find matching donors, and by creating significant public relations incentives for them to do so. These efforts provide an opportunity to capitalize on the varying strengths of multiple donors while enhancing coordination and giving donors a common stake in evaluating and improving their aggregate effectiveness in delivering environmental assistance to Eastern Europe.

4   Conclusion

Established donor institutions exhibit a common and powerful tendency to rely on favorite solutions and a preexisting stock of financial mechanisms when confronting new problems. Organizational inertia has led to predictable consequences in environmental assistance to Eastern Europe, such as neglect of important environmental problems in the region with respect to human health, resources sunk in duplicative studies not connected to concrete projects, and low rates of absorption for certain kinds of assistance which are not well matched to recipient needs or priorities. Donor institutions have tried to improve upon their efforts to disburse environmental assistance to the region. Done in isolation, however, such adaptation poses certain risks. Indeed, even if the multilateral banks get better at lending and the bilaterals improve on their ability to do technical assistance, the aggregate picture of environmental assistance and the overall fit of donor efforts with each other may still suffer critical gaps and weaknesses. The real challenge, therefore, is to create mechanisms that facilitate coordination among the many different donors and recipients in the region, and allow for synergies among projects so that the whole assistance effort becomes something greater than the sum of its parts.

Most of the environmental assistance to Eastern Europe is channeled through institutions whose mandate, structure and financial mechanisms were shaped for much broader goals than environmental assistance. "Garbage can" decision making can become an obstacle to effective environmental assistance wherever donor institutions were not designed with environmental problems in mind. This phenomenon is not unique to Eastern Europe. Indeed, environmental projects are often much harder to fund than other development projects, as they require higher levels of institution building, more flexible and concessional financial mechanisms, strong linkages among multiple related sectors, and more innovative ways of defining the problems to be solved. Traditional institutions for development assistance have certain blind sides when it comes to environmental projects or side effects, as NGO criticism of MDB activities continually reminds us.

Nonetheless, it is possible for established institutions to implement beneficial environmental assistance programs. Indeed, we have argued that the Environment for Europe process along with the Environmental Action Programme and the Project Preparation Committee are steps in the right direction. Although these are only tentative steps, they disrupt tendencies toward organizational inertia and introduce a more problem-driven focus for evaluating how well environmental assistance programs are doing in meeting priority needs in Eastern Europe.

Overcoming many environmental problems will require the involvement of multiple, established donors and recipients. This chapter invites further research on how politically acceptable institutional oversight mechanisms could be designed to overcome organizational inertia in the context of new problems or objectives. The chapters by Fairman and by DeSombre and Kauffman in this volume echo this theme: donor institutions will tend to run programs and projects the way they always have, even in the face of new objectives, unless given incentives to redesign projects and financial mechanisms. The Montreal Protocol seems to provide an example of where a strong secretariat successfully induced implementing agencies to meet new criteria and objectives related to incremental costs and phaseout of ozone-depleting substances. In contrast, the GEF appears much less successful in inducing its implementing agencies to run projects differently than before its creation. The Environment for Europe process and the PPC, watered down for political acceptability, will fall somewhere in between as far as their ability to improve coordination and collective priority setting among donors and recipients in Eastern Europe. The question of what makes oversight mechanisms more or less effective at promoting adaptation or learning among implementing agencies in order to escape the constraints of organizational inertia offers a fruitful agenda for further research.

Notes

The authors would like to express gratitude to the following people for their very helpful comments and suggestions: Richard Ackermann, Tim Murphy, Josue Tanaka, Robert Keohane, Marc Levy, and all of the participants in the Financial Transfers project.

1. Eastern Europe refers here to all countries in the EU's PHARE program: Albania, Bulgaria, Czech Republic, Estonia, Hungary, Latvia, Lithuania, Poland, Romania, Slovakia, and Slovenia.

2. The name is the French acronym for "Poland and Hungary Assistance for Restructuring Economies)," but the program was quickly extended to include Czechoslovakia and other countries in the region.

3. More precisely, PHARE's environmental programming was budgeted at 102.5 million ECU in 1990, 79.4 million ECU in 1991, 67.6 million ECU in 1992, 12.9 million ECU in 1993, and 74.7 million ECU in 1994. (PHARE 1995). The relatively smaller share for 1993 indicated decelerated spending provoked by low rates of disbursement for aid already committed in addition to decreased demand from the recipient countries. Conversion rates used are from the end of December 1993: $1.13=1ECU.

4. G-24 countries include: Australia, Austria, Belgium, Canada, Denmark, Finland, France, Germany, Greece, Iceland, Ireland, Italy, Japan, Luxembourg, the Netherlands, New Zealand, Norway, Portugal, Spain, Sweden, Switzerland, Turkey, the United Kingdom, and the United States.

5. 1994 data on bilateral donors are not yet available.

6. The reluctance of donors and recipients to fully disclose assistance data to multilateral sources means that G-24 estimates are incomplete.

7. To provide a frame of reference, cumulative G-24 assistance in all sectors to Central and Eastern Europe, including the contribution of the MDBs, reached over 61 billion ECU over the period 1990–93. Bilateral and EU assistance only amounted to 44 billion ECU over the same period. The volume of assistance to the environment sector was dwarfed by multisector assistance, such as macroeconomic assistance (nearly 11 billion ECU) and structural adjustment assistance (over 3.2 billion ECU); other sectors receiving considerably larger amounts of assistance included transport (nearly 2.2 billion ECU), energy (over 1.7 billion ECU), industry (over 2.7 billion CU), agriculture, and communications (almost 1.3 billion ECU each). However, environmental assistance had a much higher grant component than any of the aforementioned sectors or areas of assistance. (G-24 Coordination Unit 1994: 1–5)

8. Separate assistance programs for these two countries started only in 1993, after Czechoslovakia split into the Czech Republic and Slovakia.

9. This argument requires qualification. At first glance, it might appear that the priority of reducing local air pollution problems in Eastern Europe (primarily particulate emissions, plus high SO_2 concentrations in hot spots) diverges sharply from important international priorities, including transboundary SO_2 and NO_x emissions, ozone, or greenhouse gases. However, there are some important ways in which building capacity to deal with the local, health-related air pollution issues also builds capacity to address the international issues—for example, developing inventories of polluting sources, establishing and strengthening monitoring systems, improving the efficiency of power plants, and fitting industrial and power plants with filters which may reduce transboundary as well as local pollution.

10. Both banks have broadly similar sets of environmental procedures and face similar criticisms on their weakness. Proposed projects are categorized according to their potential environmental impacts. Those in category A may have a significant impact on the environment and require a full EIA; those in B have less significant impacts and require a partial EIA; those in C have no apparent environmental impact and do not require any EIA. The borrower is responsible for the preparation of environmental assessments and environmental audit reports. Often, the procedures give bank staff a way to address potential environmental problems in project design. Both the World Bank and EBRD procedures have been criticized for not adequately consulting with affected populations and NGOs (Wold and Zaelke 1992: 571; World Bank 1993a: 61). An internal World Bank review of its procedures in 1993 also pointed out that the environmental assessments are not adequately used to affect project design, and the documents produced by the project sponsor are of mixed quality, often lacking in an assessment of alternative measures (World Bank 1993a: 62–63). The EBRD, in turn, has been criticized for not complying with provisions of the 1989 U.S. Pelosi Amendment, which prohibits U.S. MDB directors from supporting projects that have significant environmental impacts if they (and the public) have not received an environmental assessment at least 120 days before the project is voted on (Fairman and Gutner 1993: 108). While the Pelosi Amendment's 120-day stipulation has been criticized by other EBRD shareholders as rather arbitrary, it has forced the U.S. executive director at the bank to abstain in voting on some loans.

11. For an example of early expectations that the EBRD might prove to be the first truly environmental MDB, see David Reed, "The European Bank for Reconstruction and Development: An Environmental Opportunity," *International Environmental Affairs* 2(4) (fall 1990): 317–343.

12. The World Bank usually refers to the International Bank for Reconstruction and Development (IBRD), and the International Development Association (IDA). The IDA is an affiliate that provides low-interest loans to the poorest countries; in East Central Europe, Albania is the only country eligible for IDA funds. In addition, the bank has two other affiliates: the Multilateral Investment Guarantee Agency (MIGA), and the International Finance Corporation. The IBRD, IDA, MIGA, and IFC are often referred to as the World Bank Group.

13. In fact, Yugoslavia, Czechoslovakia, and Poland were among the founding members of the Bank, although the latter two gave up their membership a few years later. Poland rejoined the Bank in 1986, Czechoslovakia in 1990. Romania has been a member since 1972, and Hungary signed on in 1982 (World Bank, June 1991).

14. The World Bank had a small environmental unit as early as 1971, but its impact was limited and it failed to mitigate the environmental degradation caused by many of the Bank's projects, which have traditionally been in environmentally sensitive areas such as agriculture, infrastructure development, power and transportation (See, for example, Rich 1994; Bramble and Porter 1992; Reddy 1993; Scott 1993). Environmental NGOs launched a campaign urging reform in the Bank's environmental policies in the early 1980s, which led to congressional pressure on the Bank to undertake significant reform. The result was Bank president Barber Conable's announcement in May 1987 of a number of new environmental policies.

15. The fact that 1994 was the fiftieth anniversary of the World Bank and IMF sparked an international debate on whether and how these institutions can make the policy changes needed to meet the challenges of the twenty-first century (see Owen 1994). A group of NGOs organized a "Fifty Years Is Enough" campaign to lobby for a vastly more limited Bank.

16. While the Bank's energy projects have been criticized by NGOs as weak in implementing end-use efficiency measures, some of the Bank's activities in Poland's electrical power reform have been commended for their incorporation of demand side measures (Environmental Defense Fund, NRDC 1994).

17. Indeed, its articles of agreement limit the Bank's public sector lending to no more than 40 percent of its total committed loans, guarantees, and equity investments.

18. Since the reunification of Germany, eastern Germany is no longer eligible for PHARE funding.

19. Poland, Hungary, the Czech Republic, Slovakia, Romania, and Bulgaria have all signed Europe agreements with the EU, which are seen as frameworks for eventual accession. Signatories view the PHARE program as a central mechanism for helping these countries meet the environmental preconditions of membership.

20. It should be noted that the World Bank was responsible for the establishment of a PIU in the Polish environment ministry, as part of its $18 million Polish environmental management loan. As with the PHARE PIUs, the Bank's PIU was used to manage other donor assistance and was staffed by officials from the ministry.

21. PHARE has also provided funding for the Regional Environmental Center.

22. The original document of over 150 pages was supplemented by a number of supporting technical reports, studies, and conferences held from 1991 to 1993. Their topics all focused on East Central Europe, included environmental health, environmental liability and privatization, municipal wastewater treatment, energy and environment, and so on.

23. One of the outcomes of the June 1991 meeting in Dobříš, Czechoslovakia, of environmental ministers from the EU, Eastern Europe, the United States, Canada and other nations, along with officials from the MDBs and UN organizations, was the creation of a task force chaired by the EU Commission. The task force's mandate was to "fill gaps in country-specific strategies and above all . . . help establish a framework for East-West collaboration that can lead to tangible and immediate actions to address the most serious environmental problems in Central and Eastern Europe within the available resources." (Environmental Action Programme for Central and Eastern Europe 1992). The World Bank, which shared the secretariat of the task force with the OECD, took the lead in drafting the EAP.

24. Recommendations for short-term problems include: installing dust collection systems and filters at lead, zinc, copper, and aluminum smelters located near large population centers; and investments that focus on replacing coal with gas in district heating plants, businesses, and households in cities where exposure to particulates exceeds an average level of 150 5g/m³ in winter months. Recommendations for long-term problems include phasing out leaded gasoline, and requiring new vehicles to meet EU standards; and resources to strengthen collection and dissemination of environmental data.

25. Interview, State Department official, October 1994.

IV

Conclusions

10

Increments for the Earth: The Politics of Environmental Aid

Barbara Connolly

The studies of environmental aid in this volume paint a rather dark picture. Constraints on the effectiveness of environmental aid seem more pronounced than windows of opportunity. As an analogy, imagine that you were to find yourself alone on a rapidly sinking boat. You make a quick survey of your assets and realize you have nothing but a leaky bucket and a package of bubble gum to help you out of the jam. Your options look grim. You can start bailing water, but the chances of staying ahead of the water's inflow with a leaky bucket are slim. You can think about creative ways to repair the leak in the boat with bubble gum, but repair work is difficult to do underwater. Alternatively, you could use the bubble gum to patch the bucket, and just maybe the gum would hold and you could bail fast enough to keep afloat until the Coast Guard arrived.

Using limited foreign aid resources to combat environmental degradation is much like this desperate effort to keep a sinking boat afloat. We do not imagine that environmental aid budgets will ever be large enough to provide wholesale solutions to large-scale environmental problems. Financial transfers alone will save neither the ozone layer nor the Brazilian rain forest. Achieving such objectives will depend on the interest and ability of affected actors to attract and make effective use of other, more substantial financial resources. We also do not suppose that aid donors have necessarily armed themselves with the most effective tools of intervention; they may be working with leaky buckets and bubble gum. Indeed, donor self-interest will limit the application of environmental aid to a particular set of problems or geographical areas, leaving many important environmental targets untouched. Despite these limitations, we contend that strategic use of environmental aid can make a difference in environmental quality in some situations by alleviating political obstacles to environmental reforms. By altering the incentives and capabilities of key players, environmental aid may change political dynamics in ways that enhance environmental protection.

Obstacles to environmental effectiveness are many and diverse. Financial transfers will seldom achieve the goals donors intend when they conflict starkly with recipient priorities. Often aid programs will fail to build the capacity within recipient countries to maintain environmental projects once external funding ceases. Multiple forms of incapacity in critical places may render aid projects nonviable from the start. Even when donors are fully conscious of the need for capacity building, their own interests and operating practices can divert them from that goal. Aid projects that appear effective when viewed in isolation may lose value when measured against the most important environmental priorities, or when other aid projects in related sectors duplicate or conflict with the approach of the first.

Despite these daunting obstacles, environmental assistance also features certain windows of opportunity: chances to augment financial resources in order to enable recipients to devote more attention to environmental problems, to build strong political coalitions in a position to protect the environment, to package a deal to make environmental protection appeal to actors to whom it otherwise would not. Not only can environmental assistance alter the incentives of key actors, it can also contribute to a redistribution of capabilities and hence political clout behind actors who exhibit environmental concern. Environmental aid may focus on removing capacity shortfalls within key areas or institutions in recipient countries. By increasing the political and financial resources of strategic coalitions within recipient countries that share donors' environmental goals, environmental aid may simultaneously boost concern and capacity. Finally, if the institutions surrounding environmental aid programs are carefully designed, they can provide incentives to donors, implementing agencies, and recipients to tailor their interventions in order better to address priority environmental needs, and perhaps to improve the fit of individual transfers with others through enhanced coordination.

This chapter will examine lessons that emerge from this volume about the constraints and opportunities for environmental assistance. Section 1 discusses which environmental problems and proposed solutions make their way to the financial transfer agenda, and why. Section 2 examines the factors that contribute to effectiveness or ineffectiveness during the implementation of environmental aid programs. Section 3 argues for the need to consider additional metrics of effectiveness when evaluating how the multiplicity of environmental aid transfers fit together as a whole. Oversight mechanisms, designed to facilitate evaluation of aid efforts and to promote adaptation by aid institutions in order to improve their environmental performance, are discussed in section 4; and the concluding section summarizes the role of environmental assistance in the broader scheme of available resources for environmental protection.

1 Setting the Financial Transfers Agenda: What Problems—and Solutions—Get Selected?

Three generalizations from our cases summarize our conclusions about how particular issues appear on the financial transfers agenda and how the agenda-setting process influences effectiveness. First, donors' wishes about what projects to fund by and large prevail, a bias which sets early parameters of the transfer's effectiveness. Second, trade-offs exist between defining problems in ways conducive to stable international agreements, and in ways that lead to environmental effectiveness. Third, symbolic politics plays an important role in the agenda-setting process.

Donor Dominance

The cases clearly show that donors dominate the financial transfer agenda. During the agenda-setting phase, donors—that is, governments, political coalitions pushing aid programs, and implementing agencies—determine which problems will receive aid, how those problems will be defined, and what solutions aid programs will seek to implement. Donor self-interest leaves systematic gaps in the agenda as well: certain priority problems from the recipient perspective, other ways of packaging the same problems, and efficient solutions that do not serve donor interests do not make it onto the agenda. Hence global commons issues receive funding from the GEF whereas local sustainable development problems do not, and retrofitting nuclear reactors in Eastern Europe merits a substantial emergency aid effort because of the risk to Western donors, whereas energy efficiency projects in Eastern Europe receive much less notice.

Environmental aid budgets in donor countries exist only because of donor self-interest. Without strong political constituencies invested in the mitigation of transboundary environmental problems or the export of goods and services made possible by aid programs, competing priorities within donor countries would eclipse environmental aid budgets. As a consequence, donor self-interest produces systematic gaps in the coverage of environmental aid. "Rich country problems"—like ozone depletion and loss of biodiversity—attract significant international funding, whereas urgent local problems with much more immediate consequences for human health in developing countries—such as local air and water pollution—receive less.[1] Of course, this does not imply that the global commons issues and regional environmental problems that excite donor interest do not themselves merit funding. It does mean, however, that the provision of financial transfers for environmental problems typically amounts to an attempt to persuade recipient countries to do something that donor countries consider a priority rather than to provide the resources that would enable recipients to take the environmental actions of highest priority to them.[2]

To the extent that recipients have any control over the financial transfer agenda, that control comes from their ability to impose costs on donor countries, or to exact a fee for their participation in a cooperative solution to an international environmental problem. Recipient countries can also engage in exercises of persuasion—or blackmail—if they can credibly threaten to withhold some action that donors consider vital. For example, China and India could credibly threaten to refuse to sign the Montreal Protocol unless they received compensation, since their domestic ozone-depleting substance production facilities meant that trade sanctions would not impair their ability to supply growing domestic markets[3] (ch. 4, above: 104–5). At the same time, developed countries know that participation by China, India, and other countries with large growing markets for ODS is absolutely critical for the long-term success of the Montreal Protocol process. Ozone depletion is a true global commons issue, where all actors are tied into the fate of the same ozone layer. Because of that, China, India, and other countries with domestic ODS production facilities were able to parlay their bargaining strength into not just the Multilateral Fund, but also a governance structure that gave them more control over the allocation of funds for concrete projects than recipients were able to manage in any of the other cases in this book. Although it remains true even in this case that donors set the agenda in the sense of providing financial transfers almost exclusively for problems affecting their self-interest, recipients may have some ability to affect financing levels through credible threats to link desired actions with aid receipts.

The Stability-Effectiveness Trade-Off

If the general rule is that donors set the financial transfer agenda, the more specific rule is that strong political coalitions of environmentalist and industry groupings within donor countries—dubbed "coalitions of the green and the greedy" by Kenneth Oye and James Maxwell (1994: 607)—tend both to justify environmental aid budgets and drive the selection of remedies for environmental problems. The cases in this book support prior research in which they identify a trade-off between stability and effectiveness (in the sense of impact on the physical environment) of global environmental regulations (Oye and Maxwell 1994). They claim that the most stable regulatory outcomes derive from "Stiglerian situations," where the public's desire to improve the environment converges with the self-interest of actors who gain tangible benefits from regulation. Where regulated actors derive concentrated benefits from global environmental regulation, they will not have incentives to undermine or renegotiate those agreements (Oye and Maxwell 1994: 593). Yet stable regulatory outcomes need not be fair, nor will they necessarily produce the best possible outcome from the standpoint of environmental protection.

Consider how these trade-offs appear in our cases. Successful adoption of the Montreal Protocol and the Multilateral Fund depended on a strong political coalition between industry and environmentalists (DeSombre 1995). Oye and Maxwell argue that "long-term economic interests in creating the market for CFC (chlorofluorocarbon) substitutes were one of the primary reasons that DuPont sought and ICI (Imperial Chemical Industries) accepted international regulation. In this case narrow material

interests interacted with broader environmental and political concerns"
(1994: 595). Elizabeth DeSombre and Joanne Kauffman point out, however, that industry support of a global regulatory solution is a "mixed
blessing." On the one hand, they acknowledge that without industry support, the Montreal Protocol likely would not have been negotiated. On
the other, they claim that industry backing influenced the types of solutions to the ozone problem that were considered, leading to technical and
chemical solutions in cases where other options might have been environmentally preferable (ch. 4, above: 91).

Similarly in the nuclear safety case, Western industry support proved
crucial for the mobilization of government aid funds and the provision of
technical assistance. Industry actors found sympathetic ears among some
environmentalists and domestic legislatures attentive to public concern
about the possible international risks from another Chernobyl. Indeed, in
terms of the sheer amount of aid provided and the political support behind an international, aid-driven solution, the nuclear safety case has to
count among the more successful. In terms of environmental effectiveness, however, we count this as one of the least successful cases, because
nuclear industry actors also biased the aid effort towards a narrow technical fix and prolonging the lifetime of existing nuclear reactors. This
approach assures industry lobbyists of ample government-funded retrofitting contracts as well as a foot in the door for commercial contracts to
expand nuclear power in the region, but it overlooks a broader consideration of energy options and an open assessment of the political trade-offs
between supply, efficiency, and risk (ch. 8, above: 257–259).

Contrast the Montreal Protocol and nuclear safety cases with Thomas
Bernauer's Rhine case, which exemplifies what Oye and Maxwell call
"Olsonian situations"—that is, situations of diffuse regulatory benefits
and concentrated costs. In Olsonian situations, the few clear losers will
tend to mobilize against regulation. "Because those regulated seek to
undercut or reverse regulation, these situations are marked by a high degree of regulatory instability" (Oye and Maxwell 1994: 594). The Rhine
case lacks industry actors with any strong self-interest in reducing chloride concentrations in the river. The agreed-upon solution, however, does
impose concentrated costs, particularly on the French potash mines
(MdPA). Although the riparian countries did eventually implement an in-

ternational solution, the more remarkable features of the Rhine story are the limited financial sums involved, the very small environmental effects of internationally financed reductions in chloride concentrations in comparison with unintentional reductions, and the extremely high transaction costs due to contentious bargaining and renegotiation of the agreement over a period of some sixty years (ch. 7, above). The costs of regulatory instability in this case seem to overwhelm the limited environmental benefits of financial transfers. The alternative to international environmental solutions that harness self-interest and hence lean toward industry interests, this case suggests, may be unstable and hence ineffective agreements.

Symbolic Politics

No discussion of which problems and solutions make it onto the agenda of financial transfers would be complete without acknowledging the importance of symbolic politics. It is all too easy to condemn a given instance of environmental aid for not having solved the problem it was nominally designed to solve. However, donors do not always provide aid in order to solve *environmental* problems. Often, aid programs are about solving *political* problems. Western assistance for nuclear safety in Eastern Europe falls far short of the amount required to shut down the most dangerous nuclear reactors and replace them with alternate capacity. GEF's initial funding will hardly make a dent in global warming, biodiversity loss, pollution of international waters, or ozone depletion. So why do donors bother?

At the risk of sounding overly cynical, we need to recognize the possibility that donor governments or some elements within them sometimes care more about the *appearance* of doing something to solve an important international environmental problem than they care about finding viable solutions to the problem itself. Simply by launching the G-7 nuclear safety initiative, Western governments were able to reassure concerned publics that the East European nuclear safety problem was under control. Creation of the GEF, even with its modest funding, met developed country NGO demands for additional aid resources to address global environmental problems. Indeed such tentative aid programs may serve to deflect public pressure for further actions by enabling politicians to claim that they have successfully dealt with urgent environmental needs. The policy challenge is to ask, given that aid levels fall far short of the mark required, and given that donor governments often subordinate environmental goals to other political priorities, how can activists inside and outside governments ensure that the amounts of aid actually provided are used to optimal effect?

2   What Makes Implementation of Financial Transfers Effective or Ineffective?

Once donors decide to provide financial transfers for a given purpose, what determines how successful the transfer will be at changing recipient behavior for greater environmental protection? Effectiveness, as we have seen, is a slippery and imprecise variable to measure (Young 1992; Wettestad and Andresen 1991; Underdal 1992; Levy 1993c). As a first cut, this section focuses on project effectiveness: "how well, relative to costs, a single financial transfer or set of transfers contributes to solving a particular environmental problem or set of problems, given the way in which problems are defined and established institutional arrangements" (ch. 1, above: 15). Analysis of environmental aid yields three principal conclusions about project effectiveness: (1) recipient concern is a major limiting factor; (2) financial transfers alone offer relatively weak opportunities for effective issue linkage, but fare better when they simultaneously build concern or capacity; and (3) project effectiveness requires carefully tailored efforts to build or employ existing capacity, both within recipients and donor institutions. Section 3 will consider parameters of aggregate effectiveness, in a situation where many donors are providing many transfers for many environmental problems, all at the same time.

Recipient Concern as the Limiting Factor in Successful Implementation
Whereas donor interests dominate the agenda-setting phase, recipient interests become much more critical determinants of effectiveness in the implementation phase. In fact, recipient political commitment is probably the key constraint on the implementation of pro-environment policy reforms. Where the problems and solutions identified by donors sharply diverge from recipient priorities, the provision of financial aid alone will

almost certainly not suffice to reverse recipient priorities or to induce the change in behavior desired by donors.

Surveying the cases in this book, variation in the level of recipient countries' political commitment to environmental reforms stands out as a major explanatory factor for the success or failure of financial transfers. For example, the cases of logging reform in Indonesia and nuclear safety in Eastern Europe share the feature that recipients retain very strong incentives to continue the behavior that donors wish to halt through provision of financial assistance. In Indonesia, the logging industry is part of the government's ruling coalition, and this privileged group obtains concentrated benefits from continuing unsustainable logging practices. Similarly, East European governments and utilities, still relatively unconstrained by public safety concerns, calculate that the cheapest way to secure ongoing energy supply is to continue operations of all nuclear reactors indefinitely. In the nuclear case, donor aid does not nearly suffice to alter this cost-benefit calculation, and considerable risk remains that at the end of the grant programs, the operational lifetime of the most dangerous nuclear reactors will have been prolonged, albeit at slightly safer levels, instead of shortened. In the logging case, the Indonesian government preferred to decline a World Bank forestry loan rather than infringe on logging interests. In both cases, attempts to employ financial transfers to persuade recipients to pay more attention to the environmental externalities of domestic practices failed.

The Rhine and GEF cases occupy middle ground, where financial transfers result in some beneficial projects, but a lack of recipient country commitment considerably undermines their effectiveness as a whole. In the Rhine case, why do riparian countries engage in such lengthy foot-dragging and renegotiation of the chloride settlement? One simple answer is that for three of the four countries—France, Switzerland, and Germany—reaching an international agreement is a very low priority. France, the primary recipient of the financing scheme and an upstream country, does not share the Netherlands' vulnerability to chloride concentrations in the Rhine. In fact, the costs of failing to reach agreement decrease as time passes and the chloride pollution issue begins to resolve itself due to exogenous economic factors. Unlike other cases in this book, the Rhine case involves decidedly rich countries, which makes it even less surprising that very small financial transfers do not suffice to reverse recipient priorities and induce rapid agreement.

Similarly, David Fairman argues that the GEF's tight focus on global environmental problems during its pilot phase seriously limited its operational effectiveness because of its divergence from recipient sustainable development priorities. By limiting the agenda so tightly to global commons problems, donors undermined the potential for voluntary cooperation. Even though the GEF's focus on global environmental problems was necessary to justify the creation of a new aid institution and additional funding in the eyes of donor governments, Fairman believes that this focus makes it very difficult for the GEF to gain the attention or allegiance of the nationally and locally focused actors whose cooperation it must have to fulfill its mandate.

At the other end of the spectrum, we have some clear success stories, all of which share the characteristic that recipient countries, or some powerful coalitions within them, have certain incentives of their own to implement the environmental reforms donors prefer. For example, think about why the Montreal Protocol case seems relatively successful. Article 5 countries—recipients of the Multilateral Fund—enjoy several incentives besides the fund to phase out ODS, especially the promise of long-run economic and technological benefits from switching over to cheaper substitute technologies and processes.[4] The most important contributions of the Multilateral Fund are to compensate Article 5 countries for the high transaction costs of capital investments in the switchover process, to educate participants about the benefits of switching to substitutes, and to provide additional financial inducements for signing the Montreal Protocol to ODS producer countries not affected by the protocol's trade sanctions. DeSombre and Kauffman acknowledge that market forces will play the most important role in pushing the global ODS phaseout; meanwhile, they credit the Multilateral Fund with helping "to ensure that the market transition will be more orderly and equitable than it would be without such assistance" (ch. 4, above: 120).

The same factor—recipient country commitment—explains the success of debt-for-nature swaps in Costa Rica and logging reform in the Philippines. As Cord Jakobeit notes, Costa Rica represented in many ways the ideal candidate for financial transfers. Costa Rica already possessed a well-established national park system, a vibrant NGO sector, and a reputation for being a leader in conservation. Precisely because of the country's prior commitment to conservation, Costa Rica became the favorite of aid donors during the first phase of debt-for-nature swaps.[5]

Similarly, in Michael Ross's story about logging reform in the Philippines, the key determinant of this aid effort's success lies in the political coalition struck between reform-minded donors and the proreform faction installed by the Aquino government in the Department of Environment and Natural Resources (DENR). In this case, the Aquino government welcomed donors' help in shackling and dismantling the powerful logging industry, which was largely excluded from the governing coalition. Logging reform became part of a broader effort by the Aquino government to consolidate its political base and to reduce the influence of former President Marcos's supporters. The Montreal Protocol, debt-for-nature, and Philippines logging cases all give a somewhat exaggerated appearance of the effectiveness of financial transfers, because of incentives that recipients already had to take actions desired by donors. In these cases, the financial transfers succeed in delivering environmental benefits not by reversing recipient priorities, but by capitalizing on preexisting environmental concern.

Manipulating Concern through Issue Linkage

The variation in success as a function of recipient country concern across our cases suggests a more generalizable argument: in cases where the success of aid efforts depends solely on environmental conditionality, or simple linkage of financial transfers with environmental reforms, overall effectiveness will be quite low. Lisa Martin has argued persuasively that heterogeneous capabilities and preferences among states may enhance prospects for international cooperation where those states can find credible ways of linking issues in order to provide welfare-enhancing benefits for both sides (Martin 1994). At the same time, Martin notes that scholars have not yet systematically explored the conditions for success of issue linkage. Environmental conditionality—providing financial assistance in exchange for specific environmental protection measures by recipients—is one form of issue linkage. Our cases, however, support little optimism about the promise of such forms of linkage. Financial transfers alone typically amount to weak incentives relative to recipients' ever-present temptations to renege on linked commitments for environmental reforms, unless they bring about some additional change in recipient concern or capacity. Attempts to link aid to environmental reforms often enjoy greater success when they generate long-term funding for environmental protection, enhance concern within strategic groups or bolster the political capacity of key actors, expand the political coalition backing reforms in recipient countries, or link multiple issues together so as to augment total resources available for environmental protection.

Why are financial transfers such a weak mechanism of linkage? Our cases suggest at least three explanations. The first two derive from inherent limitations of foreign aid programs. First, external aid never covers more than a small fraction of environmental expenditures in the recipient country, a pattern that we cannot expect to change in these days of declining foreign aid budgets. The GEF funds "incremental costs" only for measures to protect global commons. Debt-for-nature swaps in the largest recipient country, Costa Rica, financed only a small fraction of that country's conservation efforts, and Western aid for nuclear safety in Eastern Europe and the former Soviet Union, though a sizable sum in comparison to other instances of environmental aid, falls short of the region's needs by orders of magnitude. The broader point here is that the financial benefits at stake in most instances of environmental aid are not at all commensurate with the total costs recipients are asked to bear to effect policy change. Not surprisingly, therefore, recipients tend to balk at costly environmental reforms when they lack incentives beyond external aid. Arguably donors might enjoy greater success at altering recipients' priorities if they were willing to provide much greater sums of aid, but as noted above, we do not expect significantly larger aid budgets to materialize.

Related to the first point, the fact that donors envision environmental aid as temporary weakens the appeal of aid as an incentive to reverse recipient priorities, especially given that protection of natural resources usually involves an ongoing, long-term commitment of financial resources. Objectives such as forest conservation or the preservation of biodiversity, for example, require permanent sources of funding. Similarly, facilitating the transition to sustainable development in developing countries and in the former Soviet empire (not to mention the industrialized world) is certain to be a long-term effort. Mitigating climate change is likely to require enormous expenditures over an extended period of time. Short-term payoffs intended to make developing countries pay greater attention to global commons issues will only postpone developing countries' realization that without longer-term recurrent costs mechanisms to fund environmental protection measures, their individual interests may appear better served by exploitation of the environment, often at the expense of other countries.

Short-term aid directed at long-term, high-cost environmental problems can lead to serious contracting problems, because aid recipients can defect from their original commitments as soon as external financing ceases. This is the most prominent concern that Barbara Connolly and Martin List voice about nuclear safety assistance. Deficits of long-term planning, half-hearted attempts to apply conditionality, and underlying conflicts of interest among aid donors have led to a situation where after nuclear safety grants have been disbursed and marginal safety improvements have been implemented, recipients may well have fewer incentives to shut down the oldest nuclear power plants than before aid efforts began.

A third reason why linkage of financial transfers to environmental reforms often fails stems from collective action failures among donor countries. Remember that nonenvironmental incentives, such as export promotion or appeasing domestic political constituencies by creating the appearance of significant action, figure prominently in donors' decisions to provide aid. Such incentives, however, commonly lead to competition among aid donors, which undermines their ability to link the provision of aid to specified environmental reforms. As an illustration, Michael Ross notes that in Indonesia, donor governments were so pressed to find projects to appease strong "save the rainforests" movements within their own countries that they were unable to coordinate their efforts and bargain collectively with the Indonesian government for macropolicy changes. Already "deluged with aid projects for 'rainforest protection,'" the Indonesian government could afford to reject loans with conditionality aimed at reforming commercial logging policies.

Collective action problems among Western donors also figure prominently in the inability of Western governments to firmly link their aid to desired shutdowns of the oldest and most dangerous nuclear power plants in Eastern Europe and the former Soviet Union. Intense competition among Western nuclear suppliers over major retrofitting contracts and lucrative commercial contracts for the expansion of nuclear power in the East undermines the resolve of Western governments to stick to common conditionality policies. Faced with a choice between supporting domestic nuclear industries and advancing the cause of safety by applying stringent conditionality, many Western governments have voted with their aid programs for the commercial gains.

Despite the above qualifications, issue linkage can promote environmentally beneficial cooperation under certain circumstances. We should expect that simple linkage of financial transfers to environmental reforms will prove most successful when the actions recipients commit to undertake are limited in duration, as well as cost. In these cases, the value of the financial transfers to recipients may more nearly equal the cost of the actions they agree to take. For example, simple issue linkage appears quite effective in the Montreal Protocol case. In fact, China, India, Brazil, and most other developing countries demanded the creation of a funding mechanism as an explicit quid pro quo for their acceptance of the Montreal Protocol. What conditions allow simple linkage to operate successfully in this case? First, the phaseout of ODS is a transitional problem only, and the Multilateral Fund will continue operations until 2010, when Article 5 countries are obligated to complete their phaseout of ODS. Both donors and recipients can count on an identifiable end to the need for external financing. In addition, after covering the capital costs of switching to non-ODS technologies, recipient countries will realize economic benefits from the transition. Hence both donors and recipients benefit from the incremental cost mechanism.

The ultimately successful (although much-delayed and renegotiated) reduction of chloride emissions into the Rhine also owes something to the time- and cost-bounded nature of the chloride pollution problem. International financial transfers to facilitate cooperative solutions to the chloride problem amount to about $100 million—a markedly small sum, in contrast to the amounts transferred in most of the other cases we have studied. Not only were reductions in chloride emissions relatively cheap, the time period for internationally financed reduction projects was shorter than most. Riparian countries agreed to limit internationally financed reductions at MdPA to a ten-year time span (until 1998), because of expectations that MdPA's production and hence chloride emissions would decrease afterwards. Relatively low costs and the limited duration of the need for external financing made the chloride issue a reasonable candidate for linking financial transfers to pollution reductions, but even so, sharp asymmetries in concern among riparian countries invited protracted bargaining over distributional gains.

Despite the fact that financial transfers are often unable directly to create recipient country concern, they may succeed in building recipient concern by less direct pathways. First, financial transfers may lay the groundwork for more permanent funding mechanisms on which the longer-term success of many environmental protection efforts depend, thereby augmenting recipients' willingness to undertake such efforts. Jakobeit's analysis of the debt-for-nature case concludes that the greater value of this transfer lies not in the direct financial impact on recipient countries' conservation programs, but rather in its having provoked growing realization that viable long-term solutions will necessitate long-term financial support from mechanisms such as environmental trust funds. From the same vantage point, it becomes apparent that the longer-term success of funders' efforts to promote sustainable logging in the Philippines owes as much to the policy conditionality attached to forestry loans, such as charging the higher logging fees that provided the Philippines government with greater resources to monitor logging practices and enforce forestry regulations, as to the loans themselves. In this case, the creation of long-term funding mechanisms helps with enforcement problems as well; the Philippines government has less incentive to reverse its logging policy reforms once donors have completed loan disbursement.

In contrast, an important liability of Western nuclear safety assistance to Eastern Europe and the former Soviet Union is the lack of attention thus far to locating viable and attractive (to the recipient) sources of finance for alternative electricity generation capacity, thus exacerbating contracting problems. On the other hand, revolving environmental funds, which give East European governments access to self-sustaining means to subsidize environmental investments, are more likely to engender sustainable increases in those governments' environmental concern. Note that these examples demonstrate the considerable degree of overlap between concern and capacity, a subject that will receive further consideration later. That is, the examples show that building financial capacity within recipients to maintain or further donor-initiated projects can pay off in recipients giving higher priority to environmental concerns.

A second important way in which financial transfers can indirectly boost concern for the environment is by enhancing concern within strategic groups. In the nuclear safety case, for example, the present magnitude of Western assistance stands little chance of reversing the preferences of governments and utilities in Eastern Europe and the former Soviet Union for nuclear power. However, what Western nuclear safety assistance might realistically accomplish is to transmit Western notions of safety culture to a set of key actors, namely, the operators and regulators of NPPs who have the technical knowledge and the access to determine how these plants will be operated, and under what conditions they will be licensed for continuing operations. Increasing concern about safety conditions among these actors, who are well situated to become influential lobbyists within their own governments for a greater commitment of financial and political resources to the nuclear safety issue, may prove both cheaper and more effective than Western attempts to buy closures of the most dangerous NPPs.

Fairman's critique of the GEF's pilot phase hinges on the facility's failure, thus far, to cultivate a political constituency within recipient countries, or in other words to build concern in a strategically focused way. Developed countries' offer of additional aid in exchange for the GEF's focus to global commons problems created a situation of simple linkage—the exchange of funds for actions which developing countries would not otherwise undertake. Not surprisingly, considering the experience of other cases in this volume, simple linkage did not win developing countries' support for GEF operations. By narrowly limiting the GEF's mandate to global commons problems, the facility's creators failed to give developing countries an adequate stake in the success of resolving global problems. Had the GEF been more able to address problems that were both of pressing concern to developing countries and relevant to its global commons mandate, Fairman argues, GEF staff would have been better positioned to win the cooperation of environmental agencies as well as sectoral ministries (e.g., energy, forestry, agriculture) in identifying and developing GEF projects.

Fairman highlights the existence of a trade-off between the short-term effectiveness of individual environmental projects and the building of a local political constituency. Such a constituency is necessary both for effective implementation of financial transfers and for sustaining commitment to the activities financed by aid programs once external funding ceases. Following Fairman's logic to its end suggests that developed country representatives would do better over the long term to loosen their demands for the GEF's strict adherence to global problems and incremental cost criteria for every project. By extending greater benefits to developing countries and their own priorities, developed countries will be laying a firmer foundation of political support for implementing their own global commons agenda.[6]

A third way financial transfers can indirectly enhance recipient concern for the environment is by allocating funds so as to expand the political coalition backing reforms. For example, Ross explains how funders linked aid for the Philippines' balance of payments crisis, strongly desired by the Philippines central bank, with reforms in forest policies. While the logging reforms generated more controversy, the Aquino government placed high priority on securing economic assistance. Economic planners in the Philippines did not resist this linkage, since international concern about depletion of Philippine forests "had become a magnet for funders who hoped to help the government through its balance-of-payments crisis while promoting environmental conservation." In other words, the strategic manner in which funders applied cross-conditionality to loan disbursements linked two discrete political issues and thereby expanded the size and influence of the constituency favoring forestry reforms.

Finally, financial transfers may enhance concern about the environment by linking multiple issues together in such a way as to augment the amount of resources available for environmental protection. Debt-for-nature swaps, by creating a direct linkage between debt forgiveness and nature conservation, have made new and additional resources available for conservation. The existence of discounted debt titles on secondary markets actually creates a multiplier effect. As Jakobeit explains, by trading the debt titles for their face value (or some negotiated value) in local conservation funds, NGOs or bilateral donors can purchase a greater improvement in environmental quality than through a direct transfer of hard currency.

Despite its attractions, this form of linkage carries important limitations. First, the ability of the debt-for-nature linkage to generate additional resources for nature conservation depends on the vicissitudes of the debt crisis. As debt pressures ease, or as debtors perceive an increase in their chances for outright forgiveness, debt-for-nature swaps lose their appeal. Second, because the multiplier effect derives from financial rather than environmental characteristics of recipient countries, debt-for-nature swaps are limited in their reach and have not made significant inroads in countries that, from a purely environmental perspective, might be the most important targets of intervention to stop deforestation. As Jakobeit cautions, debt-for-nature swaps made conservation funds available principally to four major debtors in Latin America, while passing over tropical countries with the largest deforestation rates such as India, Indonesia, Burma, Thailand, and Colombia. Here as in so many other points during the process of financial transfers for the environment, tensions persist between the political and financial incentives which attract funders to a particular country or environmental problem and the desirability of directing aid resources to the problems or recipients of highest priority from an environmental perspective.

The Importance of Recipient Capacity: Using It or Building It

It will be no surprise to students of development that building recipient capacity or utilizing existing capacity is just as vital as recipient concern to effective and sustainable implementation of environmental improvements. As Fairman and Ross put it, "nothing in the field of development is as popular to promote or as difficult to accomplish as capacity building" (ch. 2, above: 41). Although our cases certainly do not solve the mystery of how to overcome institutional problems in project implementation, they do suggest three profitable lines of inquiry into the capacity problem. First, the experience of environmental aid teaches that recipient country capacity (or better, incapacity) comes in myriad forms, and increasing the effectiveness of aid programs depends on correctly identifying and addressing the missing form and location of capacity. Second, despite the clear need for capacity building, donors' own self-interests often divert them from this task. Third, our cases suggest that interventions to augment recipient capacity may prove especially profitable in those areas where capacity and concern are mutually reinforcing.

The concept of "capacity" encompasses a number of complex issues. Substantively, capacity comes, at a minimum, in financial, political, administrative, and technical forms; and successful interventions require targeting the missing form. For example, in the debt-for-nature case, the form of capacity that is most lacking in Costa Rica is sheer financial capacity. In the Philippines logging case, the capacity weaknesses include both financial and political deficits; funders make a difference by augmenting the political and financial capital of reformist forces within the government. In the case of environmental aid to Eastern Europe, administrative incapacity within recipient governments exacerbates the biases of donor-driven aid programs. The capacity deficits among Article 5 countries in the Montreal Protocol case include both financial and technological forms. The Multilateral Fund helps these countries meet transitional costs of phasing out ODS, and projects financed by the fund facilitate the transition in developing countries to non-ODS technology and processes. Addressing each of these different forms of incapacity will require very different aid strategies.

In addition, effective implementation of environmental aid programs depends on strategic choices about which actors on whom to focus capacity building efforts. For instance, Cord Jakobeit notes that in countries without a strong track record on conservation, a major benefit of debt-for-nature swaps has been the strengthening of local environmental NGOs, which can lobby their governments to raise conservation efforts on the political agenda and monitor the performance of conservation programs. In environmental aid to Eastern Europe, efforts to strengthen national environmental ministries through PHARE programs and the Environment for Europe process show some promise in helping recipient governments set overall priorities for environmental expenditures, while exerting pressure on donors to redirect assistance toward recipient priorities. Success in the case of logging reforms depends on building capacity within sectoral (in this case, forestry) ministries to integrate environmental concerns into broader development strategies. Strengthening municipal government institutions turns out to be key for resolving problems such as local water quality, where a dearth of financial capacity often creates decisive bottlenecks. EBRD efforts in Eastern Europe aimed at increasing the capacity of local governments to assess and collect user fees are starting to generate the necessary resources for local water quality projects. Other cases may benefit from capacity building efforts directed at the private sector, as in assisting firms to develop processes and technologies to reduce their dependence on ozone-depleting substances.

Unfortunately, identifying the most needed form or location of capacity within recipient countries does not necessarily lead to more successful capacity building efforts due to contradictory incentives among donors. In some cases, the contradictory incentives derive from capacity weaknesses within aid institutions. Barbara Connolly, Tamar Gutner, and Hildegard Bedarff show how the World Bank, PHARE, and the EBRD each exhibit distinctive aid strategies in Eastern Europe, emphasizing the incorporation of proper economic incentives within central government ministries, capacity building in national environment ministries and more regulatory approaches, and capacity building within private industry and municipalities, respectively. While each of these approaches may prove quite successful on different types of problems, the important point is that these donors cannot greatly alter their capacity building strategies to meet divergent circumstances. Environmental aid projects can benefit to the extent that capacity deficits in recipient countries are matched to aid institutions with the corresponding strengths. Where donor institutions compete with each other for projects and for a defining role in the region, however, their interests may impede recipients' efforts to capitalize on the different advantages of each institution.

Another reason why donors often run projects in ways that do not foster local capacity is that the twin goals of effectively completing single projects and building capacity often conflict. Fairman and Ross elaborate on this trade-off: "Funders are usually anxious to identify problems in recipient countries and supply the organizational skills and technologies (often originating in the donor's own country) to solve them. This may be the most efficient way to approach a discrete 'development' problem in low-income countries. But it also may have adverse consequences for any capacity-building efforts, because it tends to deprive local officials of the training and experience they will ultimately need to maintain the project" (ch. 2, above: 42). This tendency shows up strongly in environmental assistance to Eastern Europe, where donors routinely rely on short-term Western consultants because of their familiarity with practices of Western aid institutions and because of the political and economic benefits of using donor-based consultants. This practice not only fails to capitalize on cheaper local experts who understand local conditions, but in fact creates a drain on recipient capacity as short-staffed East European ministries are forced to commit large amounts of time and energy to "consultant tourism," or bringing the Western visitors up to speed with local conditions.

While our cases certainly illustrate the difficulty of capacity building, they also suggest certain key areas where capacity building will strikingly increase the effectiveness of financial transfers for the environment—namely, where capacity and concern intersect. For example, financial transfers can build a recipient country's political capacity to implement environmental reforms and increase recipient country concern at the same time by realigning political clout behind key players or coalitions within recipient countries who already want to do exactly what the donors want.[7] The case of logging reform in the Philippines offers the best example of this strategy. The World Bank and USAID benefited from the fact that the Aquino government and reformers at DENR were already committed to reforms. The loan packages increased the political capital of the reformist forces within the Philippines, allowing public officials committed to reforms to use the funders as scapegoats for politically unpopular measures. Forestry loans from the Asian Development Bank provided further political resources, permitting DENR reformers to distribute the new reforestation funds in ways that generated political support among members of Congress.

In a similar vein, Jakobeit argues that increasing the political capacity of local environmental NGOs may turn out to be the most effective way to promote environmental concern. As he puts it, "enhancing independent institutional capacity can empower local self-interest, thereby gradually reducing the dependence on continued external support" (ch. 5, above: 149). Here, too, the ability of financial transfers to affect recipient country concern turns on their success at altering coalitional politics within countries, so as to boost the political capital of those coalitions that stand behind conservation.

The earlier discussion on how financial transfers may increase recipient concern by developing long-term funding mechanisms for environmental reforms also exemplifies one of these key targets for effective environmental aid, where recipients' financial capacity and environmental concern reinforce each other. In their study of development assistance Fairman and Ross cite the frequent tendency for institutional components of development projects to collapse after external aid ends because of unplugged gaps in financial capacity. Often the recipient country cannot afford the recurrent costs of staffing and maintaining the institutions surrounding projects. "Capacity-building efforts, if they are to be sustained, must include provisions for long-term recurrent financing," they conclude. The same point holds with respect to concern: unless financial transfers for long-term environmental projects create long-term financing mechanisms, those projects will drop down the scale of recipient country concern as soon as the aid programs stop.

If aid programs are necessarily temporary and long-term financing mechanisms are essential for sustaining environmental improvements, why resort to environmental aid at all? One answer is that aid programs can help recipients provide for recurrent costs—by funding technical assistance to create long-term mechanisms, providing seed money, or incorporating conditionality leading to reforms that will yield sustainable revenues. In Eastern Europe, the PHARE program provides technical assistance to help recipient countries set up earmarked, revolving environment funds that draw domestic revenues from environmental fines, emissions charges, and the like. Additionally, PHARE is beginning to channel its own aid money through those national funds, as a source of seed money. The World Bank and USAID have taken a different approach in the Philippines, but also one that attends to the need for long-term funding sources. The conditionality attached to their forestry loans made it politically more palatable for the Philippines government to charge higher logging fees, which would generate sustainable revenues and compensate for losses incurred by the consequent decline in logging. Similarly, when the World Bank makes energy sector loans conditional on price liberalization, they are ensuring that recipient governments generate the necessary revenues to cover recurrent project costs. Because independent capacity to meet recurrent costs of many environmental projects weighs so heavily in securing recipient countries' continuing commitment to those projects, financial transfers that help to build recipients' long-term financial capacity can multiply their effectiveness by simultaneously generating higher levels of environmental concern.

3 Aggregate Effectiveness: How Do Individual Financial Transfers Add Up?

Although measuring effectiveness on a project-by-project basis is certainly useful, this angle does not fully capture whether aid programs live up to their full environmental potential. In the real world, many donors and institutions provide financial assistance for many different problems all at the same time. Hence individual environmental problems and associated financial transfers often overlap with each other—in ways that can either enhance or detract from environmental aid as a whole. Tightly constrained foreign aid budgets make it all the more crucial that donors and recipients employ limited financial resources to maximize environmental impact. At the aggregate level, then, we might gauge the effectiveness of environmental aid according to factors such as how well existing institutional structures and aid mechanisms fit environmental needs, how well donors and recipients coordinate aid efforts and develop synergies among them, whether the solutions advanced by various aid efforts in related project areas are complementary or contradictory, and whether aid flows target the most important environmental priorities.

Our case studies reveal two major classes of impediments to aggregate effectiveness. First, organizational inertia and bureaucratic incentives within the international institutions involved in financial transfers often result in aid mechanisms that do not necessarily fit environmental needs, projects with potentially unsustainable benefits, and failures of interagency coordination. Second, political and economic interests within both donors and recipients often impede attempts to improve coordination or to set aid priorities on the basis of environmental needs.

Organizational Inertia
Organizational inertia can seriously detract from the effectiveness of environmental aid. Because of the resistance of donor countries to creating new international bureaucracies, most environmental assistance will be

transferred through preexisting international institutions, such as the World Bank, UNEP, or the European Union. Given that many of the major donor institutions do not have environmental protection as their primary mission, their standard practices and existing financial mechanisms do not necessarily fit well with particular environmental problems. Out of the GEF's three implementing agencies, for example, only UNEP was created with a specifically environmental mission. In the other two implementing agencies, the World Bank and UNDP, broader and long-established development goals overshadow newer environmental tasks. When donor institutions confront new environmental issues, they display a persistent tendency to apply old, familiar solutions, rather than tailoring their interventions to the specific characteristics of the environmental problems. As Connolly, Gutner, and Bedarff argue, organizational inertia results in the use of financial mechanisms that do not always fit environmental problems, neglect of certain priority environmental problems from the recipient perspective, duplication or incompatibility of aid efforts, and rampant failures of coordination among donor agencies.

The chapters on the GEF, the Montreal Protocol, and aid to Eastern Europe all suggest that organizational inertia creates serious obstacles to effective aid provision. For example, the World Bank and the EBRD, as multilateral banks with broad missions for promoting economic development, conduct the vast majority of their environmental assistance through loan instruments. The greatest obstacle to the effectiveness of their environmental activity in Eastern Europe, Connolly, Gutner, and Bedarff argue, stems from "the difficulty of producing loans that are attractive to both the recipient countries and to the banks." The prospect of incurring additional debt does little to elevate the priority of environmental projects among many East European governments, particularly in the powerful finance ministries. Moreover, multilateral banks have a difficult time finding "bankable" environmental projects, and may shy away from environmental loans because of the intensive institution building often required to make projects financially viable. The obstacles incurred by a lending approach are not unique to Eastern Europe—Michael Ross notes that the World Bank's attempt to reform Indonesia's logging policies failed because the Indonesian government had no incentive to take on a loan with stringent conditionality attached when it had access to plentiful grant-based aid projects. The financial mechanisms employed by donor institutions thus do not always fit the environmental problems they target.

DeSombre and Kauffman suggest that trade-offs in environmental effectiveness may also creep in because of donor institutions' interest in demonstrating organizational effectiveness by disbursing aid rapidly. The projects that are under way thanks to the Multilateral Fund are the ones put forward by implementing agencies, and the agencies are most likely to start with the easiest projects to carry out rather than the most important, in order to speed disbursement. As a consequence, countries with the highest ODS consumption will probably lag behind the rest in achieving phaseout. In the absence of an effort to prioritize phaseout, "greater levels of ODS have been released into the atmosphere than would have if the mechanism concentrated first on transforming industry in the largest Article 5 ODS consumers or producers" (ch. 4, above: 122). As this example demonstrates, the measures by which we might evaluate the environmental effectiveness of a particular aid program do not necessarily coincide with the measures donor institutions rely on to evaluate their own performance.

Coordination Failures

The most significant sources of aggregate ineffectiveness in environmental aid are coordination failures, which are a time-honored tradition in the history of development assistance. According to Fairman and Ross, "most observers agree that the costs of uncoordinated aid programs outweigh their benefits. Many aid projects and programs are duplicative, while others work at cross-purposes" (ch. 2, above: 45). They note that donors may channel aid to ministries pursuing conflicting objectives, or may themselves provide incompatible equipment for a project or sector. Duplication of aid efforts or the proliferation of discrete projects frequently overtax the managerial capacity of recipient agencies. Despite widespread agreement on the liabilities of uncoordinated aid programs, these sources of ineffectiveness prove highly resistant to efforts at reform because they stem from underlying conflicts of interest among donor agencies, among recipient agencies, and between donors and recipients.

The GEF in its pilot phase abounded with coordination failures among implementing agencies. As Fairman depicts the history, the GEF's creators had envisaged close collaboration among its three implementing agencies, including development of a single, integrated work program. The original vision had the implementing agencies dividing up their tasks according to each agency's comparative advantage.[8] The World Bank would take the lead in coordinating work programs and activities with the other two implementing agencies and would supervise the project cycle. Unfortunately, disagreements among the implementing agencies undermined these plans for coordination, and the situation deteriorated into a turf war. The World Bank, UNEP, and UNDP battled with each other over relative shares of the GEF budget, authority to review each other's projects, and roles in assisting recipients with identification and preparation of project proposals. The implementing agencies effectively ceased to coordinate their GEF projects and approaches to problems, instead establishing duplicative and uncoordinated procedures for project identification and preparation at the country level and limiting efforts to review and critique each others' projects.

The tendency of different donors to pursue conflicting approaches in the same sector has especially perverse consequences in the case of nuclear safety in Eastern Europe and the former Soviet Union. Connolly and List emphasize that the various donors of nuclear safety assistance display divergent understandings of how the safety problem should be addressed, from those who would prefer to shut the oldest NPPs down as soon as possible to those who think that because certain Eastern countries rely so heavily on nuclear power, the best aid donors can do is to work on making the oldest reactors a bit safer. Competing solutions create serious problems for the assistance effort as a whole. Connolly and List argue, "given that donors differ in their view of the risk presented by NPPs in the East and the most appropriate (and realistic) solutions to the risk problem, certain donors are able to undermine the solutions of others by pursuing their own bilateral strategies" (ch. 8, above: 262). Contracting problems are already severe in this case because of the Eastern recipients' strong preference for continuing their reliance on existing NPPs; however, the inability of donor governments to coordinate their approaches to the problem makes the contracting problems irresolvable.

Among the many reasons for failure to improve coordination is the resistance of aid organizations to infringement on their institutional autonomy. The Montreal Protocol Multilateral Fund wins relatively high marks for the ability of the executive committee and secretariat to impose coordination on implementing agencies. However, DeSombre and Kauffman write that "the heavy involvement of the secretariat is not uniformly appreciated. Tensions between the World Bank and the secretariat over project evaluation have been particularly thorny and have contributed to delays in the approval process. In part, the Bank's structural constraints, internal reporting requirements, and project criteria make its acceptance of secretariat suggestions difficult. For its part, the secretariat has been heavily criticized for its 'micromanagement' on a project-by-project basis" (ch. 4, above: 117).

The same set of implementing agencies displayed more resistance to infringements on institutional autonomy during GEF's pilot phase. UNDP and UNEP torpedoed efforts to establish an agreed division of labor with the World Bank, in part because of real disagreements on specific areas in which each agency possessed a comparative advantage, but also because they refused to accept the authority of the GEF chairman, a World Bank official, to arbitrate their disputes. Their unwillingness to submit to World Bank leadership left all three agencies with uncoordinated procedures and approaches to problems in the field.

The commercial motives which often make financial transfers attractive to donors in the first place also discourage enhanced coordination. Fairman and Ross explain that coordination through multilateral agencies for the purpose of rationalizing aid allocation may reduce profit margins on project contracts or eliminate projects or programs, reducing the demand for some donors' exports. Donors who expect to generate more export contracts or political advantages through bilateral negotiations may resist coordination. Connolly and List's study of nuclear safety assistance provides a classic example of how donor competition over export benefits erodes coordinated solutions to an environmental problem. Recall the disagreements between French EdF and American Westinghouse officials over upgrade operations at the Kozloduy NPP in Bulgaria, during which EdF chairman Gilles Ménage accused Westinghouse of disrupting safety efforts and seeking competitive advantage by giving Bulgaria an opportunity to sidestep commitments for plant closure while still receiving assistance. The same battles over anticipated commercial benefits from nuclear contracts underlie the inability of Western governments to impose a common policy template on their nuclear safety assistance to the region as a whole.

Conflicts of interest between donors and recipients impair coordination, simply because recipient priorities often do not match donor priorities. Sustainable development objectives favored by developing countries overlap to some degree with the efforts to preserve the global commons that are the pet of developed countries, but the two are not identical. Bilateral donors such as Austria and Finland currently provide environmental assistance to Eastern Europe for transboundary air and water pollution projects that will result in environmental benefits for the donor, but their interest in bilateral aid would likely decrease if coordination and rationalization of the overall aid effort impelled them to redirect their aid to local air and water quality projects within recipient countries.

Even recipient countries have certain incentives to resist coordination, in part because of divergent priorities from donors. When donors do not coordinate their approach, the recipient may be able to play funders against each other in order to achieve its own ends. In Eastern Europe, recipients have been known deliberately to conceal information about on-going projects, in an effort to maximize aid receipts and increase their own autonomy over the use of those receipts. Connolly and List give vivid illustrations of how recipient governments and utilities in Eastern Europe and the former Soviet Union have been able to manipulate Western donors of nuclear safety assistance with conflicting approaches in order to escape donor-imposed constraints on their decisions about how long to continue operating nuclear power plants.

Unfortunately for the prospects of enhancing the effectiveness of financial transfers for environmental purposes, coordination failures stem from the nature of the messy Olsonian situations described by Oye and Maxwell. The benefits of a coordinated approach are diffuse, whereas coordination imposes concentrated costs to the autonomy of donor agencies, to the green and greedy coalitions within donor countries which lobby for and benefit from environmental aid, and even to government ministries or particular sectors within recipient countries. Because of the resistance of these actors, efforts to improve coordination in environmental aid will continue to be a major struggle.

Conflicts of Interest and Environmental Priority Setting

Even the most successful cases of environmental aid suffer from failures to prioritize aid targets. Bernauer's Rhine case flunks the test of aggregate effectiveness, in the sense that chloride pollution in the Rhine was never a high-priority environmental problem, despite the enormous amount of diplomatic attention focused on it. The use of finite diplomatic and financial resources for low-priority environmental problems may divert resources from more pressing problems. Debt-for-nature swaps and environmental assistance to Eastern Europe carry the same liability: a lack of attention to prioritizing environmental expenditures. Debt-for-nature swaps have attracted conservation funds primarily for the main Latin American debtor countries rather than hot spots of deforestation. Similarly, the first effort at setting environmental priorities for Eastern Europe as a whole, in the guise of the Environmental Action Programme for Central and Eastern Europe, did not occur until three years after initiation of environmental assistance to the region. The Action Programme called attention to gaps in the coverage of assistance for certain high priority environmental problems with respect to human health. Even now, clear conflicts persist between the priorities advanced in that document and the self-interested reasons why Western donors have created aid programs in the first place. In addition, donors have had difficulty in tailoring their standard practices and financial mechanisms to recipient priorities.

The histories of debt-for-nature swaps and environmental assistance to Eastern Europe also yield insight into precisely why failures of prioritization occur. The particular linkage drawn between the debt crisis and nature conservation in debt-for-nature swaps made additional financial resources available that would *not* have been available had NGOs and bilateral debt-holders targeted only countries with the highest deforestation rates. Making this financial transfer materialize at all dictated a strategy of focusing on countries with large accumulations of deeply discounted debt on secondary markets, so that purchasers of the debt titles could capitalize on multiplier effects. The same trade-off between prioritizing environmental targets and the ability to generate financial resources exists in the case of Eastern Europe. In the latter example, recipient priorities do not match with donor priorities or the characteristic approaches of donor institutions in key areas. If recipients somehow succeeded in restricting assistance projects only to environmental "hot spots," or local priorities, the willingness of Western governments to underwrite environmental aid programs to the region would almost certainly decline.

These cases suggest that although prioritization of environmental expenditures within the parameters of donor self-interest makes sense for enhancing aid effectiveness, prioritization efforts that divert projects from donors' self-interested priorities may in fact produce downward pressure on environmental aid budgets. In Eastern Europe, argue Connolly, Gutner, and Bedarff, "Bilateral donors . . . stand to lose major environmental and export benefits from a technocratic approach that directs donor aid to areas of recipient priority"—or that seeks to match the cheapest and most locally appropriate technology to recipient problems rather than that technology which donors most wish to sell. Western nuclear firms that lobbied forcefully for governmental aid programs to improve nuclear safety in Eastern Europe through technological improvements would lose much-anticipated commercial benefits if the effort were redirected toward energy efficiency or replacing NPPs with other kinds of plants. Similarly in the GEF case, limitations on the degree of overlap between global commons problems and local sustainable development issues create disincentives for donors to assist developing countries with establishing and implementing their own priorities. The principal message of these examples is that prioritization efforts often fall short because actors have legitimate differences in what they consider to be the most important environmental priorities, and because environmental priorities often conflict with other interests which draw donors into providing environmental aid.

4   Promoting Learning for Enhanced Effectiveness through Institutional Oversight

Given that environmental assistance is so often channeled through preexisting institutions, or institutions that were not originally equipped for environmental mandates, enhancing the effectiveness of that assistance will require either new institutions or, more likely, attempts to increase flexibility and adaptability within donor agencies so as to improve the fit of organizational practices and available financial mechanisms with existing problems. In addition, the pervasiveness of coordination failures and the deficiencies that result argue for a need to provide donors, implementing agencies, and recipients with both incentives and greater capacity to improve coordination of environmental aid. Because such efforts run up against the entrenched interests of aid institutions protecting their own autonomy, donors protecting political, environmental and commercial interests, and recipients seeking to safeguard their own priorities from donor influence, they will engender significant resistance. How, then, do we prod donors, recipients, and especially aid institutions to learn or adapt in a way that improves their environmental performance?

Of course, donor governments and implementing agencies already invest considerable effort in evaluating their environmental programs and projects. Even if each individual government or agency improves its ability to implement its own aid priorities, however, the risk remains that the aggregate results of environmental aid may not improve substantially, due to the gaps, biases, and conflicting approaches among various aid efforts. For this reason, more collaborative forms of evaluation among aid participants deserve equal attention. Our cases highlight the importance of creating institutional oversight mechanisms to monitor the performance of implementing agencies, of aid efforts as a whole, and to push beyond business-as-usual approaches. The idea behind institutional oversight is to address the profoundly political causes of organizational inertia, failures of coordination, and absence of prioritization by providing donors, recipients, and implementing agencies with incentives and capacity to do things they would not otherwise do, to increase the effectiveness of environmental aid.

The cases in this book describe three different variants of institutional oversight mechanisms. The strongest form is illustrated by the Montreal Protocol Multilateral Fund, and the weakest by the Global Environment Facility in its pilot phase. The oversight mechanisms that have grown out of the Environment for Europe process lie between these two extremes. The Montreal Protocol Multilateral Fund is controlled by an executive committee with equal representation by developed and developing countries and an independent secretariat. DeSombre and Kauffman argue that the strong institutional oversight embodied in the ExCom and secretariat explains why the Multilateral Fund, which could have been employed merely as a bribe to obtain the signatures of developing countries on the Montreal Protocol, instead has been used to fund relatively sound and well-supervised ODS phaseout projects.

The four implementing agencies of the Multilateral Fund do not specialize in different types of projects according to comparative advantage, but instead compete with each other to do similar projects. One might therefore have thought that the turf wars that developed among these implementing agencies in the GEF's pilot phase would be equally probable in implementation of the Multilateral Fund. However, the ExCom imposes discipline on the implementing agencies by insisting on coordinated work programs and sectoral approaches to organizing projects. The secretariat is heavily involved in the work of implementing agencies. "Although almost all projects brought before the ExCom are ultimately approved, many are withdrawn, postponed, or modified at the suggestion of the secretariat before they are brought forward" (ch. 4, above: 121).

It would not be unusual for donor agencies to exert careful control over project implementation. What is unusual is for implementing agencies themselves to be subjected to such intrusive (if environmentally beneficial) oversight. One factor which seems to explain this anomaly is the unusually strong bargaining position of developing countries during negotiations on the Fund's governance arrangements, due to the long-term importance of their participation in the Montreal Protocol. Developing countries' allergies to existing development agencies, particularly the World Bank, led them to use their bargaining power to secure an unusually strong oversight position within the ExCom, which ensures that implementing agencies will have to tailor their usual practices in order to produce quality ODS phaseout projects.

In its pilot phase, the GEF had the weakest form of institutional oversight, permitting the outbreak of turf wars among implementing agencies. During this initial period, GEF creators assigned primary oversight responsibilities to the World Bank and created a rather weak oversight mechanism on top of implementing agencies in the form of a participants' assembly. As we have seen, UNEP and UNDP refused to acknowledge the authority of the GEF chairman, a World Bank official, to mediate interagency disputes. The resultant failure of the agencies to coordinate projects, approaches, or evaluation procedures left organizational inertia virtually unchecked.

Negotiations over GEF restructuring produced changes in the facility's governance structure which will most likely lead to more effective supervision of implementing agencies. In exchange for agreeing to designate the GEF as the interim financing mechanism for the global climate change and biodiversity conventions signed at UNCED, developing countries demanded more control in the GEF's oversight structure. As part of the agreement, parties to the two conventions assumed direct responsibility for deciding GEF eligibility criteria and program priorities. A GEF council was created as the facility's executive body, with fourteen developed country representatives, sixteen developing country representatives, and two representatives of former Soviet bloc countries. GEF participants agreed to transform the GEF chairman into an appointed CEO, who would mediate interagency conflicts with the support of a functionally independent secretariat. This change was engineered to give the CEO greater independence from the World Bank and greater institutional resources than the GEF chairman had possessed in the pilot phase (ch. 3, above: 66–69).

The new governance structure incorporates a shift toward greater control both for developing countries and for the CEO. Developing countries' struggle to win balanced governance for the GEF was in effect a fight to set a precedent for greater control over institutional oversight mechanisms, in order to direct the coordination of aid efforts and the adaptation of implementing agencies in ways that would better serve developing country priorities. In many respects, the newly restructured GEF looks much more like the Montreal Protocol Multilateral Fund. It is too soon, however, to evaluate its effectiveness at imposing coordination and modifying the practices of implementing agencies.

The oversight mechanisms that have grown out of the Environment for Europe process in the case of environmental assistance to Eastern Europe occupy a middle ground between the Multilateral Fund's strong oversight and the much weaker oversight mechanism of the GEF's pilot phase. We might think of the Environment for Europe process as a "Law Merchant" (Milgrom, North, and Weingast 1990) solution to governance needs, rooted in the idea that increasing the transparency of aid efforts will generate improvements in their environmental performance. Biannual meetings of European environment ministers and representatives of aid institutions have drawn high-level attention to environmental needs in Eastern Europe.

As of mid-1995, the most important single outcome of the Environment for Europe process is the Environmental Action Programme for Central and Eastern Europe, a document that outlines a coherent rationale for setting environmental priorities in the region, and suggests how to find cost-effective solutions. Authored by a few motivated individuals primarily within the World Bank, the EAP seeks to persuade the entire field of donors, along with recipients, that the most sensible approach for establishing priorities is to focus on human health in recipient countries. The EAP, although certainly not a grassroots effort at priority setting, has made significant strides in advancing a common agenda and a coherent approach for aid efforts, which benefit the cause of improved coordination.

The new Project Preparation Committee (PPC), designed to bring together multiple donors on project funding in an effort to create synergies among aid efforts and to improve absorption of available environmental assistance, also grew out of the Environment for Europe process. The PPC is much weaker than the governance mechanisms which sit atop implementing agencies in the Multilateral Fund or the restructured GEF, consisting only of four staff people with no authority to modify or reject projects run by donor institutions. The PPC, watered down from the originally proposed concept in order to make it more politically acceptable to donors, will rely for its effectiveness on its ability to match donors interested in co-financing projects, to create public relations benefits for enhanced donor coordination, and to publicize the benefits of co-financing. Unlike the Multilateral Fund and GEF versions of institutional oversight, the PPC is exclusively a donors' club. Western donors still tend to dominate the oversight of environmental assistance to Eastern Europe. The many sources of ineffectiveness that stem from donors' dominance of the financial transfers agenda, such as predictable gaps in coverage and failures of coordination, will most likely remain unchecked as long as recipients fail to gain more say in the governance of aid programs. However, the PPC does provide real, if modest, incentives for donors to improve coordination, and the EAP both calls attention to recipient priorities and attendant gaps in assistance, and potentially improves the capacity of recipients to pressure donors to direct their efforts toward those gaps.

A critical insight that comes out of these three examples of institutional oversight mechanisms is that designing oversight structures is not merely a technocratic task, but a deeply political one. The bargaining among various actors to control governance structures for these three different aid efforts amounts to a struggle to control the direction of organizational learning or adaptation, and to define the criteria by which aid effectiveness will be gauged. Developing countries have achieved their most significant gains toward that end in the Montreal Protocol process.

5   Beyond Financial Transfers

Even in the best of all possible worlds, environmental aid programs will not provide sufficient resources to solve the vast majority of environmental problems. One reason has to do with symbolic politics. The dollar (or mark, or yen) figure for demonstrating that governments are doing something about an environmental risk may be much lower than the dollar figure to actually effect environmental change.[9] Symbolic politics notwithstanding, even ample government aid budgets cannot be expected to cover the full costs of preserving rain forests, revamping energy production in Eastern Europe, or coping with climate change.

Resolution of many global and regional environmental problems will necessitate both far greater and more permanent sources of funding than environmental aid can provide. The authors of this volume do not pretend to have examined more than the tip of the iceberg of such funding possibilities. In the future, modifications of international trade law to take into account the environmental externalities of current trading practices, changes in government regulation to encourage private actors to make environmental investments, the dedication of larger portions of domestic governmental budgets to environmental expenditures, tradeable permits, and environmental trust funds created to support international conventions all may yield significantly greater funding sources for environmental protection. Exploring the problems and pitfalls of any of these options certainly deserves a separate study.

Nonetheless, the research represented in this volume was born of the conviction that even very small amounts of environmental aid can significantly affect the politics behind environmental protection in beneficial ways, and that because aid resources are so limited, it is all the more important to ensure that they are used wisely. To what performance standard can we hold financial transfers for the environment? Ultimately, we expect that where financial transfers turn out to be effective in improving the environment, they do so by intervening in the politics of environmental protection, altering the incentives and capabilities of key players. Effective environmental aid manipulates the political dynamics in any of the following four ways.

First, financial transfers lead to environmental improvements by altering the incentives of key players in recipient countries in ways that promote environmental concern. In some cases environmental aid can build recipient concern directly, where provision of relatively modest resources will accomplish the desired environmental goal. More often, environmental aid builds concern in a more indirect fashion. Aid transfers can lay the groundwork for more permanent funding mechanisms which may produce sustainable increases in recipients' commitment to environmental reforms. Environmental aid can also cultivate political constituencies for donors' environmental objectives, enhancing the concern of strategic groups who are in a strong position to lobby for reforms within their own countries. Donors may target aid disbursement so as to expand and strengthen the political coalition backing specific environmental actions, or to augment the total amount of financial resources available for environmental goals.

Second, financial transfers may translate into environmental improvements when they change the capabilities of key players. Effective environmental aid programs will carefully target capacity-building efforts to specific bottlenecks, which could be political, administrative, financial, or technical capacity weaknesses within national governments, sectoral ministries, local governments, NGOs, or private industry, depending on the characteristics of the environmental problem. Opportunities to magnify the effectiveness of financial transfers spring up in areas where recipient concern and capacity merge. Here donors can rearrange political dynamics by augmenting the political and financial resources of strategic actors within recipient countries who already share donors' environmental goals.

Our studies of environmental aid show that contracting problems pose fewer obstacles to effective financial transfers than either concern or capacity. Nonetheless, in a few cases, classic contracting problems do appear. The ten-year grace period for ODS phaseout in Article 5 countries built into the Montreal Protocol does create some risk that several years in the future, recipients of the Multilateral Fund will possess larger ODS production facilities despite the fund's activities, and will then be in a position to demand larger financial concessions before implementing phaseout. Efforts to improve nuclear safety are hampered by contracting problems, because donors will likely complete aid disbursements prior to targeted dates for nuclear plant shutdowns, leaving them with little leverage if Eastern governments decide later not to shut down NPPs before the end of their design lifetimes. As the earlier discussion of issue linkage showed, green conditionality alone—the linkage of financial transfers to environmental reforms—tends to perform poorly because of an inability to provide a sufficient counterweight to recipients' incentives to renege on their commitments. Rather than directly resolving contracting problems, financial transfers do better to circumvent those difficulties through strategies to build concern or capacity.

Finally, effective environmental aid efforts will incorporate mechanisms for self-evaluation, at the aggregate level as well as within individual agencies. Because donors dominate the agenda—which problems will receive financial transfers and which solutions aid programs will push—improving the effectiveness of environmental assistance often necessitates efforts to alter the incentives and capabilities of donors and implementing agencies. The governance structure that participants design to supervise aid transfers can contribute as much value as the aid itself, because these oversight mechanisms can direct adaptation and capacity-building efforts within donors and implementing agencies to counter powerful tendencies toward organizational inertia, coordination failures, and policy fragmentation. Control of institutional oversight mechanisms amounts to the ability to set directions for adaptation and to define the standards of effectiveness against which financial transfers are measured.

It is thus through selective intervention in the political process, by altering the incentives and capabilities of key participants on both the donor and recipient ends, that financial transfers can catalyze additional resource flows for environmental protection and ensure that the recipients of those resources are able to use them well. Only astute political leadership, willing to adapt the policies of states and international organizations in light of experience, can make environmental aid effective.

Notes

The author would like to express deep thanks to the following people for the comments and insights they contributed to this chapter: Elizabeth DeSombre, David Fairman, Tamar Gutner, Robert Keohane, Ron Mitchell, and Michael Ross.

1. For a polemic critique of the tendency of Western environmentalists to focus on "rich country problems" to the detriment of local environmental problems with much more serious and immediate health consequences in the developing world, see Easterbrook 1994.

2. Thanks to Ron Mitchell for clarifying this point.

3. Provisions banning exports of ODS and products containing ODS to nonparties were incorporated into the Montreal Protocol. These trade restrictions were designed to force countries reliant on ODS imports to sign the Montreal Protocol in order to preserve their inflow of ODS until their final phaseout. (Benedick 1991). However, countries with their own domestic production facilities for ODS would not be seriously affected by these trade measures.

4. For a more comprehensive listing of these incentives, see ch. 4, above: 119–121.

5. Even with Costa Rica's favored recipient status, Jakobeit cautions, debt-for-nature swaps were only able to finance a small portion of the country's conservation efforts and national park system, and the external aid did not suffice to eradicate the root causes of deforestation and loss of biodiversity (ch. 5, above: 139–140).

6. The GEF story supports lessons from negotiation theory, which suggest that issue linkage will be most helpful in promoting cooperation if the linkage creates additional value, or offers new benefits to both sides. Lawrence Susskind points out that issue linkage in this sense is quite distinct from conditionality, or one-sided trades, which may create leverage for one side in a negotiation but do not create additional value. One-sided trades are less stable, or less conducive to sustained cooperation (Susskind 1994: 91–92). The GEF's incremental cost criteria,

which effectively limit payments to developing countries to actions they would not otherwise undertake, represents an exercise in conditionality rather than the creation of additional value for both sides. Critiques of the GEF's pilot phase suggest that a broader, more mutually beneficial form of issue linkage could improve the GEF's political prospects for successful project implementation.

7. Thanks to Bill Clark for stating this point so clearly.

8. The World Bank would be the GEF's financial trustee and administrator. In addition to its coordination and supervision responsibilities, the Bank would identify and fund investment projects. UNDP was to coordinate GEF activities at the country level, to administer technical assistance and capacity-building projects, and to promote the GEF's cause and encourage submission of project proposals among various recipients. UNEP had the task of ensuring that GEF's policies and strategies did not conflict with existing and proposed global conventions, as well as supporting the independent Scientific and Technical Advisory Panel (ch. 3, above: 60–61).

9. I owe this point to Bill Clark.

Appendix: The Scope of Global Environmental Financing—Cases in Context

Wendy E. Franz

Introduction

The 1992 United Nations Conference on Environment and Development highlighted many ways to accomplish the complex task of funding for global environmental protection. The financial chapter of Agenda 21 pointed to the need for a variety of institutions and mechanisms, including multilateral funds, United Nations agencies, regional banks, nongovernmental development organizations, and bilateral aid, to deal with the nature and scope of environmental challenges. Financing is fast becoming one of the greatest challenges to comprehensive and worldwide environmental protection. This volume explores the effectiveness of several financial transfer institutions for the environment, some global in scope and membership (the Global Environment Facility, the Montreal Protocol Fund, and debt-for-nature swaps), others focused on regional activities and problems (nuclear safety in Eastern Europe, logging reform in the tropics, and river pollution). They cover a time period stretching from the early 1970s to the present.

The sum total of the efforts at environmental protection covered in this book reached nearly $5.5 billion in December 1994 (see table A.1). In addition, a balance of $82 million remained in the Montreal Protocol Fund as of that date, and another $742 million has been authorized, but not disbursed, by the Global Environment Facility. While the magnitude of resources provided through institutions for environmental protection is, as Robert Keohane notes in the introduction, only part of the story of effective environmental protection, ample financial contributions are a critical first step in the effort to protect the global environment. The

Table A.1
Case studies of financial transfers for environmental protection

Case	Time period	Cumulative transfers or disbursements as of December 1994 $ US millions (to nearest million)
Eastern European nuclear safety[a]	1992–	874
Environmental assistance to Eastern Europe[b]	1990–	2080
Montreal Protocol Fund[c]	1991–2010	195
Global Environment Facility[d]	1990–	95
Debt-for-nature swaps[e]	1987–1993	983
Rhine River[f]	1972–1998	127
Forests reform—Philippines and Indonesia[g]	1985–	1154
Sum Total		**5478**

a. Total commitments from Western Governments by the end of 1993 for nuclear safety in Eastern Europe and the former Soviet Union. Includes bilateral aid and multilateral funds (including the European Union's PHARE and TACIS and the EBRD's nuclear safety account).
b. Totals for 1990–1994. Based on $788 million from World Bank, 47 million ECU from EBRD (conversion of $1.24 = 1 ECU), $374 million from PHARE, and $860 million in bilateral assistance (G-24).
c. Total cash disbursements to UNDP, UNEP, UNIDO, World Bank, and bilateral aid as of 8 December 1994. Pledged contributions to the fund reached $393 million. Income to the fund totaled $308 million. Other expenditures (secretariat 1991–1994, programme support 1991–1994, and cash advance to the ozone secretariat) totaled $10 million. The balance in the fund as of 8 December 1994 was $82 million. Data source: Executive Committee of the Multilateral Fund for the Implementation of the Montreal Protocol, Contributions and Fund Disbursements (report from the Treasurer), *UNEP/OzL.Pro/ExCom/15/4/Rev.1*, 9 December 1994.
d. Total disbursed is as of 31 March 1994. The total authorized as of August 1994 was $742 million. Total of funds committed reached $413 million. Data source: GEF, July 1994 participants' meeting, annex 4. *Quarterly Operational Report* August 1994.
e. First and second generation swaps. Figure includes: $47 million from NGOs, figure for cost. (Cost = expenditures by environmental agency to acquire the sovereign debt), and DM 947 million ($582.2, $1 = DM 1.6) from Germany in debt forgiveness, and $353.3 million from the United States Enterprise for the Ameri-

Table A.1
(continued)

cas Initiative in debt forgiveness. Data source: *International Monetary Fund, World Debt Tables 1993:* 116, and Cord Jakobeit.

f. The Rhine River plan was agreed to in principle in 1972. The first plan for action was agreed to in 1976. France reneged, and it was not until 1985 that the flow of funds and compliance with the agreement took place. Figure based on FF 632 million, converted $1 = FF 5.

g. External financial support to National Forestry Action Plans as of December 1992. Includes $854 million to the Philippines (grants and loans) and $300 million to Indonesia (grants and loans). Data source: Tropical Forests Action Programme, *TFAP Update No. 28,* prepared by the Coordinating Unit of the Tropical Forests Action Programme, Forestry Department, FAO.

purpose of this appendix is to put the financial scope and magnitude of the financial transfers discussed in this volume in the context of global environmental transfers for the environment, and to identify trends in environmental financing.

Financing for environmental protection has emerged as an essential component in official development assistance at both the multilateral and bilateral levels. The effort to integrate environmental concerns with development assistance is still new. Multilateral institutions and national governments have, over the course of the last seven to ten years, adopted criteria to distinguish international aid for the environment from other development assistance. Program and project goals are being reconsidered in light of the need for global environmental protection.

The World Bank, for example, recruited its first environmental adviser in 1969 and established an office of environmental affairs shortly thereafter, but did not bring environmental concerns into the mainstream of its operations until 1987. A year later, the Bank created its environment department, established four regional environmental divisions, and increased the number of environment division staff members. The Bank adopted a set of criteria to distinguish environmental lending and first reported on environmental aspects of its operations in fiscal 1990 (World Bank 1990: 11–12). In 1989, at the Paris Summit, the Group of Seven industrialized nations requested that the Organization for Economic Cooperation and Development develop a comprehensive set of environmental indicators to use in evaluating the state of the environment. The

Development Assistance Committee adopted definitions for the identification of environmental aid in 1991. Subsequently, the OECD has launched a new program of environmental performance reviews to evaluate national progress on domestic environmental objectives and international commitments. These reviews include data for financial contributions in addition to treaty compliance and international cooperation. A set of pilot reviews was prepared in 1992/93, and member countries are expected to be reviewed every five years (OECD 1994a).

If member countries have not yet distinguished environmental financing from traditional development assistance and multilateral institutional support, these reports will take important steps in identifying national contributions to the cause of global environmental protection. Several countries have already made efforts to define environmental assistance in such a way as to make it identifiable. For example: Canada's International Development Agency adopted its environment and development policy in 1986 (Ministry of Supply and Services, Canada 1991); Sweden's Riksdag decided in 1988 that Swedish development assistance should incorporate environmental goals (Government of Sweden 1991); the 1998 DANIDA action plan focuses on enhancing the environmental aspects of Danish development cooperation (UN 1992b). The percentage of official development assistance projects with environmental protection as their primary goal has increased considerably in the last five years. This trend is expected to continue. Many governments have also undertaken environmental finance initiatives that fall outside the realm of official development assistance, such as government-business partnerships and support for the work of nongovernmental organizations.

It is important to note that data for environmental financing before 1990 are rare. As the preceding discussion suggests, environmental protection has not been systematically considered as part of development programs until recently; the careful articulation of the connection between sustainable development and environmental protection that emerged at the Rio Summit in 1992 was emblematic of the changes that have taken place slowly over the course of the past fifteen years, accelerating during the past five years. The Stockholm Conference on the Human Environment held in 1972 was more concerned with the physical environment than with the complexities of the social aspects of environment and development. Several important environmental initiatives grew out of the Stockholm Conference, including the United Nations Environment Program. Transfers under those auspices, however, have been quite modest. Recent experience suggests that nations and multilateral institutions are now making the connection explicit and are rapidly increasing the transfers of financial resources from rich to poor. The percentage of development assistance that is going to environmental objectives is growing, although it varies by institution and nation. The web of global environmental finance is being woven through multilateral, national bilateral, and nonstate initiatives. Each will be considered in turn.

Multilateral Financial Transfers

Significant financial contributions are being made by the World Bank and regional development banks. Agencies of the United Nations (e.g., UNEP) are components of international efforts to protect the environment, although they do not register as contributors of large financial flows.

The *World Bank* is a large contributor of resources for environmental protection (beyond those connected to the Global Environment Facility or the Montreal Protocol Fund). The World Bank distinguishes environmental lending using the following criteria. A project is deemed to be "primarily environmental" if "either the costs of environmental protection measures or the environmental benefits accruing from the project exceeded 50 percent of the total costs or benefits." A project is said to be "significantly environmental" if "environmental protection costs or environmental benefits exceed 10 percent of total project costs or benefits" (World Bank 1993a). By July 1994, the World Bank portfolio contained 118 primarily environmental projects entailing total commitments on the order of $9 billion in loans and credits (World Bank 1994e: 4). One hundred three of 118 project commitments were made between 1990 and 1994. The contributions of the World Bank, the Inter-American Development Bank, and the Asian Development Bank are shown in table A.2. In addition to the primarily environmental World Bank projects identified in the table, the Bank reports another forty-three projects with significant environmental components in fiscal year (FY) 1992, 29 such projects in FY 1993, and thirty-one in FY 1994 (World Bank 1992c, 1993a, 1994e).

Table A.2
Environmental lending by multilateral development banks
$ US millions (number of projects)

Year	World Bank[a]	Asian Development Bank[b]	Inter-American Development Bank[c]
1986	50 (1)		
1987	200 (2)		
1988	350 (6)		
1989	250 (6)		
1990	1,200 (18)		
1991	900 (12)		
1992	1,500 (25)		
1993	2,000 (23)	1,592	1,200 (13)
1994	2,400 (25)		

a. Figures for fiscal years—primarily environmental projects defined as projects where the "costs of environmental protection measures or the environmental benefits accruing from the project exceeded 50% of the total costs or benefits." Data sources: *World Bank and the Environment, Fiscal 1991; World Bank and the Environment, Fiscal 1992; World Bank and the Environment, Fiscal 1993; Making Development Sustainable: The World Bank Group and the Environment, Fiscal 1994.* Figures do not include the Global Environment Facility.
b. Figures for fiscal year. Data source: Asian Development Bank, *Annual Report 1993: 59.*
c. 1993 loans for thirteen projects that will "benefit the environment and help conserve natural resources." Data source: Inter-American Development Bank, *Annual Report 1993: 22.*

World Bank projects with primarily environmental objectives constituted 7 percent of total bank lending in 1991.

The *Asian Development Bank* began to integrate environmental concerns into aspects of Bank operations in 1990 and 1991. By 1992, the ADB had not yet reported on the dollar value of its environmental projects, but its annual report noted that the "number of loans with environmental orientation continued to rise," and that the declarations of UNCED were expected to significantly influence the policies, programs, and projects of institutions like itself (ADB 1992). In 1993, the ADB reported that nearly $1.6 billion had been spent for environmental loans (ADB 1993).

Table A.3
World Bank environmental projects by region[a]
$ US millions (to nearest million)

	1993	1994	Totals	Percent of total
Africa	266	166	432	9
Asia	830	995	1,825	39
Europe	276	184	460	10
Middle East	207	0	207	4
Latin America, Caribbean	681	1,062	1,743	37
Total	2,260	2,407	4,667	

a. Primarily environmental projects defined as projects where the "costs of environmental protection measures or the environmental benefits accruing from the project exceeded 50% of the total costs or benefits." Calcualted using data from: *World Bank and the Environment, Fiscal 1993; Making Development Sustainable: The World Bank Group and the Environment: Fiscal 1994.*

The financial resources of *UNEP* are small relative to the role that it has played on the international stage. Although UNEP has, since its creation in 1972, demonstrated its ability to provide environmental leadership (particularly under Mostafa Tolba), its total expenditures are not as large as even the smallest transfer cases considered in this book. UNEP expenditures from its Environment Fund Programme (program and program-support costs) reached $59 million in 1990, $87 million in 1991, and $86 million in 1992 (UNEP 1992: 146). Both UNEP and the UNDP are, however, implementing agencies for the MPF.

As important as where the finances are coming from, and in what quantities, is where the money is going. Several generalizations can be made about the regional allocation of World Bank, Montreal Protocol Fund, and Global Environment Facility projects (see tables A.3, A.4, and A.5). Nearly half of all GEF projects (44 percent) were allocated to Asia, while Latin America and Africa were allocated 25 percent and 18 percent respectively. The World Bank (in 1993 and 1994) gave 39 percent of its environmental lending projects to Asian countries and 37 percent to Latin American countries. The Montreal Protocol Fund had 35 percent of project allocations in Africa, 31 percent in Latin America, and 22 percent in Asia.

Table A.4
Montreal Protocol Fund transfers by region[a]
$ US millions (to nearest million)

As of 22 February 1994	Totals	Percent of total
Africa	38	35
Asia, Pacific	24	22
Europe, Middle East	9	8
Latin America, Caribbean	33	31
Interregional, global	4	4
Totals	108	

a. Sum total of approved funding as of 22 February 1994. Figures for UNEP, UNDP projects, as well as World Bank Activities under way and completed. Data source: UNEP/OzL.Pro/ExCom/12/7.

Bilateral Financial Transfers

The members of the Development Assistance Committee of the OECD[1] provide official development assistance (not specifically environmental aid) to developing countries through multilateral mechanisms in large amounts: $13.7 billion in 1991, $15.2 billion in 1992, and $16.8 billion in 1993. However, larger amounts are reserved by these nations for bilateral distribution: $40.7 billion in 1991, $42.3 billion in 1992, $38.5 billion in 1993 (DAC: 65). Bilateral environmental development assistance for nine OECD countries between 1989 and 1993 is shown in table A.6. Ever larger proportions of national bilateral official development assistance are contributing to environmental protection. Bilateral environmental assistance can be viewed in two ways: total value and value as a percentage of official development aid. Among OECD countries, Japan and Norway are top contributors of environmental aid. *Norway* is highly geared to environmental aid, which befits a country whose prime minister, Gro Harlem Brundtland, chaired the World Commission on Environment and Development. Eighteen percent of Norway's official development aid budgets between 1991 and 1993 goes to environment related activities, a proportion unmatched by any other OECD country (OECD 1993c). Norway's official development aid for the environment is

Table A.5
Global Environment Facility transfers by region[a]
$ US millions (to nearest million)

	Totals (12/91 to 5/93)	Percent of total
Africa	44	18
Asia, Pacific	105	44
Europe, Middle East	19	8
Latin America, Caribbean	60	25
Interregional, global	10	4
Totals	238	

a. Total approved investment projects plus total approved technical assistance projects from December 1991 to December 1992 and total approved investment projects plus total approved technical assistance projects from January to May 1993. Data source: *Report by the Chairman to the May 1993 Participant's Meeting, Part I, Main Report: 5–7*).

supplemented by new and additional programs for global issues, which include contributions to multilateral institutions and programs for Eastern Europe. *Japan* leads OECD countries in total environment assistance, as total grants, loans, and technical assistance reached nearly 2.7 billion in 1992. Another $106 million was distributed to multilateral institutions by Japan, bringing the total of environment sector activities to $2.8 billion. This constituted 16.9 percent of Japan's total official development assistance (ODA) in 1992. In 1991, environment sector activities represented 7.2 percent of the total, 12.4 percent in 1990, and 9.8 percent in 1989 (Japan, Ministry of Foreign Affairs, 1994). In addition to official development assistance, Japan has provided technical assistance to manage and dispose of radioactive waste and other hazardous substances ($100 million) in 1993. Debt-for-nature swap contributions from the banking sector reached $1 million, and a 1992 Nature Conservation Fund worth $3 million was established by Keidanren (Federation of Economic Organizations) for cooperation with developing countries in their efforts toward nature conservation (OECD 1994a).

Other OECD countries have made contributions to global environmental protection, both through environmental official development assistance (see table A.6), and through programs initiated with business,

Table A.6
Bilateral environmental financial transfers
$ US millions (to nearest million)

	Canada[a]	Germany[b]	Japan[c]	Netherlands[d]	Norway[e]	Portugal[f]	Switzerland[g]
1989			1248				
1990			1605				
1991			1049				
1992		512	2697		221	30	15.0
1993	312*	603		101			17.0

*since 1986

	United Kingdom[h]	United States[i]
1989		
1990		
1991		680
1992		682
1993	161	638

a. Value of sixty-seven projects undertaken since 1986 by the Canadian International Development Agency. Another forty-six projects worth $311 million included environmental elements. Data source: Ministry of Supply and Services, Canada 1991: 71–72.

b. Financial means for development projects dealing with environmental and resource protection. Conversion from DM using average of daily figures for 1991, 1.659 DM = $1. Data source: OECD 1993a. For 1992, figure refers to pledged amount.

c. Data source: *Japan's ODA: Annual Report 1993:* 180. Ministry of Financial Assistance, Tokyo, March 1994, pp. 189. Includes grants, loans, and technical assistance. Calculated using $1 = ¥ 100.

d. Data source: Bilateral environmental assistance of $159 million for environmental policies in developing countries. The remaining $59 million went for multilateral environmental contributions, and the MILIEV program, a grant program for Dutch companies. Data source: OECD 1995a. *Environmental Performance Reviews: the Netherlands,* 1995. On average in 1993, 100 Gld = $53; figures calculated using this conversion.

e. Norway's Net ODA in 1992 was $1230 million (Data source: DAC, *Development Cooperation: Efforts and Policies of the Members of the Development Assistance Committee,* 1993 report: 81). The OECD 1993c *Environmental Performance Reviews: Norway,* 1993 reports that 18 percent of Norway's official aid is allocated for the environment in 1992, p. 129. The figure was calculated using these two sources.

f. Portugal's net ODA in 1992 was $300 million. (Data source: DAC, *Development Cooperation: Efforts and Policies of the Members of the Development As-*

Table A.6
(continued)

sistance Committee, 1993 report: 81). The OECD 1993d *Environmental Performance Reviews Portugal,* 1993 reports that 10 percent of Portugal's official aid was allocated for the environment in 1992, p. 124. The $30 million figure was calculated using these two sources.

g. Data source: Raymond Clemencon, Institute of Global Conflict and Cooperation, University of Chicago.

h. Finances devoted to helping recipient countries deal with environmental problems. Roughly one-third had environmental protection as its main objective. This bilateral aid was mostly geared toward current or former Commonwealth countries. Data source: OECD 1994b *Environmental Performance Reviews: the United Kingdom,* 1994. On average in 1993, £ 1 = $1.28; figures calculated using this conversion.

i. USAID environment and energy obligations, fiscal years. Data source: USAID's Bureau for Management, 26 September 1994 data.

nongovernmental organizations, and other government agencies. Highlights of initiatives not shown in table A.6 are below.

• *Canada* provides support to domestic and international NGOs for environmental activities through the Environment and Development Support Program at $500,000 per year, initiated in 1989 (Ministry of Supply and Services, Canada 1991).

• *Germany* began making environmental assessments of development projects in 1980, earlier than many of the other countries surveyed. In addition to the $603 million contributed to development projects dealing with environmental and resource protection, Germany supports international and German NGOs in providing environmental assistance. Further, the Germans have contributed to debt-for-nature swaps ($9 million as of 1992) and provided additional debt relief contingent on measures taken to protect and preserve the environment for a number of developing countries (OECD 1993a).

• Although figures for *Iceland* are not available, the OECD *Environmental Performance Reviews: Iceland* suggests that much of the bilateral aid distributed by the Icelandic Aid Agency (ICEIDA) is for programs related to natural resources (research, education, and training in fisheries and geothermal energy) (OECD 1993b).

• In 1990–1993, the *Netherlands* committed roughly $1 million for environmental protection activities in Central and Eastern Europe. In 1994, a total of about $6 million was provided to Central and Eastern European countries. The Dutch Special Programme on the Environment, charged with integrating environmental concerns into development cooperation policy, includes an environmental and economic self-sufficiency program, called MILIEV, which provides grants to Dutch companies for up to 55 percent of investment in projects that help to improve the environment in developing countries. The Netherlands has contributed to debt-for-nature swaps in Costa Rica and Tunisia. In addition to the 1994 environmental development assistance noted in table A.6, the Netherlands allocated $47 million for environmental policies in developing countries through bilateral and multilateral mechanisms, and MILIEV, in 1991, and $60 million in 1992 (OECD 1995a).

• In the *United States,* it is the United States Agency for International Development that administers US bilateral aid. Within USAID, the Office of Environment and Natural Resources in the Bureau for Research and Development advances environment concerns in global development programs (USAID 1993). Environment and energy obligations were roughly 15 percent of total development assistance in 1991 (USAID 1993). Environment and energy project obligations were projected to reach $679 million in 1994. The U.S. Enterprise for the America's Initiative had provided $353 million for debt forgiveness in seven developing countries by 1993. The United States supports several government-business partnerships aimed at promoting environmental exports and international business participation in environmental protection.

Bilateral aid is disbursed worldwide, but a few comments may be made about the regional distribution. The OECD reports that Poland is "among the non-OECD countries that have received the largest amount of funding through *grants to promote environmental protection*" (OECD 1993d). By November 1993, total environmental assistance had reached $230 million for 236 projects, and debt-for-nature swaps had reached $120 million per year. Worldwide, the United States is expected to spend $679 million expected to be spent on environment and energy projects in 1994. Of this amount, an estimated $273 million would be dedicated to Asia/Near East, $239 million to the Europe/New Independent States, $61.2 million to Africa, $41.6 million to Latin America and the Caribbean, $58.1 million to global obligations.

Nongovernmental Organizations

As discussed in Cord Jakobeit's chapter on debt-for-nature swaps, the activities of nongovernmental organizations were critical to the execution of the transfer mechanism. The $49.3 million that was spent on debt-for-nature swaps between 1987 and 1993 was contributed by eleven NGOs and three governments (IMF 1993: 116).[2] The universe of NGOs whose concerns and financial resources are directed to environmental protection is vast. Over 1,400 NGOs were accredited to UNCED, a number unprecedented in the history of UN conferences. Accounting for all NGO contributions to the cause of the environment is not a task to be undertaken here. There are, however, several environmental groups that are notable not only for their visibility, but for the size of the financial resources that they have been able to garner for their operations. Nonetheless, these resources do not compare to those provided by some of the other sources discussed here. The budget of the World Wide Fund for Nature reached nearly $270 million in 1991/1992. However, the budgets of other large environmental NGOs (Greenpeace, International Union for the Conservation of Nature, Friends of the Earth, and the Sierra Club) do not exceed $50 million each. As such, the resources of a large number of environmental NGOs taken together will not match the magnitude of other sources.

Conclusion

Multilateral development banks, United Nations agencies, the Global Environment Facility, national and regional aid programs, and nongovernmental organizations are all a part of the financing web that is being created worldwide. Although the obstacles to effective transfers for environmental protection are often severe, this appendix suggests that the magnitude of resources being provided is increasing. Governments and international institutions are integrating environmental considerations into their development programs, and projects with environmental objectives and environmental components are increasing as a percentage of overall development assistance.

Notes

1. Australia, Austria, Belgium, Canada, Denmark, Finland, France, Germany, Ireland, Italy, Japan, Luxembourg, the Netherlands, New Zealand, Norway, Portugal, Spain, Sweden, Switzerland, United Kingdom, United States. (Only three

OECD countries are not members of this committee: Greece, Iceland, and Turkey.)

2. Conservation International, Conservation Trust of Puerto Rico, Debt for Development Coalition, National Park Foundation of Costa Rica, Monteverde Conservation League, Missouri Botanical Gardens, Nigerian Conservation Foundation, Rainforest Alliance, Smithsonian Institute, The Nature Conservancy, World Wildlife Fund, Sweden, Norway, and USAID.

Glossary

ADB   Asian Development Bank

CFCs   chlorofluorocarbons

CI   Conservation International

DAC   Development Assistance Committee

DENR   Department of Environment and Natural Resources (The Philippines)

EAI   Enterprise of the Americas Initiative

EAP   Environmental Action Programme for Central and Eastern Europe

EBRD   European Bank for Reconstruction and Development

EdF   Electricité de France

EIB   European Investment Bank

ECU   European Currency Unit

EU   European Union

FAO   Food and Agricultural Organization

FCCC   Framework Convention on Climate Change

FMB   Forest Management Bureau

GEF   Global Environment Facility

HFCs   hydrochlorofluorocarbons

IAEA   International Atomic Energy Agency

IAs   implementing agencies

IBRD   International Bank for Reconstruction and Development (World Bank)

ICI   Imperial Chemical Industries

ICPR   International Commission for the Protection of the Rhine Against Pollution

IDA   International Development Association

IDB   Inter-American Development Bank

IEA   International Energy Agency

IE/PAC   Industry and Environment Programme Activity Centre

IFC   International Finance Corporation

IMF   International Monetary Fund

ITTA   International Tropical Timber Agreement

ITTO   International Tropical Timber Organization

IUCN   International Union for the Conservation of Nature (World Conservation Union)

MDBs   multilateral development banks

MdPA   Mines de Potasse d'Alsace

MIGA   Multilateral Investment Guarantee Agency

MPMF   Montreal Protocol Multilateral Fund

NGOs   nongovernmental organizations

NPF   National Parks Foundation

NPPs   nuclear power plants

ODP   ozone-depleting potential

ODS   ozone-depleting substances

OECD   Organization for Economic Cooperation and Development

OORG   Ozone Operations Resource Group

PHARE   Poland and Hungary Assistance for Restructuring Economies (European Union program)

PPC   Project Preparation Committee (of the EAP)

PPP   polluter pays principle

PWRs   pressurized-water reactors

SEC   Slovak Energy Company

STAP   Scientific and technical advisory panel (of the GEF)

TACIS   Technical Assistance to the Commonwealth of Independent States (European Union program)

TFAP   Tropical Forestry Action Plan

TNC   The Nature Conservancy

UNCED   United Nations Conference on Environment and Development

UNDP   United Nations Development Program

UNEP   United Nations Environmental Program

UNIDO   United Nations Industrial Development Organization

USAID   United States Agency for International Development

WANO   World Association of Nuclear Operators

WHO   World Health Organization

WWF   World Wildlife Fund

References

Ackermann, Richard. 1991. "Environment in East Central Europe: Despair or Hope?" *Transition,* World Bank/CECSE 2, no. 4 (4 April): 9–11.

Aegis. 1994. "European Bank Reserves Judgment on Mochovce." London: East West Environment, 2.

Alcamo, Joseph, ed. 1992. *Coping with Crisis in Eastern Europe's Environment.* New York: Parthenon Publishing Group.

Allison, Graham. 1971. *Essence of Decision: Explaining the Cuban Missile Crisis.* New York: Little, Brown.

Andrews, Richard N. L. 1993. "Environmental Policy in the Czech and Slovak Republic." In Anna Vari, Pal Tamas, eds. *Environment and Democratic Transition: Policy and Politics in Central and East Central Europe.* Boston: Kluwer Academic Publishers, 5–48.

Ascher, William. 1994. *Communities and Sustainable Forestry in Developing Countries.* Report of the Center for Tropical Conservation, Duke University, May.

ADB (Asian Development Bank). 1991. *Annual Report.*

ADB. 1992. *Annual Report.* Manila.

ADB. 1993. *Annual Report.* Manila.

Ball, Nicole. 1992. *Pressing for Peace: Can Aid Induce Reform?* Washington, DC: Overseas Development Council.

Baumgartl, Bernd. 1994. "Only a Guideline: The Environmental Action Program for Central and Eastern Europe." *Environmental Impact Assessment Review* 14: 147–156.

Bautista, Germelino M. 1992. *The Forestry Revenue System in the Philippines: Its Concept and History.* Natural Resources Management Program Policy Studies, Report 92–20. Manila: Louis Berger International, October.

Benedick, Richard Elliot. 1991. *Ozone Diplomacy: New Directions in Safeguarding the Planet.* Cambridge, MA: Harvard University Press.

Berg, Eliot. 1991. "Comment." In V. Thomas, A. Chibber, M. Dailami, J. de Melo, eds., *Restructuring Economics in Distress.* New York: Oxford University Press.

Bernauer, Thomas. 1995. "The Effect of International Environmental Institutions: How We Might Learn More." *International Organization* 49:2: 351–377.

BMZ (Bundesministerium für Wirtschaftliche Zusammenarbeit). 1990. *Achter Bericht zur Entwicklungspolitik der Bundesregierung.* Bundestagsdrucksache 11/7313 (May) Bonn.

Boado, Eufresina L. 1988. "Incentive Policies and Forest Use in the Philippines." In *Public Policies and the Misuse of Forest Resources.* Cambridge: Cambridge University Press.

Böhmer, Jochen. 1994a. "Can the International Debt Strategy Use the Concept of Corporate Compositions?" *Intereconomics* 29(1): 37–42.

Böhmer, Jochen. 1994b. Telephone interview. Bundesministerium für Wirtschaftliche Zusammenarbeit (German Ministry for Economic Cooperation), 11 January.

Bowles, I., and G. Prickett 1993. *An Analysis of the GEF's Pilot Phase Approach to Biodiversity and Global Warming* [draft]. Washington, DC: Conservation International and Natural Resources Defense Council.

Boyle, Stewart. 1993. Letter to Mark Tomlinson, EBRD. London: International Institute for Energy Conservation (31 December).

Bramble, Barbara J. 1986. *Third World Debt and Natural Resources Conservation.* Washington, DC: National Wildlife Federation, International Program.

Bramble, Barbara J. 1989. *External Debt, Democratization, and Natural Resources in Developing Countries: The Case of Brazil.* Washington, DC: National Wildlife Federation.

Bramble, Barbara, and Gareth Porter. 1992. "Non-Governmental Organizations and the Making of U.S. International Environmental Policy." In Andrew Hurrell and Benedict Kingsbury, eds., *The International Politics of the Environment.* New York: Oxford University Press.

Brand, Monica L. 1990. *Debt-for-Nature Swaps: Moving from Peril to Promise.* Washington, DC: U.S. Agency for International Development (LAC/DP).

Brechin, Steven. 1996. *International Organizations and Trees for People.* Baltimore: Johns Hopkins University Press.

Broad, Robin, and John Cavanagh. 1993. *Plundering Paradise: The Struggle for the Environment in the Philippines.* Berkeley: University of California Press.

Bruijnzeel, L. A., and W. R. S. Critchley. 1994. *Environmental Impacts of Logging Moist Tropical Forests.* IHP Humid Tropics Programme Series: No. 7. Paris: UNESCO, Division of Water Sciences, International Hydrology Programme.

Cagin, Seth, and Philip Dray. 1993. *Between Earth and Sky: How CFCs Changed Our World and Endangered the Ozone Layer.* New York: Pantheon Books.

Carrière, Jean. 1991. "The Crisis in Costa Rica: An Ecological Perspective." In David Goodman and Michael Redclift, eds., *Environment and Development in Latin America. The Politics of Sustainability.* Manchester: Manchester University Press.

Cassen, Robert, and Associates. 1986. *Does Aid Work? Report to an Intergovernmental Task Force.* Oxford: Clarendon Press.

CCE (Commission des Communautées Européennes, Service de l'environnement et de la protection des consommateurs). 1973. *Etude de differentes pollutions constatées dans le Bassin Rhenan.* Brussels, September, ENV/45/74.

Central Bureau of Statistics (Indonesia). 1992. *Indonesian Foreign Trade Statistics, 1971–1991.* Jakarta.

Chabrzyk, G. 1993. "An Assessment of the Environmental Impact of the PHARE Programme." Brussels: European Parliament, Scientific and Technological Options Assessment, 26 April.

Chamberlin, Michael. 1992. "Encouraging Banks to Make Donations." In Smithsonian Institution and Natural Resources Defense Council, eds. *Debt-for-Nature Swaps: Progress and Prospects.* Unpublished conference report. Washington, DC: Smithsonian Institution and Natural Resources Defense Council.

Coase, Ronald H. 1960. "The Problem of Social Cost." *Journal of Law and Economics* 3(October): 1–44.

Cody, Betsy A. 1988. *Debt-for-Nature Swaps in Developing Countries: An Overview of Recent Conservation Efforts.* CRS Report for Congress 88–647 ENR, 26 September. Washington, DC: Congressional Research Service, Library of Congress.

Cohen, Michael D., James D. March, and Johon P. Olsen. 1972. "A Garbage Can Model of Organizational Choice." *Administrative Science Quarterly* 17: 1–25.

Colchester, Marcus, and Larry Lohmann. 1990. *The Tropical Forestry Action Plan: What Progress?* Penang, Malaysia: World Rainforest Movement and the Ecologist.

Commission of the European Communities. 1992. PHARE General Guidelines 1993–97. Revised Draft. Brussels, 16 November.

Commission of the European Communities. 1993. "Existing and Planned Efforts by the Community to Assist the Central and Eastern European Countries in Improving the Safety of their Nuclear Reactors." Report by the services of the Commission. Brussels, 21 December.

Communiqué. 1972. *Communiqué of the Ministerial Conference on the Pollution of the Rhine.* The Hague, 25 and 26 October.

Cox, R., and H. Jacobson. 1973. *The Anatomy of Influence: Decision Making in International Organizations.* New Haven: Yale University Press.

Curtis, Randall K. 1993. *Interview with Randall K. Curtis.* The Nature Conservancy, Washington, DC, 15 December.

DAC (Development Assistance Committee). 1993. *Development Cooperation: Efforts and Policies of the Members of the Development Assistance Committee.* Paris: OECD.

Danida. 1991. *Effectiveness of Multilateral Agencies at Country Level: Case Study of 11 Agencies in Kenya, Nepal, Sudan, and Thailand.* Copenhagen: Danida, Ministry of Foreign Affairs.

Danish, K. 1994. "The Bhutan Trust Fund and the Potential of National Conservation Trust Funds." Unpublished paper prepared for International Politics of the Environment Conference, Woodrow Wilson School, Princeton University, 17 May.

Dawkins, Kristin. 1990. "Debt-for-Nature Swaps." In Lawrence E. Susskind, Esther Siskind, and J. William Breslin, eds., *Nine Case Studies in International Environmental Negotiations*. Cambridge, MA: The MIT-Harvard Public Disputes Program.

DeBardeleben, Joan, ed. 1991. *To Breathe Free: East Central Europe's Environmental Crisis*. Baltimore: Johns Hopkins Press.

DeSombre, Elizabeth R. 1995. "Baptists and Bootleggers for the Environment: The Origins of United States Unilateral Sanctions." *Journal of Environment and Development* 4,1 (Winter): 53–75.

Dichter, Thomas W. 1988. "The Changing World of Northern NGOs: Problems, Paradoxes, and Possibilities." In John P. Lewis, ed., *Strengthening the Poor: What Have We Learned?* New Brunswick, NJ: Transaction Books.

Dick, J., and L. Bailey. 1994. *Indonesia's Environmental Assessment Process (AMDAL)*. Halifax, Nova Scotia: Environmental Management Development in Indonesia Project, Dalhousie University.

Dieperink, Carel. 1992. *Intergovernmental Settlement of Environmental Conflict: The Case of the Chlorides Pollution of the River Rhine*. Paper prepared for the First European Conference of Sociology, Vienna, 26–29 August.

Djamaludin. 1991. "The Implementation of Indonesian Selective Cutting and Replanting (TPTI) Silviculture System for Timber Improvement in the Logged-Over Areas." Pp. 95–110 in *Fourth Round-Table Conference on Dipterocarps*. Vol. 41. Bogor, Indonesia: Southeast Asian Regional Centre for Tropical Biology.

Dogsé, Peter, and Bernd Von Droste. 1990. *Debt-For-Nature Exchanges and Biosphere Reserves. Experiences and Potential*. MAB Digest 6. Paris: UNESCO.

Dumanoski, Diane. 1990. "In Shift U.S. to Aid World Fund on Ozone." *Boston Globe* (16 June): 1.

Duncan, Alex. 1985. "Aid Effectiveness in Raising Adaptive Capacity in the Low-Income Countries." In John P. Lewis, ed., *Development Strategies Reconsidered*. New Brunswick, NJ: Transaction Books.

Dupont, Christophe. 1993. "The Rhine: A Study of Inland Water Negotiations." In Gunnar Sjöstedt, ed., *International Environmental Negotiation*. Newbury Park: Sage: 135–148.

Earth Negotiations Bulletin. 1994. "Convention on Biological Diversity COP-I Summary." Vol. 9 no.28. Available on Econet conference "biodiversity." 19 December.

Easterbrook, Gregg. 1994. "Forget PCB's. Radon. Alar." *New York Times Magazine* (11 September): 60–63.

EBRD. 1990. "Articles Establishing the European Bank for Reconstruction and Development." London.

EBRD. 1992a. Environmental Policy. London.

EBRD. 1992b. Municipal Development Operations Policy. London: EBRD. July.

EBRD. 1993. "Environments in Transition: The Environmental Bulletin of the EBRD." London.

EBRD. 1994a. Annual Report. London.

EBRD. 1994b. Environments in Transition: The Environmental Bulletin of the EBRD." London.

Environment for Europe. 1991. Environment for Europe: Conference of Environment Ministers and Representatives of International Organizations. Dobris Castle, Czechoslovakia. 21–23 June. Conference proceedings.

Environmental Action Programme for Central and Eastern Europe. 1993. Document submitted to the Ministerial Conference in Lucerne, Switzerland (28–30 April).

Environmental Action Programme for Central and Eastern Europe. 1992. Unofficial outline (25 February).

Environmental Defense Fund, Natural Resources Defense Council. 1994. "Power Failure: A Review of the World Bank's Implementation of Its New Energy Policy." Washington, DC.

Erwin, T. L. 1988. "The Tropical Forest Canopy: The Heart of Biotic Diversity." In Edward O. Wilson, ed., *Biodiversity*. Washington, DC: National Academy Press.

European Bank for Reconstruction and Development. 1993. "Nuclear Safety Account to Grant ECU 24 Million to Upgrade Kozloduy Plant in Bulgaria." Press release. London: European Bank for Reconstruction and Development (16 June).

European Parliament. 1993. *Report of the Committee on Energy, Research and Technology on Nuclear Safety in the Countries of Eastern Europe and the Commonwealth of Independent States*, A3–0396/93/Part B. Brussels, Luxembourg: European Parliament.

Fairman, D. 1994. "Report of the Independent Evaluation of the Global Environment Facility" [review essay]. *Environment* 36:6 (July/August): 25–30.

Fairman, David, and Tamar Gutner. 1993. "International Mandates to Multilateral Development Banks: Bridging the Gap Between Objectives and Performance," in Lawrence E. Susskind, William R. Moomaw and Adil Najam, eds. *Papers on International Environmental Negotiation*, vol. 3. Cambridge, MA: Program on Negotiation at Harvard Law School.

Fakultas Kehutanan of the Institute Pertanian, Bogor. 1989. *Report on Field Case Studies of Forest Concessions*. Indonesian Ministry of Forestry, UN Food and Agriculture Organization, Field document no. 1–5, Jakarta, December.

FAO (Food and Agriculture Organization). 1981. *Tropical Forest Resources*. Rome.

FAO. 1993. *Forest Resources Assessment 1990: Summary of Final Results for the Tropical World.* Prepared for the Eleventh Session of the Committee on Forestry, Rome, 8–12 March 1993.

FAO. Various. *Forest Products Yearbook.* Rome: UN Food and Agriculture Organization.

Fennesz, Andrea. 1993. "The Assistance of the G-24 Countries." Brussels: Commission of the European Communities, G-24 Coordination Unit.

First Ministerial Conference of Developing Countries on Environment and Development, 1991. "Declaration of the First Ministerial Conference of Developing Countries on Environment and Development." Beijing: First Ministerial Conference Secretariat.

Fisher, Duncan. 1992. *Paradise Deferred: Environmental Policymaking in Central and East Central Europe.* London: Royal Institute of International Affairs and Ecological Studies Institute, June.

Flavin, Christopher. 1987. "Reassessing Nuclear Power: The Fallout from Chernobyl." *Worldwatch Paper 75.* Worldwatch Institute.

Forest Management Bureau. 1993. *1992 Philippine Forestry Statistics.* Quezon City: Department of Environment and Natural Resources, Forest Management Bureau, the Philippines.

Forest Management Bureau. 1994. List of Existing/Active Timber License Agreements (as of 31 May 1994). Internal document, Forest Management Bureau, the Philippines.

French, Hilary F. 1991. "Eastern Europe's Clean Break with the Past," Worldwatch (March/April).

French, Hilary F. 1995. "Forging a New Global Partnership." In Lester R. Brown, et al., *State of the World.* New York: W. W. Norton.

FOE (Friends of the Earth). 1990. *Funding Change: Developing Countries and the Montreal Protocol.*

FOE. 1992a. "Dangerous Liaisons." London.

FOE. 1992b. "Report on the Executive Committee of the Interim Multilateral Fund," 6th meeting.

FOE. 1993. "Summary Report of the 10th Meeting of the Executive Committee," 28–30 June.

G-24 Coordination Unit. 1994. "Scoreboard of Assistance Commitments to the CEEC." Brussels: Commission of the European Communities, June.

G-24 Coordination Unit. 1994b. "G-24 Assistance Commitments." Brussels: Commission of the European Communities, 12 December.

G-24 Coordination Unit, and OECD Centre for Cooperation with European Economies in Transition. 1993. "Environmental Assistance to Central and Eastern Europe." Prepared for Environment for Europe Ministerial Conference. Lucerne, Switzerland, 28–30 April.

G-77. 1993. "Group of 77 and China Position on GEF/PA.93/2: Elements for Establishing the Restructured Global Environment Facility." G-77/93/1. Washington, DC: Global Environment Facility. Dated 23 September.

GEF (Global Environment Facility). 1991a. *Report by the Chairman to the May 1991 Participants' Meeting*. Part 2. Washington, DC: Global Environment Facility.

GEF. 1991b. *Report by the Chairman to the December 1991 Participants' Meeting*. Part 2. Washington, DC: Global Environment Facility.

GEF. 1992a. *Report by the Chairman to the April 1992 Participants' Meeting*. Part 1. Washington, DC: Global Environment Facility.

GEF. 1992b. *Report by the Chairman to the April 1992 Participants' Meeting*. Part 2. Washington, DC: Global Environment Facility.

GEF. 1992c. *The Pilot Phase and Beyond*. Working paper series no. 1. Washington, DC: Global Environment Facility.

GEF. 1992d. *Report by the Chairman to the December 1992 Participants' Meeting*. Part 1. Washington, DC: Global Environment Facility.

GEF. 1993a. *Report by the Chairman to the May 1993 Participants' Meeting*. Part 1. Washington, DC: Global Environment Facility.

GEF. 1993b. *Chairman's Summary, GEF Replenishment Meeting, Beijing, China*. Washington, DC: Global Environment Facility. Dated 25 May.

GEF. 1993c. *Elements for Establishing the Restructured GEF*. GEF/PA.93/2.

GEF. 1993d. *The GEF and the Evaluation: Learning from Experience and Looking Forward*. GEF/PA.93/7. A Background Note for the GEF Participants' Meeting, Cartagena, Colombia. Washington, DC: Global Environment Facility. Dated 2 December.

GEF. 1993e. *Report by the Chairman to the December 1993 Participants' Meeting*. Washington, DC: Global Environment Facility.

GEF. 1994a. *Draft Instrument for the Establishment of the Restructured Global Environment Facility*. GEF/PA.93/6/Rev.6. Washington, DC: Global Environment Facility. Dated [and approved] 17 March.

GEF. 1994b. *GEF Council: A Proposed Statement of Work*. GEF/C.1/2. Washington, DC: Global Environment Facility.

GEF. 1995a. *Bulletin and Quarterly Operational Summary*, no. 14.

GEF. 1995b. *Operational Strategy*. GEF secretariat draft GEF.

GEF STAP. 1992. *Criteria for Eligibility and Priorities for Selection of Global Environment Facility Projects*. Washington, DC: Global Environment Facility. Dated May.

GEF STAP. 1994a. *Review of the Global Warming Portfolio*. Washington, DC: UNEP/GEF. Available on Econet conference "gef.report."

GEF STAP. 1994b. *Summary Report of the Review of the Biodiversity Portfolio, Tranches I-V*. Washington, DC: UNEP/GEF. Available on Econet conference "gef.report." Dated April.

George, Susan. 1992. *The Debt Boomerang: How Third World Debt Harms Us All.* London: Pluto Press.

Gibson, J. Eugene, and Randall K. Curtis. 1990. "A Debt-for-Nature Blueprint." *Columbia Journal of Transnational Law* 28(2): 331–412.

Gibson, J. Eugene, and William J. Schrenk. 1991. "The Enterprise for the Americas Initiative: A Second Generation of Debt-for-Nature Exchanges—with an Overview of Other Recent Exchange Initiatives." *George Washington Journal of International Law and Economics* 25(1): 1–70.

Gill, Tom. 1958. "Diagnosis of Forestry Problems." *The Philippine Lumberman* (October–November): 5–10.

Gilpin, Kenneth N. 1994. "Mopping Up Foreign Debt." *The New York Times,* 18 April 1994: D1, D6.

Goppel, J. M. 1991. *Legal Aspects of International Water Management: the Rhine.* Paper presented at Waterscapes '91: International Conference on Water Management for a Sustainable Environment, Saskatoon, Canada, June.

Government of Sweden. 1991. *Sweden's National Report to UNCED.*

Greener, Laurie P. 1991. "Debt-For-Nature Swaps in Latin American Countries: The Enforcement Dilemma." *Connecticut Journal of International Law* 7(1): 123–180.

Greenpeace UK. June 1992. *Neglecting Alternatives: How Government Inaction Is Destroying the Ozone Layer.*

Greenpeace UK. October 1993. *The Refrigerant Revolution, Hydrocarbons: The Right Choice for China.*

Grubb, Michael et al. 1993. *The Earth Summit Agreements: A Guide and Assessment.* London: Royal Institute of International Affairs.

Gutner, Tamar. March 1993. "International Actors, Institutional Development and Policymaking in Czechoslovakia and Hungary: The Case of Environmental Policy." Presented at the Western Political Science Association Conference.

Haas, E. 1990. *When Knowledge Is Power: Three Models of Change in International Organizations.* Berkeley: University of California Press.

Haas, Peter M. 1992. "Introduction: Epistemic Communities and International Policy Coordination." *International Organization* 46(1): 1–35.

Haas, Peter M., Robert O. Keohane, and Marc A. Levy, eds. 1993. *Institutions for the Earth: Sources of Effective International Environmental Protection.* Cambridge: MIT Press.

Haggard, Stephan. 1990. "The Political Economy of the Philippine Debt Crisis." In Joan Nelson, ed., *Economic Crisis and Policy Choice: The Politics of Adjustment in the Third World.* Princeton: Princeton University Press.

Hartkopf, G. 1976. "International Protection of the Rhine." *Environmental Policy and Law* 1(4): 166–168.

Haughton, Jonathan, Darius Teter, and Joseph Stern. 1992. *Report on Forestry Taxation.* Private memorandum, 8 September.

Hawkins, Ann P. 1990. *Swapping Debt for Nature: Emergence of a New Global Order?* Ph.D. dissertation, Cornell University.

Hecht, Susanna, and Alexander Cockburn. 1990. *The Fate of the Forests: Developers, Destroyers, and Defenders of the Amazon.* New York: Harper Perennial.

Hermann, Chris, Lane Krahl, Keith Openshaw, James Seyler, and Craig Smith. 1992. *Interim Evaluation of the Natural Resources Management Program.* Report issued by Tropical Research and Development and Abt Associates, Inc., 30 September.

Herttrich, Michael, Rolf Janke, and Peter Kelm. 1994. "International Cooperation to Promote Nuclear-Reactor Safety in the Former USSR and Eastern Europe." *Green Globe Yearbook 1994.* New York: Oxford University Press, 89–101.

Hiltunen, Heidi. 1994. *Finland and Environmental Problems in Russia and Estonia.* Helsinki: The Finnish Institute of International Affairs.

Hoensch, Volker. 1993. "Betriebsorganisation in osteuropäischen Kernkraftwerken." In *Atomwirtschaft* 38 (2): 130–133.

Horta, K. 1994. "The First GEF Council Meeting." Memorandum to interested NGOs. Washington, DC: Environmental Defense Fund. Available on Econet conference "gef.forum." Dated 8 July.

Houghton, R. A. 1990. "The Future Role of Tropical Forests in Affecting the Carbon Dioxide Concentrations of the Atmosphere." *Ambio* 19: 204–209.

Houghton, R. A. 1993. *Forests and Climate.* Paper presented at the Global Forest Conference, Bandung, Indonesia, 17–20 February.

Hrynik, Tamara J. 1990. "Debt-For-Nature Swaps: Effective but Not Enforceable." *Case Western Reserve Journal of International Law* 22: 141–163.

Hurrell, Andrew. 1992. "Brazil and the International Politics of Amazonian Deforestation." In Andrew Hurrell and Benedict Kingsbury, eds. *The International Politics of the Environment: Actors, Interests, and Institutions.* Oxford: Clarendon Press.

Hutchison, Jane. 1993. "Class and State Power in the Philippines." In Kevin Hewison, Richard Robison, and Garry Rodan, eds., *Southeast Asia in the 1990s: Authoritarianism, Democracy, and Capitalism.* Sydney: Allen and Unwin.

IAWR (Internationale Arbeitsgemeinschaft der Wasserwerke im Rheineinzugsgebiet). 1988. *Salz im Rhein, Rost im Rohr.* Amsterdam.

ICPR. (International Commission for the Protection of the Rhine Against Pollution). 1974–1992 *Tätigkeitsberichte* (annual reports). Koblenz, Germany.

ICPR. 1991. *Rhein Aktuell.* Koblenz, Germany, April.

Ingram, C. Denise. 1989. *Analysis of the Revenue System for Forest Resources in Indonesia.* Indonesian Ministry of Forestry and UN Food and Agriculture Organization, Field Document No. VI-2, Jakarta, November.

Institute for East-West Studies. 1992. Beyond Assistance: Report of the IEWS Task Force on Western Assistance to Transition in the Czech and Slovak Federal

Republic, Hungary and Poland. New York, Prague: Institute for East-West Studies, European Studies Center.

International Atomic Energy Agency. 1992. *The Safety of WWER-440 Model 230 Nuclear Power Plants.* Vienna.

International Monetary Fund (IMF). 1993. World Debt Tables 1993. Washington, DC: IMF.

International Tropical Timber Council. 1990. *Draft Annual Report for 1989.* ITTC (8)/2,2 April.

International Tropical Timber Council. 1992. *Draft Annual Report for 1991.* ITTC (12)/2, 15 March.

Israel, Arturo. 1987. *Institutional Development.* Baltimore: Johns Hopkins University Press.

Jay, Keith, and Constantine Michalopoulos. 1989. "Interaction Between Donors and Recipients." In Anne Krueger, Constantine Michalopoulos, and Vernon Ruttan, eds., *Aid and Development.* Baltimore: Johns Hopkins University Press.

Johnson, Nels, and Bruce Cabarle. 1993. *Surviving the Cut: Natural Forest Management in the Humid Tropics.* Washington, DC: World Resources Institute.

Johnson, S. 1992. *The Earth Summit: The United Nations Conference on Environment and Development.* London: Graham and Trotman/Martinus Nijhoff.

Jordan, A. 1994a. "Financing the UNCED Agenda." *Environment* 36(3): 16–20ff.

Jordan, A. 1994b. "Paying the Incremental Costs of Global Environmental Protection: The Evolving Role of GEF." *Environment* 36(6): 12–20ff.

Kabala, Stanley J. 1993. "Environmental Affairs and the Emergence of Pluralism in Poland." In Anna Vari and Pal Tamas, eds., *Environment and Democratic Transition: Policy and Politics in Central and East Central Europe.* Boston: Kluwer Academic Publishers.

Kahler, Miles. 1992. "External Influence, Conditionality, and the Politics of Adjustment." In Stephan Haggard and Robert Kaufman, eds., *The Politics of Economic Adjustment.* Princeton: Princeton University Press.

Kampffmeyer, Thomas. 1987. *Towards a Solution of the Debt Crisis—Applying the Concept of Corporate Compositions with Creditors.* Occasional Papers of the German Development Institute (GDI), No. 89. Berlin: German Development Institute.

Kauffman, Joanne. 1994. "Domestic and International Linkages in Global Environmental Politics: A Case Study of the Montreal Protocol," Paper presented at the International Studies Association meeting, Washington, DC, March.

King, K. 1994. *Issues to be Addressed by the Program for Measuring Incremental Costs for the Environment.* GEF Working Paper No. 8. Washington, DC: Global Environment Facility.

Kiss, Alexandre. 1985. "The Protection of the Rhine Against Pollution." *Natural Resources Journal* 25(3): 613–637.

Kjorven, O. 1992. *Facing the Challenge of Change: The World Bank and the Global Environment Facility.* Report 1992/3. Lysaker, Norway: Fridtjof Nansen Institute.

Klinger, Janeen. 1994. "Debt-for-Nature Swaps and the Limits to International Cooperation on Behalf of the Environment." *Environmental Politics* 3(2): 229–246.

Kloss, Dirk. 1994. *Umweltschutz und Schuldentausch. Neue Wege der Umweltschutzfinanzierung am Beispiel lateinamerikanischer Tropenwälder.* Frankfurt: Vervuert.

Koch, Matthias, and Michael Grubb. 1993. "Agenda 21." Chapter 9 of Grubb et al. 1993: 97–157.

Korten, Frances F. 1994. "Questioning the Call for Environmental Loans: A Critical Examination of Forestry Lending in the Philippines." *World Development* 22: 971–979.

Kraemer, Moritz, and Jörg Hartmann. 1993. "Policy Responses to Tropical Deforestation: Are Debt-for-Nature Swaps Appropriate?" *Journal of Environment and Development* 2(2): 41–65.

Krasner, S. 1985. *Structural Conflict: The Third World against Global Liberalism.* Berkeley: University of California Press.

Kummer, David M. 1992. *Deforestation in the Postwar Philippines.* Chicago: University of Chicago Press.

Kurylo, M. J., and NASA Panel for Data Evaluation. 1988. "Present State of Knowledge of the Upper Atmosphere 1988: An Assessment Report," NASA Reference Publication 1208, Washington, DC: U.S. Government Printing Office, cited in World Meteorological Organization/National Aeronautics and Space Administration, 1997, "Report of the Ozone Trends Panel."

Lammers, J. G. 1989. "The Rhine: Legal Aspects of the Management of a Transboundary River." In Wil D. Verwey, ed., *Nature Management and Sustainable Development.* Amsterdam: Foundation for Nature Management and Sustainable Development, 440–457.

Le Monde. 1993. "PHARE, le programme de soutien aux reformes," 10 April.

Le Prestre, P. 1989. *The World Bank and the Environmental Challenge.* Cranbury, N.J.: Associated University Presses.

Lean, Geoffrey, and Teresa Poule. 1993. "China's Kitchens in New Cold War," *Sunday Independent,* London, 11 July.

LeMarquand, David G. 1977. *International Rivers: The Politics of Cooperation.* Vancouver: University of British Columbia, Westwater Research Centre.

Levy, Marc. 1993a. "East-West Environmental Politics After 1989: The Case of Air Pollution." In Robert O. Keohane, Joseph S. Nye, and Stanley Hoffmann, eds. *After the Cold War: International Institutions and State Strategies in Europe, 1989–1991.* Cambridge, MA: Harvard University Press.

Levy, Marc A. 1993b. "European Acid Rain: The Power of Tote-Board Diplomacy." In Haas, Keohane, and Levy: 75–132.

Levy, Marc A. 1993c. "The Effectiveness of International Environmental Institutions: What We Think We Know and How We Might Learn More." Paper presented at the annual convention of the International Studies Association, Acapulco, Mexico, 23–27 March.

Lewin, Elisabeth et al. 1994. *Evaluación independiente del programa "Manejo sostenible de recursos naturales en Costa Rica."* Informe final, Enero. San José, Costa Rica: ACLA, ACOSA, OC-SPN.

Lewis, John P. 1993. *Pro-Poor Aid Conditionality.* Washington, DC: Overseas Development Council.

Livernash, Robert. 1992. "The Growing Influence of NGOs in the Developing World." *Environment* 34(5): 12–43.

Lovejoy, Thomas E. 1984. "Aid Debtor Nations' Ecology." *New York Times* (4 October): A31.

Lovejoy, Thomas E. 1994. *Interview with Thomas E. Lovejoy.* Washington, DC: Smithsonian Institution, 15 June.

MacNeill, Jim, Pieter Winsemius, and Taizo Yakushiji. 1991. *Beyond Interdependence: The Meshing of the World's Economy and the Earth's Ecology.* New York: Oxford University Press.

Mahony, Rhona. 1992. "Debt-for-Nature Swaps: Who Really Benefits?" *The Ecologist* 22(3): 97–103.

Manning, Chris. 1971. "The Timber Boom, with Special Reference to East Kalimantan." *Bulletin of Indonesian Economic Studies* 8: 30–60.

Martin, Lisa L. 1994. "Heterogeneity, Linkage and Commons Problems." *Journal of Theoretical Politics* 6(4).

Mazmanian, D., and P. Sabatier. 1983. *Implementation and Public Policy.* Glenview, IL: Scott, Foresman.

McNeil, M. 1994. "The Global Environment Facility: Negotiations to Establish a Voting Mechanism for Phase 2." Unpublished paper prepared for STM 221, John F. Kennedy School of Government, Harvard University, Cambridge, MA. Dated 26 January.

MdPA (Mines de Potasse d'Alsace). 1988. *La protection du Rhin.* Mulhouse, France: MdPA.

Melzer, A. 1993. "On the Flow of Bilateral Assistance of G-24 Countries to Hungary in the Field of Environment." Report to the Expert Group preparing the Lucerne Conference of European Ministers of Environment. Environment for Europe, Lucerne, Switzerland, February.

Milgrom, Paul, and John Roberts. 1992. *Economics, Organization, and Management.* Englewood Cliffs, NJ: Prentice-Hall.

Milgrom, Paul, Douglass North, and Barry Weingast. 1990. "The Role of Institutions in the Revival of Trade." *Economics and Politics* 2: 1–23.

Ministry of Foreign Affairs, Japan, 1994. *Japan's ODA 1993*. Tokyo: Ministry of Foreign Affairs.

Ministry of Forestry, Indonesia. 1994. *Forestry Sector Study, Mid Term Report, Forestry Sector Policy Analysis,* Working paper no. 2, Utilisation of the Reforestation Fund. Asian Development Bank Project Preparation Technical Assistance T.A. No. 1781-INO.

Ministry of Supply and Services, Canada. 1991. *Canada's National Report to UNCED.*

Miranda, Marie Lynn, Olga Marta Corrales, Michael Regan, and William Ascher. 1992. "Forestry Institutions." In Narendra P. Sharma, ed., *Managing the World's Forests: Looking for Balance Between Conservation and Development.* Dubuque, IA: Kendall/Hunt Publishing.

MIRENEM (Ministerio de Recursos Naturales). 1993. *Estrategia de Financiamiento del Sistema Nacional de Areas de Conservación.* Avance, Enero. San José, Costa Rica: MIRENEM.

Moldan, Bedrich, and Jerald L. Schnoor. 1992. "Czechoslovakia: Examining a Critically Ill Environment." *Environment Science and Technology* 26(1): 16.

Molina, M. J., and F. S. Rowland. 1974. "Stratospheric Sink for Chlorofluoromethanes: Chlorine Atom-Catalyzed Destruction of Ozone. *Nature* 249: 810–812.

Moltke, Konrad von. 1988. "Debt for Nature: An Overview." In World Wildlife Fund/National Wildlife Federation, eds., *Debt for Nature: An Opportunity.* Washington, DC: World Wildlife Fund/National Wildlife Federation.

Montreal Protocol on Substances that Deplete the Ozone Layer (1987), London Amendments (1990), and Copenhagen Amendments (1992).

Mosley, Paul. 1987. *Conditionality as Bargaining Process: Structural Adjustment Lending, 1980–1986.* Essays in International Finance No. 168, October. Princeton: Princeton University, Department of Economics, International Finance Section.

Mosley, Paul. 1992. "The Philippines." In Paul Mosley, Jane Harrigan, and John Toye, eds., *Aid and Power: The World Bank and Policy-Based Lending,* vol. 2. London: Routledge.

Mosley, Paul, Jane Harrigan, and John Toye. 1991. *Aid and Power: The World Bank and Policy-Based Lending,* vol. 1. London: Routledge.

Myers, Norman. 1988. "Threatened Biotas: 'Hot Spots' in Tropical Forests." *The Environmentalist* 8(3): 187–208.

Myers, Norman. 1992. "The Anatomy of Environmental Action: The Case of Tropical Deforestation." In Andrew Hurrell and Benedict Kingsbury, eds., *The International Politics of the Environment.* Oxford: Clarendon Press.

NAC (Ninth Conference of Heads of State or Government of Nonaligned Countries) 1989. "Environment." NAC/9/EC/Doc.8/Rev.4. Dated 7 September. Cited in Porter and Brown (1991: 127).

National Statistical Coordinating Board. Various. *Philippine Statistical Yearbook.* Manila: National Statistical Coordinating Board.

Nelson, Joan M. 1989. "The Politics of Long-Haul Economic Reform." In Joan Nelson, ed., *Fragile Coalitions: The Politics of Economic Adjustment.* New Brunswick, NJ: Transaction Books.

Nelson, Joan M., and Stephanie J. Eglinton. 1992. *Encouraging Democracy: What Role for Conditioned Aid?* Washington, DC: Overseas Development Council.

Nelson, Joan M., and Stephanie J. Eglinton. 1993. *Global Goals, Contentious Means: Issues of Multiple Aid Conditionality.* Washington, DC: Overseas Development Council.

Oberndörfer, Dieter. 1989. *Schutz der tropischen Regenwälder durch Entschuldung.* Schriftenreihe des Bundeskanzleramtes Band 5. Munich: Beck.

OECD (Organization for Economic Cooperation and Development). 1985. *Twenty-Five Years of Development Cooperation.* Paris.

OECD. 1992a. *DAC Principles for Effective Aid.* Paris.

OECD. 1992b. *Development Co-operation, 1992 Report.* Paris.

OECD. 1993a. *Environmental Performance Reviews: Germany.* Paris.

OECD. 1993b. *Environmental Performance Reviews: Iceland.* Paris.

OECD. 1993c. *Environmental Performance Reviews: Norway.* Paris.

OECD. 1993d. *Environmental Performance Reviews: Portugal.* Paris.

OECD. 1994a. *Environmental Performance Reviews: Japan.* Paris.

OECD. 1994b. *Environmental Performance Reviews: United Kingdom.* Paris.

OECD. 1995a. *Environmental Performance Reviews: Netherlands.* Paris.

OECD. 1995b. *Environmental Performance Reviews: Poland.* Paris.

Official Journal of the European Communities. 1993. "Cooperation with the Countries of Central and Eastern Europe and the Former Soviet Union," *Official Journal of the European Communities* 36 309: 183.

Ooi Jin Bee. 1990. "The Tropical Rain Forest: Patterns of Exploitation and Trade." *Singapore Journal of Tropical Geography* 2: 117–142.

Orlebar, Edward. 1994. "Business and the Environment: Call for 'Carbon Bond'." *Financial Times* (1 June): 20.

Owen, Henry. 1994. "The World Bank: Is 50 Years Enough?" *Foreign Affairs* (September/October): 97–108.

Oye, Kenneth A., and James H. Maxwell. 1994. "Self-Interest and Environmental Management." *Journal of Theoretical Politics* 6(4).

Oye, Kenneth A., and James H. Maxwell, 1995, "Self-Interest and Environmental Management." In Robert O. Keohane and Elinore Ostrom, eds., *Local Commons and Global Interdependence: Heterogeneity and Cooperation in Two Domains.* Newbury Park, CA: Sage.

Ozone Secretariat. 1993. *Handbook for the Montreal Protocol on Substances that Deplete the Ozone Layer,* 3d ed., August.

Page, Diana. 1990. "Debt-for-Nature Swaps: Experiences Gained, Lessons Learned." *International Environmental Affairs* 1, no. 4: 278–288.

Panayotou, Theodore. 1992. *Transferable Development Rights as an Instrument of Conservation.* Unpublished paper. Cambridge, MA: Harvard University, Kennedy School of Government.

Parson, Edward A., Peter M. Haas, and Marc A. Levy. 1992. "A Summary of the Major Documents Signed at the Earth Summit and the Global Forum." *Environment* 34(8): 13–36.

Parson, Edward. 1992. *Learning from Past Successes: Protecting the Ozone Layer,* Ph.D. dissertation. Cambridge, MA: Harvard University, Kennedy School of Government.

Patterson, Alan. 1990. "Debt-For-Nature Swaps and the Need for Alternatives." *Environment* 32(10): 5–32.

Paul, Samuel. 1990. *Institutional Reforms in Sector Adjustment Operations: The World Bank's Experience.* Washington, DC: World Bank.

Pearce, Fred. 1994. "Are Sarawak's Forests Sustainable?" *New Scientist* 144(26 November): 28–32.

Pearson Commission (Commission on International Development). 1969. *Partners in Development.* New York: Praeger.

Petroleum Economist. 1993. "Nuclear Power Reforms Gather Momentum: Nuclear Safety Contracts in Central and Eastern Europe." *Petroleum Economist* 60(7): 9.

PHARE Advisory Unit. 1991. The PHARE Compendium 1990–1991. Brussels: Commission of the European Communities.

PHARE Advisory Unit. 1992. PHARE 1992 Summaries of Programmes Approved, vol 2. Brussels: Commission of the European Communities.

PHARE Information Service. 1993. "PHARE & TACIS Environmental Activities." Brussels: Commission of the European Communities (January).

PHARE. 1995. "Environment to the Year 2000: Progress and Strategy Paper." Brussels: European Commission.

Poore, D., P. Burgess, J. Palmer, S. Rietbergen, and T. Synnott. 1989. *No Timber without Trees: Sustainability in the Tropical Forest.* London: Earthscan Publications.

Pressman, Jeffrey, and Aaron Wildavsky. 1984. *Implementation,* 3d ed. Berkeley: University of California Press.

Preston, Robert, and Jimmy Burns. 1993. "EBRD Spends More on Itself than It Hands Out in Loans." *Financial Times* (13 April): 1.

Princen, Thomas, and Matthias Finger. 1994. *Environmental NGOs in World Politics: Linking the Local and the Global.* London: Routledge.

Radke, Christian, and Barbara Unmüssig. 1989. "Debt-for-Nature Swaps." *VDW Intern,* no. 80 (October): 11–30.

Reddy, Amulya. 1993. "Has the World Bank Greened?" In *Green Globe Yearbook:* 65–73.

Reed, David, ed. 1992. *Structural Adjustment and the Environment.* Boulder, CO: Westview Press.

Reed, David., ed. 1993. *The Global Environment Facility: Sharing Responsibility for the Biosphere,* vol. 2. Washington, DC: World Wildlife Fund.

Regional Environmental Center. 1993. "Environmental Funds in Central and Eastern Europe." Prepared for a seminar "Financing Environment" in connection with the Regional Environmental Center's General Assembly meeting. Budapest, 2 December.

Remigio, Amador A., Jr. 1992. "Philippine Forest Resource Policy in the Marcos and Aquino Governments: A Comparative Assessment." Paper presented at a conference on "The Political Ecology of South East Asia's Forests," 23–24 March, London.

Repetto, Robert. 1989. "Balance-Sheet Erosion: How to Account for the Loss of Natural Resources." *International Environmental Affairs* 1(2): 103–137.

Repetto, Robert, and Malcolm Gillis, eds. 1988. *Public Policies and the Misuse of Forest Resources.* Cambridge: Cambridge University Press.

Resor, Jamie. 1993. *Interview with Jamie Resor.* World Wildlife Fund, Washington, DC, 13 December.

Resor, Jamie, and Barry Spergel. 1992. *Conservation Trust Funds: Examples from Guatemala, Bhutan, and the Philippines.* January. Washington, DC: World Wildlife Fund.

Revilla, J. A. V. 1994. *The National Forest Inventory Project and Its Links with Sustainable Forest Management (Update on the NFI Project Including a Preliminary Estimate of Long-Term Timber Supply from the Natural Forests).* Indonesian Ministry of Forestry and UN Food and Agriculture Organization, field document no. 38, May.

Reyes, Luis J., Jr. 1955. "Are We Overcutting Our Forests?" *Philippine Lumberman* (February-March): 5, 33.

Rich, B. 1985. "The Multilateral Development Banks, Environmental Policy and the United States." *Ecology Law Quarterly* (12): 681–745.

Rich, B. 1990. "The Emperor's New Clothes: The World Bank and Environmental Reform." *World Policy Journal* (spring): 305–329.

Rich, Bruce. 1994. *Mortgaging the Earth. The World Bank, Environmental Impoverishment, and the Crisis of Development.* Boston: Beacon Press.

Ricupero, R. 1993. "Chronicle of a Negotiation: The Financial Chapter of Agenda 21 at the Earth Summit." *Colorado Journal of International Environmental Law and Policy* 4: 81–101.

Roan, Sharon. 1989. *Ozone Crisis: The Fifteen-Year Evolution of a Sudden Global Emergency.* New York: John Wiley and Sons.

Robison, Richard. 1986. *Indonesia: The Rise of Capital.* Sydney: Allen and Unwin.

Romy, Isabelle. 1990. *Les Pollutions Transfrontières des Eaux: L'Exemple du Rhin. Mozens d'action des lésés.* Lausanne: Payot.

Rondinelli, Dennis. 1986. "Administration of Integrated Rural Development Policy: The Politics of Agrarian Reform in Developing Countries." In Atul Kohli, ed., *The State and Development in the Third World.* Princeton: Princeton University Press.

Round, Robin. Nd. *At the Crossroads—The Multilateral Fund of the Montreal Protocol: A Report for Friends of the Earth International.* London: Friends of the Earth, International.

Rubin, Steven M. 1993. *Interview with Steve Rubin.* Conservation International, Washington, DC, 14 December.

Rubin, Steven M., Jonathan Shatz, and Colleen Deegan. 1994. "International Conservation Finance: Using Debt Swaps and Trust Funds to Foster Conservation of Biodiversity." *The Journal of Social, Political and Economic Studies* 19(1): 21–43.

Sarkar, Amin. 1994. "Debt Relief for Environment: Experience and Issues." *Journal of Environment & Development* 3(1): 123–136.

Schelling, Thomas C. 1955. "American Foreign Assistance." *World Politics* 7(4 July): 606–626.

Schelling, Thomas C. 1957. "American Aid and Economic Development: Some Critical Issues." In *International Stability and Progress: United States Interests and Instruments.* New York: American Assembly.

Schreiber, Helmut. 1990. "The Threat from Environmental Destruction in East Central Europe." *Journal of International Affairs* 44 (Winter): 359–391.

Schreiber, Henry, and Laurence Dufond. 1992. "Les MdPA et la nappe phréatique." *Dossier: Mines et carrières-industrie minérale,* April: 67–71.

Schwabach, Aaron. 1989. "The Sandoz Spill: The Failure of International Law to Protect the Rhine from Pollution." *Ecology Law Quarterly* 16: 443–471.

Schwarz, Adam. 1994. *A Nation in Waiting: Indonesia in the 1990s.* Boulder, CO: Westview Press.

Scott, Dennis J. 1993. "Half-Hearted Cleanup at the World Bank." *Legal Times:* 44–45.

Sebenius, James K. 1983. "Negotiation Arithmetic: Adding and Subtracting Issues and Parties." *International Organization* 37(2): 281–316.

Sher, Michael S. 1993. "Can Lawyers Save the Rain Forest? Enforcing the Second Generation of Debt-for-Nature Swaps." *Harvard Environmental Law Review* 17(1): 151–224.

Sikkink, K. 1993. "Human Rights, Principles Issue Networks and Sovereignty in Latin America." *International Organization* 47(3): 411–441.

Sjoberg, H. 1994. *From Idea to Reality: The Creation of the Global Environment Facility.* Working paper no. 10. Washington, DC: Global Environment Facility.

Smith, David, and John Coppendale. 1993. "Assessment of G-24 Assistance to the Countries of Central and Eastern Europe in the Environment Sector." Final Report. Herts, United Kingdom: PA Consulting Group.

Susskind, Lawrence E. 1994. *Environmental Diplomacy: Negotiating More Effective Global Agreements.* New York and Oxford: Oxford University Press.

Sutter, Harald. 1989. *Forest Resources and Land Use in Indonesia.* Indonesian Ministry of Forestry and UN Food and Agriculture Organization, field document no. I-1, Jakarta, October.

Swedish Nuclear Inspectorate. 1992. Beskrivning av arbetsläeget per." (24 August): 1.

Swedish Nuclear Inspectorate. 1993. "Kärnkraftsäkerhet i Ignalina—ett led i ett regionalt samarbetsprogram." (23 December): 5.

Taake, Hans-Helmut. 1994. "Die Integration der Entwicklungsländer in die Weltwirtschaft." *Europa Archiv* 49(8): 223–230.

Tammes, Gerrit J. 1990. "Debt-for-Nature Conversion: What Limits Their Further Growth?" *International Environmental Affairs* 2(2): 153–159.

Thalwitz, W. 1991. Letter responding to Tolba 1991, and attached memo from Thalwitz to World Bank Senior Vice President for Operations M. Qureshi. Washington, DC: World Bank. Unpublished. Both dated 12 April.

Tolba, Mostafa. 1991. Letter to World Bank president Barber Conable on the Global Environment Facility. Nairobi, Kenya: United Nations Environment Programme. Unpublished. Dated 24 January.

Tolba, Mostafa. 1994. Presentation at Harvard University Science Center, November.

Tropical Forests Action Program. 1993. *TFAP Update No. 28.* Prepared by the TFAP Coordinating Unit.

Ugalde, Alvaro. 1994. *Interview with Alvaro Ugalde.* United Nations Development Program and former director of the National Park Service, San José, Costa Rica, 17 February.

Ullsten, Ola, Salleh Mohammed Nor, and Montague Yudelman. 1990. *Report of the Independent Review.* Official Review of the Tropical Forestry Action Plan, Kuala Lumpur, Malaysia, May.

Umaña, Alvaro, and Katrina Brandon. 1992. "Inventing Institutions for Conservation: Lessons from Costa Rica." In Sheldon Annis et al., eds., *Poverty, Natural Resources, and Public Policy in Central America.* New Brunswick, NJ: Transaction Publishers.

Underdal, Arild. "The Concept of Regime 'Effectiveness'." *Cooperation and Conflict* 27 (September 1992).

UNEP 1994. *Toward UNEP's GEF Strategy.* Report of a Workshop held in Nairobi on 22–23 March 1994. Nairobi. Unpublished. Dated 3 May.

UNEP, UNDP, and the World Bank. 1993. *Report of the Independent Evaluation of the Global Environment Facility Pilot Phase.* Washington, DC: Global Environment Facility. Dated 23 November.

United Nations. 1992a. *Nations of the Earth Report—United Nations Conference on Environment and Development: National Report Summaries* vol. 1, Geneva: UNCED.

United Nations. 1992b. *Nations of the Earth Report—United Nations Conference on Environment and Development: National Report Summaries,* vol. 2, Geneva: UNCED.

United Nations. 1992c. *Nations of the Earth Report—United Nations Conference on Environment and Development: National Report Summaries* vol. 3, Geneva: UNCED.

United Nations Environment Programme. 1992. *Annual Report of the Executive Director.*

United Nations Environment Programme (UNEP):

UNEP Economic Options Committee. 1995. "1994 Report of the Economic Options Committee for the 1995 Assessment of the Montreal Protocol on Substances that Deplete the Ozone Layer."

United Nations Environment Program, Executive Committee of the Multilateral Fund for the Montreal Protocol:

UNEP/OzL.Pro/ExCom/3/8/Rev.1, April 1991. "Implementation Guidelines and Criteria for Project Selection and Criteria for Project Selection."

UNEP/OzL.Pro/ExCom/3/18, 6 May 1991. "Draft Report of the Third Meeting of the Executive Committee of the Interim Multilateral Fund."

UNEP/OzL.Pro/ExCom/7/20, 26 May 1992. "Institutional Strengthening."

UNEP/OzL.Pro/ExCom/10/39, 28 May 1993, "(Draft) Incremental Cost Policies and Operational Interpretation."

UNEP/OzL.Pro/ExCom/10/35/Rev.1. "Status Report on Country Programmes."

UNEP/OzL.Pro/ExCom/10/40, June 1993. "Consolidated Report on Progress to the 10th Meeting."

NEP/OzL.Pro/ExCom/12/36, 21 March 1994. "Modification in the Indicative List of Categories of Incremental costs Under the Montreal Protocol (Submitted by the Government of India)."

UNEP/OzL.Pro/ExCom/12/Inf.6, 24 March 1994. "Policies, Procedures, Guidelines and Criteria (as of November 1993)."

UNEP/OzL.Pro/ExCom/12/37, 31 March 1994. "Report of the Twelfth Meeting of the Executive Committee of the Multilateral Fund for the Implementation of the Montreal Protocol."

UNEP/OzL.Pro/ExCom/15/4Rev1. "Draft Report of the Fifteenth Meeting of the Executive Committee of the Multilateral Fund for Implementation of the Montreal Protocol."

UNEP/OzL.Pro/ExCom/17/5, 28 June 1995. "Report from the Treasurer on the Status of Contributions and Disbursements."

UNEP/OzL.Pro/ExCom/17/9, 30 June 1995. "Consolidated Progress Report."

United Nations Environment Program. Open Ended Working Group of the Parties to the Montreal Protocol:

UNEP/OzL.Pro/WGII(2)/7, 5 March 1990. "Open Ended Working Group of the Parties to the Montreal Protocol."

UNEP/OzL.Pro/WG.IV/6, 29 May 1990. "Draft Decisions on the Financial Mechanism for the Implementation of the Montreal Protocol (Proposal by the Executive Director)," London.

UNEP/OzL.Pro.2/3, 29 June 1990. "Appendices to Decision II/8 ("Financial Mechanism") Adopted by the Second Meeting of the Parties, London, 20–29 June 1990," annex IV.

UNEP/OzL.Pro4/15, 25, November 1992, "Report of the Fourth Meeting of the Parties to the Montreal Protocol on Substances That Deplete the Ozone Layer, Copenhagen, 23–25 November 1992."

Uphoff, Norman. 1986. *Local Institutional Development*. West Hartford, CT: Kumarian Press.

U.S. Department of Commerce. 1977. "Forest Resources Production." *Survey on Business Opportunities in Indonesia*.

U.S. Department of the Treasury. 1993. *The Operation of the Enterprise for the Americas Facility*. Report to Congress (January). Washington, DC.

U.S. General Accounting Office. 1994. Environmental Issues in Central and Eastern Europe: U.S. Efforts to Help Resolve Institutional and Financial Problems. Washington, DC: General Accounting Office, May.

U.S. House of Representatives. 1992. "Informal Consultative Meeting with Industry." In *Hearings of the Committee on Science, Space and Technology*, July 11, 1992, Washington, DC: U.S. Government Printing Office: 127–138.

"U.S. Instructions to the Meeting in Geneva," May 1990, mimeograph on file with authors.

USAID (U.S. Agency for International Development). 1993. Office of Environment and Natural Resources: 1993–1994 Directory. Washington, DC.

USCAN (U.S. Climate Action Network). 1995. "Update on May Council Meeting." Available on econet conference "gef.forum." Dated 22 May.

Victor, D., and J. Salt. 1994. "From Rio to Berlin: Managing Climate Change." *Environment* 36(10): 6–15ff.

Vitug, Marites Dañguilan. 1993. *Power from the Forest: The Politics of Logging*. Manila: Philippine Center for Investigative Journalism.

Vollmer, U. 1993. "Forest Policy." In Laslo Pancel, ed., *Tropical Forestry Handbook,* vol. 2. Berlin: Springer-Verlag.

WALHI. 1991. *Sustainability and Economic Rent in the Forestry Sector.* Jakarta.

Wallace, David Rains. 1992. *The Quetzal and the Macaw. The Story of Costa Rica's National Parks.* San Francisco, CA: Sierra Club Books.

Waterbury, John. 1993. *Exposed to Innumerable Delusions: Public Enterprise and State Power in Egypt, India, Mexico, and Turkey.* Cambridge: Cambridge University Press.

Watson, R. T., and Ozone Trends Panel, Parther, M. J., and Ad Hoc Theory Panel, and Kurylo, M. J., and NASA Panel for Data Evaluation 1988. "Present State of Knowledge of the Upper Atmosphere 1988: An Assessment Report." NASA Reference Publication 1208, Washington, DC: US Government Printing Office, cited in World Meteorological Organization/National Aeronautics and Space Administration, 1997, "Report of the Ozone Trends Panel."

WCED. 1987. *Our Common Future.* Report of the World Commission on Environment and Development. Oxford: Oxford University Press.

Weber, Steven. 1994. "Origins of the European Bank for Reconstruction and Development." *International Organization* 48(1): 1–38.

Wegner, Rodger. 1993. "Non-Governmental Organizations." *Intereconomics* 28(6): 285–294.

Weiss, Mary. 1992. "The Enterprise for the Americas Initiative: An Instructive Model for International Funding for the Environment." *New York University Journal of International Law and Politics* 24(2): 921–955.

Weisskopf, Michael. 1990. "U.S. Intends to Oppose Ozone Plan." *Washington Post* (9 May): 1.

Wells, M. 1994. "The Global Environment Facility and Prospects for Biodiversity Conservation." *International Environmental Affairs* 6,1: 69–80.

Wettestad, Jorgen and Steinar Andresen. 1991. *The Effectiveness of International Resource Cooperation: Some Preliminary Findings.* Lysaker, Norway: The Fridtjof Nansen Institute.

White House, "Statement by Chief of Staff," 15 June 1990, mimeograph on file with authors.

Williams, M., and P. Petesch. 1993. *Sustaining the Earth: Role of Multilateral Development Institutions.* Washington, DC: Overseas Development Council.

Williamson, Oliver E. 1985. *The Economic Institutions of Capitalism: Firms, Markets, Relational Contracting.* New York: The Free Press.

Wilson, James Q. 1980. "Conclusion." In James Q. Wilson, ed., *The Politics of Regulation.* New York: Basic Books.

Winterbottom, Robert. 1990. *Taking Stock: The Tropical Forestry Action Plan after Five Years.* Washington, DC: World Resources Institute.

Wold, Chris A., Durwood Zaelke. 1992. "Promoting Sustainable Development and Democracy in Central and Eastern Europe: The Role of the European Bank for Reconstruction and Development." *American University Journal of International Law and Policy.*

Wolf, A., and D. Reed. 1994. *Incremental Cost Analysis in Addressing Global Environmental Problems*. Washington, DC: Worldwide Fund for Nature.

World Bank. 1989. *Philippines: Environment and Natural Resource Management Study*. Washington, DC: World Bank.

World Bank. 1990a. *Funding for the Global Environment: The Global Environment Facility*. Discussion paper. Washington, DC: World Bank.

World Bank. 1990b. *The World Bank and the Environment: Fiscal 1990*. Washington, DC: World Bank.

World Bank. 1991. *The World Bank and the Environment: Fiscal 1991*. Washington, DC: World Bank.

World Bank. June 1991. "The World Bank Group's Support for Economic Transformation in Central and Eastern Europe." Washington, DC: World Bank.

World Bank. 1991a. *Establishment of the Global Environment Facility*. World Bank Resolution No. 91–5. Washington, DC: World Bank.

World Bank. 1991b. *Forestry: The World Bank's Experience*. Report of the Operations Evaluation Department. Washington, DC: World Bank.

World Bank. 1992a. *World Development Report 1992*. New York: Oxford University Press.

World Bank. 1992b. World Bank Annual Report 1992. Washington, DC: World Bank.

World Bank. 1992c. *The World Bank and the Environment. Fiscal 1992*. Washington, DC: World Bank.

World Bank. 1992d. "Environmental Policy Issues." Background note, *Environment for Europe*, April.

World Bank. 1993a. *The World Bank and the Environment. Fiscal 1993*. Washington, DC: World Bank.

World Bank. 1993b. World Bank Annual Report 1993. Washington, DC: World Bank.

World Bank. 1993c. *World Debt Tables, 1993–94*. Vol. 1: Analysis and Summary Tables. Washington, DC: World Bank.

World Bank. 1994a. Annual Report. Washington, DC: World Bank.

World Bank. 1994b. "Central Europe Department Projects Related to Energy/Environment." Department document (August/September).

World Bank. 1994c. *Conditional Lending Experience in World Bank-Financed Forestry Projects*. Report of the Operations Evaluation Department, 27 December (draft).

World Bank. 1994d. *The Evolution of Environmental Concerns in Adjustment Lending: A Review*. Environment Working Paper No. 65, Environment Department, March 1994, authored by Jeremy Warford, Adelaida Schwab, Wilfrido Cruz, and Stein Hansen. Washington, DC: World Bank.

World Bank. 1994e. *Making Development Sustainable.* Washington, DC: World Bank.

World Bank. 1994f. *Review of Implementation of the Forest Sector Policy.* Report by the Agriculture and Natural Resources Department and the Environment Department. Washington, DC: World Bank.

World Bank. 1994g. *World Debt Tables, 1994–95.* Vol. 1: Analysis and Summary Tables. Washington, DC: World Bank.

World Bank ENVGC. 1994. *Facing the Global Environment Challenge: A Progress Report on World Bank Global Environment Operations.* June–July. Washington, DC: Global Environment Coordination Division, World Bank.

World Bank Global Environmental Coordination Division. 1994. *Implementation Performance Review of Bank Implemented Montreal Protocol Investment Operations,* Washington, DC: World Bank.

World Bank, International Energy Agency, and European Bank for Reconstruction and Development. 1993. *Nuclear Energy and Electric Power in Armenia, Bulgaria, Lithuania, Russia, Slovakia and Ukraine: Strategies and Financing— Summary Report,* June. Washington, DC: World Bank.

World Development Report 1992. 1992. New York: Oxford University Press for the World Bank.

WRI (World Resources Institute). 1989. *Natural Endowments: Financing Resource Conservation for Development.* Washington, DC: WRI.

WRI (World Resources Institute). 1992. *World Resources 1992–93.* New York: Oxford University Press.

WRI (World Resources Institute), ed. 1993. *Biodiversity Prospecting: Using Genetic Resources for Sustainable Development.* Washington, DC: WRI.

WRI (World Resources Institute). 1994. *World Resources 1994–95.* New York: Oxford University Press.

Wright, Michael. 1992. "Reaching Beyond Parks." In Smithsonian Institution and Natural Resources Defense Council, eds., *Debt-for-Nature Swaps: Progress and Prospects.* Unpublished conference report. Washington, DC: Smithsonian Institution and Natural Resources Defense Council.

Young, Oran R. 1992. "The Effectiveness of International Institutions: Hard Cases and Critical Variables." In James N. Rosenau and Ernst-Otto Czempiel, eds., *Governance without Government: Change and Order in World Politics.* New York: Cambridge University Press.

Index

Nuclear reactors, 5, 329–330
  accidents in, 233–234
  closure of, 256, 340, 342, 352, 363
  energy exportation and, 262–263
  operations of, 271–272
  replacing, 269–270
  safety of, 234–235, 236–239, 243,
    245–247, 275–276, 277, 341–
    342

ODP. *See* Ozone-depleting potential
ODS. *See* Ozone-depleting substances
OECD. *See* Organization for Economic Cooperation and
  Development
Olkusz, 284–285
OORG. *See* Ozone Operations Resource Group
Organization for Economic Cooperation and Development (OECD), 35,
  374, 375, 377, 378
  aid coordination by, 47–48
  environmental issues, 369–370
  polluter pays principle of, 220–221
Ozone-depleting potential (ODP), 113
Ozone-depleting substances (ODS),
  89, 93, 123, 330–331, 351
  developing countries and, 90–91,
    121–122
  and Montreal Protocol, 96, 97, 104–
    105, 358, 363
  North-South issues and, 94–95
  phasing out, 103, 108–109, 112,
    113–114, 118–119, 124–125, 336,
    340, 345, 364n3
Ozone depletion, 4, 11, 26, 55, 85n4,
  95, 102, 330–331
  impacts of, 92–94
  Montreal Protocol and, 85n5, 89,
    96–97
  program funding mechanisms and,
    103–104
  reducing, 7, 19–20, 119–120
Ozone Operations Resource Group
  (OORG), 117

PA. *See* Participants assembly
Pacific islands, 193
PA Consulting Group, 289
Paks plant, 249, 278n2
Panama, 159
Paraguay, 142
Paras, Jerome, 181
Paris Club, 140–141
Participants assembly (PA) (GEF), 67,
  68
Peru, 35
PHARE. *See* Poland and Hungary As-
  sistance for Restructuring
  Economies
Philippine Central Bank, 181
Philippines, 7, 10, 21, 35, 37, 45, 113
  conditionality in, 195, 196–197n9
  debt-for-nature swaps in, 134, 150,
    159
  forestry program loans in, 190–191,
    192, 197n10, 347, 348
  logging policies in, 26, 168, 174,
    177–184, 336, 337, 343, 345
PIUs. *See* Program implementation
  units
Poland, 35, 285, 302, 303, 322nn13,
  15, 19, 20, 378
  debt-for-nature swaps in, 135, 141
  environmental assistance for, 287–
    288, 291, 298, 301, 305, 307,
    308, 310
Poland and Hungary Assistance for
  Restructuring Economies (PHARE),
  320n2, 322nn18, 19, 346
  environmental assistance, 286, 287,
    289, 290, 294, 300–301, 302–
    311, 312, 313, 320n3, 322n21,
    345, 348
  institution building, 292, 293
  nuclear safety assistance, 248–249,
    250, 260, 263, 266, 268, 278nn7,
    8, 279n13
Policy dialogue, 37–38
  on environment, 39–41
Political institutions, 12–13

Politics, 15, 343
  of chloride pollution, 208–210, 219–
    223, 226–229, 332–333
  symbolic, 333–334
Polluter pays principle (PPP), 220–221
Pollution. *See also* Air pollution; Water pollution
  in Eastern Europe, 281, 284–285
Potassium. *See also* Mines de Potasse
    d'Alsace
  and chloride pollution, 207
PPC. *See* Project Preparation
    Committee
PPP. *See* Polluter pays principle
Preston, Lewis, 66
Private good substitutes, 206, 224–
    225, 228
Problem solving, 16–17
Program implementation units (PIUs),
    306–307, 322n20
Project Preparation Committee (PPC),
    319
  operations of, 315–318, 360–361
Property rights, 203–205, 230n.3,
    231n9

Rainforest Action Network, 65
Ramos, Fidel, 180
Reforestation, 180–181, 182–183
Regional Environmental Center (Budapest), 310, 322n21
Reyes, Luis, 178
Rhine Action Program, 214
Rhine River, 15, 201, 210, 219–220,
    221, 222
  chemical pollution of, 205–206,
    213–214
  chloride pollution of, 6, 21–22, 202–
    203, 205, 207–209, 211, 215,
    218, 225, 229–230nn1, 2, 332–
    333, 335–336, 340–341, 355
Rielly, William, 156
Romania, 240, 278n3, 284, 291, 298,
    303, 322n19
Ruhr Valley, 201

Rural sector, 42, 181, 182, 184
Russia, 35, 236, 253, 262, 298
  nuclear safety in, 240, 246–247,
    249, 264, 268, 278n4

Salt waste, storage and elimination of,
    211, 212–213, 215, 226, 231n9
Sandoz, 213
Sarawak, 172
Scandinavia, 233. *See also various
    countries*
Scientific and technical advisory panel
    (STAP), 61, 62, 67
SEC. *See* Slovak Energy Company
Self-interest, 25
Siemens-KWU, 242, 244
Skoda, 253, 278n1
Slawkow, 284–285
Slippery slope problem, 261, 269,
    275–276
Slovak Energy Company (SEC), 263
Slovakia, 263, 322n19
  environmental assistance for, 287,
    291, 307, 310, 328n8
  nuclear power in, 241, 244, 253
Slovenia, 298, 310
Southeast Asia, 158
Soviet Union, former, 4, 69, 108, 162,
    296. *See also various countries*
  energy use and planning in, 257–
    258, 272–273, 279nn14, 15
  and nuclear reactors, 233–234, 236–
    238, 240–242, 243, 340
  nuclear safety in, 239, 247–254,
    258–259, 264, 275–277, 338,
    341–342, 352, 354
STAP. *See* Scientific and technical advisory panel
"State of the Environment in Europe,"
    309
Stockholm Conference on the Human
    Environment, 370–371
Structural adjustment programs, 132,
    135, 139, 189–190
Suharto, 191